Laser-Molecule Interaction

Laser-Molecule Interaction

Laser Physics and Molecular Nonlinear Optics

J. R. LALANNE

A. DUCASSE

S. KIELICH

Foreword by
B. Couillaud

Translated by
L. Orrit

A Wiley-Interscience Publication

JOHN WILEY & SONS, INC.

New York • Chichester • Brisbane • Toronto • Singapore

This text is printed on acid-free paper.

This is a translation of
Interaction Laser Molecule: Physique du laser et optique non linéaire moléculaire
by Jean René Lalanne, André Ducasse, and Stanislaw Kielich

Translated with the assistance of the French Ministry of Culture

Library of Congress Cataloging in Publication Data:

Lalanne, J. R., 1940–
Laser molecule interaction : laser physics and molecular nonlinear optics / by J. R. Lalanne, A. Ducasse, and S. Kielich ; foreword by B. Couillaud ; translated by L. Orrit.
p. cm.
"A Wiley-Interscience publication."
Includes bibliographical references and index.
ISBN 0-471-12066-9
1. Nonlinear optics. 2. Lasers. I. Ducasse, A. II. Kielich, Stanisław. III. Title.
QC446.2.L35 1996
535′.2—dc20 95-24308

Printed in the United States of America

10 9 8 7 6 5 4 3 2 1

Contents

PART TWO
THE LASER WAVE AND ITS PROPERTIES

Foreword

While as far as we know, lasers have been with us for only a few decades, nonlinear optics has been part of our world since the beginning. The domain has never been as lively, exciting, and promising as it is now. Some of the most fundamental concepts, such as, the requirement of a population inversion for lasers are shaken, and the first experimental demonstrations are on their way. A larger number of techniques have now been perfected to the point of becoming the essential laboratory tools as well as the key components of a growing field of applications ranging from telecommunications to medical diagnostics and therapeutics.

The studies of nonlinear optics have really started with the discovery of the laser, and while most of the individuals who were part of the pioneering efforts are still with us, at least three new generations of researchers and engineers have joined over the years. Every day adds another family of new young investigators eager to push the limits, exercise their creativity, and add their contribution to the field. The industry that has developed around the discoveries in the field is finally reaching maturity, and the demand for highly qualified engineers is growing every day.

The authors of this book recognized these needs two decades ago. Their interest and motivation to educate at all levels ranging from undergraduate to seasoned engineers has been constant since then. This interest is illustrated by the numerous lectures, topic schools, and teaching materials that they have developed and continuously improved during their long collaboration. Their book *Laser Molecule Interaction* reflects the passion to educate demonstrated by Jean-René, André, and Stanislaw. It represents the results of numerous, sometimes extremely challenging discussions on the basic concepts and of better ways to introduce them. The content of this book is also the result of a constant challenge and critique by the two generations of students that have been exposed to the material presented. As a former colleague and occasional participant in the discussions and lectures, I am delighted to finally see that the authors have decided to broaden their audience by publishing the results of their long effort in an original book.

If this book allows a better and faster understanding of sometimes difficult concepts, triggers motivation, and develops an increased interest of the reader to the field of nonlinear optics, as I am firmly convinced that it will, the authors will have reached their longstanding goal. I wish them success, and I also hope that the reader will find as much interest and fun reading this book as I have experienced while sharing many years of my professional life with the authors.

BERNARD COUILLAUD

Coherent Inc.
March 1995

Preface

Light-matter interaction is now a privileged field of physics. The invention of the *laser* in 1959 has greatly contributed to the highly spectacular development of this field during the past 30 years. In particular, the first observation of coherent generation of a harmonic optical wave by P. A. Franken in 1961 gave rise to a novel optics—*nonlinear optics*—the unceasing development of which has achieved remarkable proportions in the domain of fundamental research as well as in that of practical applications. Thus, there is no exaggeration in the statement that the applications of the laser and nonlinear optics extend to practically all the domains of science and technology (in 1992, the laser market has exceeded the ''gigadollar'' level). This remarkable diversity of applications has resulted in an unceasing growth of the number of scientific publications and international conferences; many researchers in a great variety of specialities have become directly involved in nonlinear optics—often like Monsieur Jourdain,* speaking in prose unwittingly!

In 1970, when one of us (SK) had been invited as Associated Professor to Bordeaux University 1, we profited by our activities in fundamental research, often led jointly and always directed toward the laser and its interaction with matter, to promote a line of training in laser science and nonlinear optics. Thus, in the course of 25 years, we have contributed to the education and updating of about a thousand teachers (of the universities and other higher education establishments) as well as researchers and engineers. It is just this experience in teaching, summarized in the publication of a monograph by one of us (JRL) as the result of courses read at the Commissariat à l'Energie Atomique, that we now propose to transmit more widely by way of the present book, characterized as follows:

- This book is primarily intended for students of the second and third cycles of universities and schools of engineering. More exactly, we have written it essentially for the use of students preparing doctoral theses in *Chemical Physics*. Certainly, too, it will also prove useful for students of physics and junior researchers as an introductory handbook on the laser and its interaction with matter and, obviously, for wider circles of readers, many

*Molière, in *Le Bourgeois gentilhomme*.

of whom have already visited our laboratories, with the aim of obtaining first-hand knowledge of the laser, its operation, and possibilities.

- It is by no means conceived as an exhaustive presentation of nonlinear optics; in fact, there exist about a dozen books in English, written for physicists, which the reader will find cited in our bibliography. However, they consistently deal *either* with the laser *or* with nonlinear optics. We, on the other hand, have preferred *to deal with these two subjects jointly* since they are inseparable in practice. We have thus decided to restrict ourselves to the presentation of only certain examples characteristic of the very numerous nonlinear ''effects'' induced by lasers. Moreover, with regard to the specific requirements of the readers for whom our book is destined, we have decided to give priority to examples of *laser-molecule interaction*, which we chose for the title of our book.
- The book is intended to be *self-consistent*. Correct mastery of laser-matter interaction presupposes the acquisition of knowledge in three fields of physics; thus, the reader will be presented with the fundamentals of quantum mechanics, tensor calculus, and statistical physics as they intervene throughout our book and with a necessarily limited list of works of reference on these subjects which, if needed, may help the general reader to achieve a better assessment of laser technique.
- The few works of reference cited in each chapter fall into two categories: (1) basic works on the subject dealt with in the respective chapter, and (2) introductory articles in published widely read reviews such as *La Recherche*, *Physics Today*, or *Laser Focus World*. The references cited in the second category will direct the reader wishing for further specialization to the many original scientific papers produced in the course of all these years, the perusal of which is, in many cases, by no means easy.
- Our book contains about *35 original problems together with their solutions*, thereby providing the student with a handy tool for testing his or her newly acquired knowledge and also constituting a valuable addition to the examples discussed in the text. The book is conceived as follows:

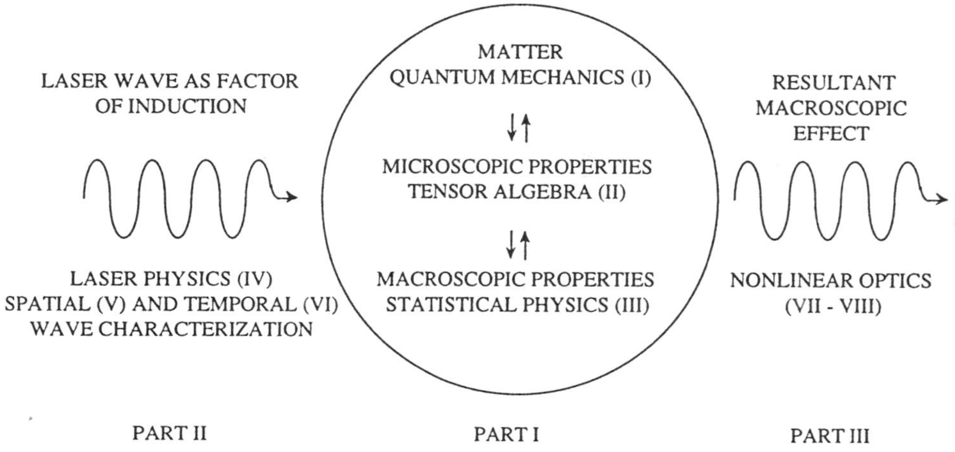

The roman numbers in parentheses denote the chapters. The three parts are of equal length.

Part I provides a partial review of the description of microsystems and an introduction to their nonlinear optical properties. Chapter I gives quantum mechanical description of the microsystem in the absence of a laser wave, first isolated then in interaction with its environment. Chapter II contains an introduction to nonlinear optics starting from a *tensorial description* of the optical properties of microsystems. Susceptibility tensors are discussed as to their structure and reduction. The tensorial expressions for electric n-poles and the fields radiated by them are presented. Chapter III is devoted to the transition from the microscopic to the macroscopic. The fundamentals of *statistical physics* are reviewed briefly. The calculation of the mean values of tensorial quantities at thermal equilibrum is performed on a variety of examples leading to the description of spherically symmetric (isotropic) macroscopic systems. Thus, at the end of Part I, the reader may have become familiar with the tensorial description of the optical properties of matter and acquired mastery of the mathematical and physical concepts and procedures that will enable him to start on the study of laser waves and their properties in applications to macroscopic systems.

Part II, which comprises Chapters IV, V, and VI, deals with the *laser*. In Chapter IV, its operation and the properties of fundamental wave emitted are discussed on the basis of Einstein's phenomenological approach of wave-matter amplifying interaction. Chapter V deals with the *spatial properties* of the laser wave related to the geometry of the cavity containing the amplifying medium, as well as the surface and angular energy concentrations one can obtain using the transformations undergone by laser beams when traveling through. Chapter VI gives a presentation of *different types of laser operation* (continuous-wave and pulsed) and of the *time structure* of the waves. The parameter *time* is given the attention it deserves with regard to its very important role in laser physics, particularly in the time-resolved spectroscopy and in physics of high intensities. We expect Part II may provide the reader with sufficient experimental mastery of optical sources—by no means a matter of just "pressing the right switch" but often requiring optimization in the course of experiment.

In Part III the results already established are combined to give a basic treatment of laser-molecule interaction. Chapter VII is devoted to the application of the density operator formalism to *stationary resonant laser-molecule interaction* in a semiclassical approach in which the laser wave is treated classically whereas the material system is dealt with in accordance with quantum mechanics. Expressions for the nonlinear polarization, susceptibility, and light refractive index are derived and their applications in Lamb laser theory, saturated absorption spectroscopy, and laser manipulation of microscopic systems are discussed. Finally, Chapter VIII is entirely devoted to a presentation of *nth-order susceptibility*. Its properties are made apparent in some examples of nonlinear interaction involving weak and strong as well as stationary and nonstationary energy transfer, such as harmonics generation and scattering, optical Kerr effect, degenerate four-wave mixing, and photon echo.

Inevitably, our book is the result of a compromise: In our attempts to adhere throughout to the level of principles, we deliberately limited the subject matter and sacrificed bulk in order to achieve better understanding. We hope that our book, which is of the nature of an introduction, may stimulate the reader to supplementary, more highly specialized studies.

We wish to express our indebtedness to B. Couillaud, our longtime friend and associate; to all our colleagues teaching within the curriculum on lasers; to F. Boisard, who enabled us to set up this line of teaching; and to B. Veyret and R. Perrier, who have helped us on many occasions. Also, we wish to thank N. Robineau who prepared the manuscript for the printers. The authors would also like to gratefully acknowledge K. Flatau, their friend, for his help in the preparation of Part I of this book.

J. R. Lalanne
A. Ducasse
S. Kielich

Bordeaux, March 1993

Acknowledgments

We are highly indebted to Mr. B. Couillaud, who wrote the Foreword, and Mrs. L. Orrit, née Volker, who translated this book with the appreciated help of M. Orrit.

J. R. L.
A. D.

Bordeaux, March 1995

Notation

Amplification coefficient	in field	a
	in intensity	a^2
Directional coefficients		$c_{i\alpha}$
Complex conjugate		cc
Population rate		λ
Relaxation constant		γ
Time constant of relaxation		τ
Determinant		$\mid \; \mid$
Matrix elements	relaxation	γ
	population	λ
Amplifier medium thickness		d
Cavity factor		h
Degeneracy factor		g'
Linear gain	in field	g
	in intensity	$G = 2g$
Cavity length		L
Coherence length		Δx_c
Matrix	generally	(M)
	adjoint	(M^{adj})
	inverse	(M^{-1})
	Hermitian	$(M^{\dagger})$
	unitary	$(M^{\dagger}) = (M^{-1})$
Operator	generally	$\mathcal{O}$
	vectorial	$\boldsymbol{\mathcal{O}}$
	relaxation	Γ
	population	Λ
Curvature radius		R
Reflectivity	in field	r
	in intensity	r^2

Representation	interaction	denoted by I
	Schrödinger	without notation
Kronecker delta		δ_{ij}
Coherence time		Δt_c
Tensors	generally	[t]
	Kronecker	$[\delta_{ij}]$
	Levi-Civita	$[\epsilon_{ijk}]$
Trace of a matrix or tensor		Tr
Transmission	in field	T
	in intensity	T^2
Vectors	generally	**V**
	row and column	{ }

Universal Constants

Quantity	Notation	Magnitude (in SI units)
Charge of the electron	e	1.60218×10^{-19} C
Boltzmann constant	k	1.38066×10^{-23} Jk^{-1}
Dirac constant	$\hbar$	1.05457×10^{-34} Js
Planck constant	h	6.62618×10^{-34} Js
Mass of the electron at rest	m_e	9.10939×10^{-31} kg
Permeability of vacuum	$\mu_0 = 1/\epsilon_0 c^2$	$4\pi \times 10^{-7}$ Hm^{-1}
Permittivity of vacuum	ϵ_0	8.85419×10^{-12} Fm^{-1}
Bohr radius	$a_0 = 4\pi\epsilon_0\hbar^2/m_e e^2$	5.29177×10^{-11} m
Light velocity in vacuum	c	2.997925×10^8 m s^{-1}

Prefixes Denoting Order of Magnitude

10^{-3}	milli	m	10^3	kilo	k
10^{-6}	micro	μ	10^6	mega	M
10^{-9}	nano	n	10^9	giga	G
10^{-12}	pico	p	10^{12}	tera	T
10^{-15}	femto	f	10^{15}	peta	P
10^{-18}	atto	a	10^{18}	exa	E

Greek Alphabet

Alpha	A	α	Theta	Θ	θ	Sigma	Σ	σ
Beta	B	β	Lambda	Λ	λ	Tau	T	τ
Gamma	Γ	γ	Mu	M	μ	Phi	ϕ	φ
Delta	Δ	δ	Nu	N	ν	Chi	X	χ
Epsilon	E	ϵ	Xi	Ξ	ξ	Psi	Ψ	ψ
Zeta	Z	ζ	Pi	Π	π	Omega	Ω	ω
Eta	H	η	Rho	p	ρ	Upsilon	Y	υ

Throughout, only the capital letters Γ, Δ, θ, Λ, Σ, ϕ, and ψ will be used.

PART ONE

Description of Microscopic Systems and of the Optical Properties of Matter

CHAPTER I

Description of Microscopic Systems and of Associated Physical Quantities: A Review of Quantum Physics

An optical wave propagating in a material interacts with the microscopic elements of the latter, i.e., *with the electrons and the nuclei of its constituent atoms and molecules*. We need *quantum physics* for a correct description of this microscopic world.We shall devote the present, introductory chapter to a brief review of the essential fundamentals of *quantum physics* which we shall be applying throughout our book. The reader aiming at a fuller and more profound description is invited to consult the literature cited in Section I.6. Moreover, we shall introduce the reader to *the use of certain mathematical tools*, such as the time development operator, the interaction picture, and others, to which we shall have recourse later on.

I.1 Postulates of quantum physics used in the description and their consequences

Quantum physics is founded on an axiomatic which we consider to involve nine postulates [for references, see Cohen-Tannoudji et al. (1976)]. We shall be needing the postulates P_1, P_2, P_3, P_4, P_6, and P_7 only. They fall into the following three categories:

I.1.1 Postulates of description

Postulate P_1 will serve in the description of the *microscopic systems,* whereas Postulates P_3 and P_7 will enable us to describe the *physical quantity* .

I.1.1.1 Postulate P_1 of description of the system

P_1 At a given moment of time t the state of a system is represented by a ket vector (which we refer to simply as a ket and denote by $| \;>$), belonging to a particular vectorial space (Hilbert space, denoted by E), also referred to as the *space of states* of the system. The ket is a time-dependent vector which we write in the form $|\psi>$.

The above symbolism is due to Dirac [see Cohen-Tannoudji et al. (1976)]. To the space E is associated a dual space E*—the space of linear funtionals of the kets, each element of which is called a *bra vector* (or simply a bra) and is denoted by $< \;|$. It can be shown that an *antilinear correspondence* holds between the bra and the ket, in the form

$$\begin{aligned} |\psi> &= \lambda_1 |\psi_1> + \lambda_2 |\psi_2> \\ <\psi| &= \lambda_1^* <\psi_1| + \lambda_2^* <\psi_2| \end{aligned} \qquad (I.1.1)$$

imposing a metrics on the space of states. The product (bra × ket) defines a *scalar product*; its essential properties, which we shall be applying in what follows, are assembled below:

$$<\psi|\psi> \text{ real; positive; zero if and only if } |\psi> = 0 \qquad (I.1.2)$$

$$<\psi'|\psi> = <\psi|\psi'>^* \qquad (I.1.3)$$

$$<\psi'|\lambda_1 \psi_1 + \lambda_2 \psi_2> = \lambda_1 <\psi'|\psi_1> + \lambda_2 <\psi'|\psi_2> \qquad (I.1.4)$$

$$<\lambda_1 \psi'_1 + \lambda_2 \psi'_2 |\psi> = \lambda_1^* <\psi'_1|\psi> + \lambda_2^* <\psi'_2|\psi> \qquad (I.1.5)$$

(λ_1 and λ_2 are two arbitrary complex numbers).

Let $|u_n>$ represent an orthonormal basis, assumed as discrete, of space E. We can write

$$|\psi> = c_n |u_n>$$

On left multiplication by $<u_m|$ and taking orthogonality of the basis into account, we obtain

$$<u_m|\psi> = c_n <u_m|u_n> = c_n \delta_{mn} = c_m$$

where δ_{mn} is the Kronecker delta.

Thus, the mth component of a ket $| \psi >$, within the basis $| u_n >$, is given by $< u_m | \psi >$. The set of these numbers is disposed in the form of a *column vector*, symbolized by { }. An identical procedure, carried out on a $< \psi |$ and the dual basis $< u_n |$, leads to the components $c_m^* = < \psi | u_m >$, disposed in the form of a *row vector,* symbolized as { }.

I.1.1.2 Postulates P_2 and P_7 of description of a physical quantity

P_2 Any measurable physical quantity A can be described by a Hermitian operator A acting within E whose orthonormal set of eigenvectors $| u_n >$ constitutes a basis in E.

P_7 In the Schrödinger picture, A is obtained from A (**r**, **p**, t) on replacing **r** and **p** by

$$\mathbf{r} \rightarrow \boldsymbol{r}, \qquad \mathbf{p} \rightarrow -i\hbar\nabla$$

where **p** is the generalized momentum of the system.

This postulate sets up a correspondence between the quantity and the operator. As the case may be, the expression for A will have to be transformed prior to application of the principle of correspondence in order to render it Hermitian (the notion of hermiticity will be defined later on.)

An operator acts so as to transform a vector. To a ket vector belonging to E, it associates another ket, as follows:

$$| \psi' > = A | \psi > \qquad (I.1.6)$$

Now, it will be remembered that we have already had to deal with a class of mathematical entities acting in this way. We have *matrices* in mind. Thus, operators can be represented in matricial form. We resume this point briefly in Inset I.1.

We define a Hermitian operator $A^\dagger$ by the relation

$$A^\dagger = A$$

We now proceed to explain why the operators representative of physical quantities are, in quantum physics, endowed with properties of this kind.

I.1.2 Postulates P_3 and P_4 of measure and their consequences

P_3 Measurement of a physical quantity A can lead only to an eigenvalue of the corresponding operator A.

The equation in eigenvalues of an operator A, assumed Hermitian, reads $A | u_n > = a_n | u_n >$. We carry out the double operation (†) defined above. This gives $< u_n | A^\dagger = a_n^* < u_n |$. On multiplying the two preceding equations by $< u_n |$ and $| u_n >$, respectively, we get

$$< u_n | A | u_n > = a_n < u_n | u_n >$$

$$< u_n | A^\dagger | u_n > = a_n^* < u_n | u_n > \tag{I.1.7}$$

The equality $A = A^\dagger$ implies that $a_n = a_n^*$, i.e., a *real eigenvalue.*

Thus, hermiticity of an operator ensures the principal physical characteristic of any measurement, which is that it shall result in a real number.

P_4 When measuring a physical quantity A on a system in a state represented by a vector $|\psi(\mathbf{r},t)>$, the probability $P(a_n)$ of obtaining, as a result, the discrete nondegenerate eigenvalue a_n is given by

$$P(a_n) = < \psi | u_n \times u_n | \psi > \text{ (no summation over the index n)} \tag{I.1.8}$$

where $| u_n >$ is the eigenvector of A associated with the eigenvalue a_n.

(Cases involving continuous and degenerate eigenvalues can be expressed in a similar way.)

An eigenvalue is referred to as nondegenerate if it is in one-to-one correspondence with an eigenvector.

The postulates P_2 and P_7 have as consequences the following two highly important results:

— On expanding $| \psi >$ on the orthonormal basis $| u_n >$, we obtain

$$P(a_n) = c_m^* c_{m'} < u_m | u_n \times u_n | u_{m'} > = c_m^* c_{m'} \delta_{mn} \delta_{nm'} = | c_n |^2$$

The probability of obtaining a_n is equal to the squared modulus of the coefficient at $| u_n >$ in the expansion of $| \psi >$ over the basis of the eigenvectors of A.

— The quantum mechanical mean value obtained from a measurement of A is, in the nondegenerate case, given by $< a > = P(a_n)\, a_n = | c_n |^2 a_n$. Thus

$$< \psi | A | \psi > = c_m^* c_{m'} < u_m | A | u_{m'} > = c_m^* c_{m'} a_{m'} < u_m | u_{m'} >$$

$$= c_m^* c_{m'} a_{m'} \delta_{mm'} = | c_m |^2 a_m$$

where n and m are two summation indices. We thus get

$$< a > = < \psi | A | \psi > \tag{I.1.9}$$

Inset I.1 **_How to represent an operator A by a matrix $(M)_A$_**

Let $|u_n>$ denote the basis, assumed as discrete, orthonormal and finite (of order n) of the eigenvectors of A, i.e., $A|u_m> = a_m|u_m>$ where a_m is a complex number. Consider another operator B acting on $|\psi>$, thus $|\psi'> = B|\psi>$. On applying the convention of summation over recurring subscripts we obtain

$$|\psi'> = c'_m|u_m> = c_m B|u_m>$$

And on left multiplication by $< u_{m'}|$ we obtain

$$c'_m < u_{m'}|u_m> = c'_m \delta_{m'm} = c'_{m'} = c_m < u_{m'}|B|u_m> = c_m B^m_{m'}$$

We arrive at the equivalent representations:

$$\begin{pmatrix} c'_1 \\ \vdots \\ c'_m \\ \vdots \\ c'_n \end{pmatrix} = \begin{pmatrix} B^1_1 & \cdots & B^n_1 \\ \cdots & \cdots & \cdots \\ B^1_{m'} \cdots & \boxed{B^m_{m'}} & \cdots B^n_{m'} \\ \cdots & \cdots & \cdots \\ B^1_n & \cdots & B^n_n \end{pmatrix} \begin{pmatrix} c_1 \\ \vdots \\ c_m \\ \vdots \\ c_n \end{pmatrix} \quad \left(\begin{matrix}\text{matrix} \\ \text{notation}\end{matrix}\right)$$

column vector of dimension n — square matrix of dimension n × n — column vector of dimension n

$$|\psi'> = B|\psi> \quad \text{(Dirac notation)}$$

The square table of n^2 numbers $B^m_{m'}$ provides the *matrix representation of the operator B within the basis selected.*

Note: If $|u_m>$ is an eigenvector of B, thus $B|u_m> = b_m|u_m>$, then $B^m_{m'} = < u_{m'}|B|u_m> = b_m < u_{m'}|u_m> = b_m\delta_{m'm}$. Thus, the matrix $(B^m_{m'})$ is diagonal and its elements are given by the eigenvalues of B.

We refer to some useful definitions:

The adjoint matrix (M^{adj}) is obtained on transposing the elements of (M) [each element is replaced by the one symmetric to it with respect to the principal diagonal of (M), and subsequently each element is replaced by its minor].

The minor of a matrix element is defined as the cofactor of an element of (M) multiplied by $(-1)^{r+c}$, where r is the row index and c the column index of the respective element.

The cofactor of a matrix element is given by the determinant constructed with the elements remaining on removal of all the elements of the row and the column corresponding to the element under consideration.

The inverse matrix (M^{-1}), under the condition that $|M| \neq 0$, is defined as

$$(M^{-1}) = (\mathbb{1}) / (M) = (M^{adj}) / |M|$$

Hermitian matrices are such that $(M^{\dagger}) = (M)$.

The operation, denoted by †, defines transposition followed by conjugation.

The unitary matrix is defined as $(M^{\dagger}) = (M^{-1})$.

A unitary matrix conserves the norm of the vectors which it transforms. Thus, if (M) is unitary, and with respect to $|\psi'\rangle = (M)|\psi\rangle$, we obtain

$$\langle \psi' | \psi' \rangle = \langle \psi | (M^{\dagger})(M) | \psi \rangle = \langle \psi | (M^{-1})(M) | \psi \rangle$$
$$= \langle \psi | \mathbb{1} | \psi \rangle = \langle \psi | \psi \rangle$$

I.1.3 Postulate P_6 of evolution

P_6 The evolution in time of a ket $|\psi(\mathbf{r},t)\rangle$ is governed by the Schrödinger equation:

$$i\hbar |\dot{\psi}(r, t)\rangle = H \, |\psi(r, t)\rangle \qquad \text{(I.1.10)}$$

H is the Hamiltonian operator, obtained by applying P_7 to the Hamiltonian representing the energy of the system.

We are now in a position to express the following consequence of the postulate P_6, to which we shall repeatedly be having recourse if the quantum mechanical system is in a stationary state, i.e., if its energy is time-independent, the postulate P_3 causes the energy (E_n) measured to be an eigenvalue of the operator *H*. We have

$$H \, | u_n(\mathbf{r}, t)\rangle = E_n \, | u_n(\mathbf{r}, t)\rangle \qquad \text{(I.1.11)}$$

where $| u_n(\mathbf{r}, t)\rangle$ is the respective eigenvector. On applying the postulate P_6 to the state $| u_n(\mathbf{r}, t)\rangle$ and with regard to Eq. I.1.11 we get $i\hbar | \dot{u}_n(\mathbf{r}, t)\rangle = E_n \, | u_n(\mathbf{r}, t)\rangle$ which, on integration, gives

$$| u_n(\mathbf{r}, t)\rangle = | u_n^0(\mathbf{r})\rangle \exp(-i E_n t / \hbar) \qquad \text{(I.1.12)}$$

Stationary states are characterized by a space-time separation of the vectors representing them.

I.2 Some "tools" of quantum physics

Having gained some practice in dealing with operators and their matrix representations, we now proceed additionally to a concise study of a particular operator, a picture, and a formalism which will be of use in our further work.

I.2.1 Time-development operator

I.2.1.1 Its definition

The time-development operator expressed as $U(\mathrm{t}, \mathrm{t}_0)$ and applied to a vector describes the evolution in time of the latter during the interval of time from the initial moment t_0 to some moment of time t. The process can be described by the following formula:

$$|\psi(\mathbf{r}, \mathrm{t})> = U(\mathrm{t}, \mathrm{t}_0)|\psi(\mathbf{r}, \mathrm{t}_0)> \tag{I.2.1}$$

The time-development operator possesses the following six properties:

(a) $U(\mathrm{t}_0, \mathrm{t}_0) = \mathbb{1}$ (I.2.2)

(b) It fulfills the differential equation:

$$\mathrm{i}h\,\dot{U}(\mathrm{t}, \mathrm{t}_0) = HU(\mathrm{t}, \mathrm{t}_0) \tag{I.2.3}$$

On insertion of Eq. I.2.1 into Eq. I.1.10, we obtain

$$\mathrm{i}\hbar\,\partial\left(U(\mathrm{t}, \mathrm{t}_0)|\psi(\mathbf{r}, \mathrm{t}_0)>\right)/\partial\mathrm{t} = HU(\mathrm{t}, \mathrm{t}_0)|\psi(\mathbf{r}, \mathrm{t}_0)>$$

i.e.,

$$\mathrm{i}\hbar\,\dot{U}(\mathrm{t}, \mathrm{t}_0)|\psi(\mathbf{r}, \mathrm{t}_0)> = HU(\mathrm{t}, \mathrm{t}_0)|\psi(\mathbf{r}, \mathrm{t}_0)>$$

This equation, which holds for arbitrary times t_0, leads to the relation I.2.3.

(c) Integral form of the time-development operator.

Integrating Eq. I.2.3 gives

$$U(\mathrm{t}, \mathrm{t}_0) = (1/\mathrm{i}\hbar)\int_{\mathrm{t}_0}^{\mathrm{t}} HU(\mathrm{t}', \mathrm{t}_0)\,\mathrm{dt}' + C$$

where C is a constant operator. At $\mathrm{t} = \mathrm{t}_0$, we have $C = U(\mathrm{t}_0, \mathrm{t}_0) = \mathbb{1}$. Hence

$$U(\mathrm{t}, \mathrm{t}_0) = (1/\mathrm{i}\hbar)\int_{\mathrm{t}_0}^{\mathrm{t}} HU(\mathrm{t}', \mathrm{t}_0)\,\mathrm{dt}' + \mathbb{1} \tag{I.2.4}$$

(d) The inverse time-development operator $[U(\mathrm{t}, \mathrm{t}_0)]^{-1}$ is defined as

$$[U(t, t_0)]^{-1} = U(t_0, t) \tag{I.2.5}$$

In fact, on neglecting spatial dependence, we have

$$|\psi(t)\rangle = U(t, t_0)|\psi(t_0)\rangle$$

$$|\psi(t_0)\rangle = U(t_0, t')|\psi(t')\rangle$$

whence

$$|\psi(t)\rangle = U(t, t_0)\, U(t_0, t')|\psi(t')\rangle$$

and we find

$$U(t, t') = U(t, t_0)\, U(t_0, t')$$

On putting t' = t we get

$$U(t, t) = 1 = U(t, t_0)\, U(t_0, t)$$

which enables us to prove Eq. I.2.5.

(e) The time-development operator has to be unitary (see the preceding inset) in order to keep unchanged the norm of the vectors it acts on. Thus

$$U(t', t) = U^{-1}(t, t') = U^{\dagger}(t, t') \tag{I.2.6}$$

(f) Form of the evolution operator for conservative systems.

Integration of Eq. I.2.3 in the case of time-independent H leads to

$$U(t, t_0) = \exp[-iH(t - t_0)/\hbar] \tag{I.2.7}$$

and

$$U^{-1}(t, t_0) = U^{\dagger}(t, t_0) = \exp[-iH(t - t_0)/\hbar] \tag{I.2.8}$$

I.2.2 Interaction picture

When formulating postulate P_7, we touched on a particular picture widely used in quantum physics. We have in mind the Schrödinger picture. Since the evolution of the microscopic system is governed by Eq. I.1.10 [with the corresponding time-development operator $U(t,t_0)$, given by Eq. I.2.3], we immediately take note of the fundamental role played by operator H. For the case of laser-matter interaction we shall show (see Chapter VII) that the Hamiltonian takes the form of $H = H_0 + H'$, where H_0 describes the nonperturbed system whereas H' accounts for the perturbation induced by the laser. In general, $H' \ll H_0$, so that the relevant part (H') of the Hamiltonian H is "masked" by the part H_0. This, at least, is the situation we get on applying the Schrödinger picture.

There is much to be gained by going over to another picture involving a renormalization privileging H' at the expense of H_0. This is achieved by having recourse to the interaction picture. The change of picture proceeds in accordance with the following formula:

$$U_0^{-1}(t, t_0) \times U(t, t_0) = U^I(t, t_0) \qquad \text{(I.2.9)}$$

where $U(t,t_0)$ is the time development operator in the Schrödinger picture, whereas $U_O(t,t_0)$ is that part of the time development that alone is dependent on the unperturbed Hamiltonian H_O, in Schrödinger picture. $U^I(t,t_0)$ is now the time-development operator in interaction picture (denoted by a superscript I). In what follows, a quantity with *no* index I will be assumed to belong to the Schrödinger picture.

Let us now search for the equation of evolution fulfilled by $U^I(t,t_0)$.
Applying Eq. I.2.9 we get, after some calculations,

$$i\hbar \dot{U}^I(t, t_0) = U_0^{-1}(t, t_0) H' U(t, t_0) \qquad \text{(I.2.10)}$$

On left multiplication by $U_0(t, t_0)$, Eq. I.2.9 gives

$$U(t, t_0) = U_0(t, t_0)\ U^I(t, t_0) \qquad \text{(I.2.11)}$$

whereas on inserting Eq. I.2.11 into Eq. I.2.10 we arrive at

$$i\hbar \dot{U}^I(t, t_0) = U_0^{-1}(t, t_0) H'\ U_0(t, t_0)\ U^I(t, t_0)$$

On defining the perturbation Hamiltonian H'^I in the interaction picture as

$$H'^I = U_0^{-1}(t, t_0) H'\ U_0(t, t_0)$$

we obtain:

$$i\hbar \dot{U}^I(t, t_0) = H'^I U^I(t, t_0) \qquad \text{(I.2.12)}$$

The equation we have just derived is obviously quite similar to Eq. I.2.3; however,

(a) it is expressed in the interaction picture, and

(b) its Hamiltonian no longer contains the term H_0^I, irrelevant to the description of the interaction alone.

I.2.3 Dyson formalism

Dyson's is a perturbation calculus involving a restricted series expansion which we shall use in Chapter VII. It originates in the integration of Eq. I.2.12:

$$U^I(t, t_0) = (1 / i\hbar) \int_{t_0}^{t} H'^I(t_1)\, U^I(t_1, t_0)\, dt_1 + \mathbb{1} \qquad \text{(I.2.13)}$$

In the zeroth order of expansion of $U^I(t,t_0)$ in the perturbation $H'^I(t)$, Eq. I.2.13 gives $U_0^I(t_1, t_0) = \mathbb{1}$.

In the first order of the expansion we get

$$U_1^{\mathrm{I}}(t, t_0) \approx \mathbb{1} + (1/i\hbar) \int_{t_0}^{t} H'^{\,\mathrm{I}}(t_1)\, dt_1 \tag{I.2.14}$$

and in the second order we obtain

$$U_2^{\mathrm{I}}(t, t_0) \approx \mathbb{1} + (1/i\hbar) \int_{t_0}^{t} H'^{\,\mathrm{I}}(t_1)\, dt_1 + \left[1/(i\hbar)^2\right] \int_{t_0}^{t} H'^{\,\mathrm{I}}(t_1)\, dt_1 \int_{t_0}^{t_1} H'^{\,\mathrm{I}}(t_2)\, dt_2$$

In general, proceeding to higher orders we obtain

$$U^{\mathrm{I}}(t, t_0) = \mathbb{1} + \sum_{n=1}^{\infty} U_n^{\mathrm{I}}(t, t_0) \tag{I.2.15}$$

where

$$U_n^{\mathrm{I}}(t, t_0) = \left[1/(i\hbar)^n\right] \int_{t_0}^{t} \cdots \int_{t_0}^{t_{n-1}} H'^{\mathrm{I}}(t_1)\, H'^{\mathrm{I}}(t_2) \cdots H'^{\mathrm{I}}(t_n)\, dt_1 \cdots dt_n \tag{I.2.16}$$

The above equation contains an n-fold integral over n times. This expansion of the time-development operator in interaction picture is just what is meant by *Dyson formalism*. It will be seen that it is particularly well adapted to the calculation of microscopic optical susceptibilities of order n.

We shall now make use of the concepts already discussed to set the framework for the quantum description of microscopic systems and shall extend our discussion of the tools of quantum physics and their use.

I.3 Framework of the description

An ineluctable difficulty we encounter is that a measurement performed *on a set* of quantum systems is affected by *a twofold incertitude*. The first difficulty, as we have seen, is of quantum origin: Except for special cases of systems in an eigenstate of the operator corresponding to the quantity measured, the result of the measurement is of probabilistic nature since we obtain a quantum mean value of eigenvalues, as given by Eq. I.1.9.

The second difficulty is statistical in nature and stems from the circumstance that the systems are rarely in one and the same state. This leads to a *statistical average* which combines with the *quantum mean value* mentioned above. With the exception of some recent, very remarkable experiments applying lasers and which perhaps will permit us to isolate *a single* quantum system (see Section VII.3.3), almost all experiments on laser-matter interaction bear *on sets* of atoms and molecules. We shall now explain why and how the vector of state description proposed by the postulate P_1 is of little use in the latter case *(mixed case)*, whereas it retains its applicability in the *pure case* (that of a *single* system). The problem posed by these difficulties will lead us to define and apply a special operator—

I.3.1 Description by the vector of state

I.3.1.1 Pure case

Let us consider N identical systems, denoted by (1.) Each of them is described (neglecting the space-time dependence of the vector) by $|\psi(1)> = c_n^{(1)}|u_n>$, where $|u_n>$ is an orthonormal basis in E space.

The probability of finding the system in state $|u_n>$ is given by Eq. I.1.8, which reads $P^{(1)}(a_n) = |c_n^{(1)}|^2$.

The statistical mean probability, per system, of finding the set of N systems in the state $|u_n>$ is

$$N\,P^{(1)}(a_n)/N = |c_n^{(1)}|^2 \tag{I.3.1}$$

I.3.1.2 Mixed case

Let us consider N systems distributed in i classes with respective statistical probabilities $p_i = N_i / N$. Obviously

$$N = \sum_i N_i \quad \text{and} \quad p = \sum_i p_i = 1$$

The probability $P(a_n)$ for a single system is

$$P(a_n) = \sum_i N_i\,P^{(i)}(a_n)/N = \sum_i p_i\,P^{(i)}(a_n) \tag{I.3.2}$$

It should be noted that $P(a_n)$ is a quadratic function of c_n. Here resides the main difficulty of representing a statistical mean state by a vector of state. To convince ourselves that this is so, let us try to construct, by way of a linear combination, a "mean" vector of the description of a mixed state with two species, denoted 1 and 2.

Suppose we described the overall state of the system by the vector

$$|\psi> = \lambda_1|\psi_1> + \lambda_2|\psi_2> \tag{I.3.3}$$

in order to try to identify $|\lambda_1|^2$ and $|\lambda_2|^2$ with p_1 and p_2. The probability $P(a_n)$ is given by Eq. I.1.8. Inserting Eq. I.3.3 into Eq. I.1.8, we obtain

$$\begin{aligned} P(a_n) = <\psi|u_n \times u_n|\psi> &= p_1\,P^{(1)}(a_n) + p_2\,P^{(2)}(a_n) \\ &+ \lambda_1^*\lambda_2\,c_n^{(1)*}\,c_n^{(2)} + \lambda_1\lambda_2^*\,c_n^{(1)}\,c_n^{(2)*} \end{aligned} \tag{I.3.4}$$

and find that the "rectangular" terms cause the probability of obtaining the statistical mean state defined by Eq. I.3.3 to differ from $\left(p_1\,P^{(1)}(a_n) + p_2\,P^{(2)}(a_n)\right)$, the

the *density operator*. It is characterized by the following properties.statistical average of the probabilities of obtaining the states $|\psi_1>$ and $|\psi_2>$.

There is thus no hope of representing a statistical mixed state by a single mean vector. However, we can deal with each class independently and obtain the required result by performing, *a posteriori*, a statistical calculation on the set of classes; but such calculations are usually bulky and by no means easy. We are thus led to define another procedure.

I.3.2 Description by the density operator

I.3.2.1 Pure case

(a) Definition of the density operator

Let us consider a physical quantity B to which is associated an operator B, represented on the orthonormal basis $|u_n>$ by matrix elements B_n^p, given by $B_n^p = < u_n | B \; | u_p >$.

The mean value $< b >$ of the measurement on B, relative to state $| \psi >$, reads (see Eq. I.1.9)

$$< b > = < \psi | B | \psi > = c_n^* c_p B_n^p \qquad \text{(I.3.5)}$$

We define the density operator as

$$\rho = | \psi > < \psi | \qquad \text{(I.3.6)}$$

It can readily be shown that the product of a ket (a column vector) and a bra (a row vector) takes the form of a matrix.

Let us calculate the elements ρ_p^n of the matrix (ρ.)

$$\rho_p^n = < u_p \big| \psi \times \psi \big| u_n > = c_p c_n^* \qquad \text{(I.3.7)}$$

On insertion of Eq. I.3.7 into Eq. I.3.5, we obtain

$$< b > = \rho_p^n B_n^p = < u_p | \rho | \Big(\sum_n \big| u_n \times u_n \big| \Big) | B \; | u_p >$$

Now, $\sum_n \big| u_n \times u_n \big|$ is the unit operator $\mathbb{1}$. This important identity is referred to as the completeness relation. We obtain

$$< b > = < u_p | \rho B \; | u_p > = \mathrm{Tr}\,(\rho B) \qquad \text{(I.3.8)}$$

Hence, the mean value $< b >$ turns out to be the *trace* (Tr—the sum of the diagonal elements) of the product operator ρB.

We now shall formulate some properties of the density operator.

(b) Its properties

— The trace of the density operator equals 1.

In fact, assuming $|\psi>$ as normalized, we have

$$\mathrm{Tr}(\rho) = <u_n|\psi \times \psi|u_n> = \sum_n |c_n|^2 = 1$$

— ρ is Hermitian.

To probe this, we calculate the matrix element ρ_n^p.

$$\rho_n^p = <u_n|\rho|u_p> = <u_n|\psi \times \psi|u_p> = [<u_p|\psi \times \psi|u_n>]^* = [\rho_p^n]^*$$

— To obtain the equation of evolution of ρ, we calculate the time derivative of ρ:

$$\dot{\rho} = |\dot{\psi} \times \psi| + |\psi \times \dot{\psi}|$$

On expressing $|\dot{\psi}>$ and $<\dot{\psi}|$ with Eq. I.1.10 and its conjugate transpose we arrive at

$$i\hbar \dot{\rho} = [H, \rho] \qquad \text{(I.3.9)}$$

This is the *Liouville equation*, widely applied in laser-matter interaction. $[H, \rho]$ is referred to as the *commutator* of the Hamiltonian and density operators:

$$[H, \rho] = H\rho - \rho H \qquad \text{(I.3.10)}$$

Dealing with the pure case, we shall now express the probability of finding b_q (the qth eigenvalue of B) as the result when measuring the physical quantity B on a system of N entities being in the same state $|\psi^{(k)}>$. Equation I.1.8 leads to

$$P_{b_q}^{(k)} = <\psi^{(k)}|u_q \times u_q|\psi^{(k)}> \quad (\textit{no} \text{ summation over the index q})$$

Above, $|u_q \times u_q| = P_q$ is the *projector operator* [see Cohen-Tannoudji et al. (1976)]. We get

$$P_{b_q}^{(k)} = <\psi^{(k)}|P_q|\psi^{(k)}>$$

Equation I.1.9 shows that $P_{b_q}^{(k)}$ is the quantum mean value of P_q. With regard to Eq. I.3.8, it takes the form

$$P_{b_q}^{(k)} = \mathrm{Tr}(\rho^{(k)} P_q) \qquad \text{(I.3.11)}$$

I.3.2.2 Mixed case

The above expression derived for $P_{b_q}^{(k)}$ is *linear* in ρ^{k}, whereas the equivalent expression I.3.4 was *quadratic* in $|\psi>$. This makes a great difference, strongly favoring the density operator picture in the treatment of the mixed case. Obviously, for a mixture with many states $|\psi^{(k)}>$ and statistical probabilities p_k such that $\sum_k p_k = 1$, we can write (see Eq. I.3.2)

$$P_{b_q} = p_k P_{b_q}^{(k)} = \mathrm{Tr}\left(p_k \rho^{(k)} P_q\right) = \mathrm{Tr}\left(\rho P_q\right) \qquad \text{(I.3.12)}$$

with

$$\rho = P_k \rho^{(k)} \qquad \text{(I.3.13)}$$

The latter equation defines a *mean density operator*, enabling us to treat the pure and mixed case identically. This is a considerable advantage. It justifies our choice of the density operator formalism for the treatment of the three selected descriptions (see Eq. I.4). Nonetheless, in order to make the reader more familiar with the application of the vector of state picture, we shall propose some problems, the solutions of which will require this manner of treatment, to be discussed in Part III (Chapter VIII) of our book.

I.3.2.3 Physical meaning of the elements of the mean density operator

(a) Diagonal elements

These are of the form

$$\rho_n^n = p_k \rho_n^{n\,(k)} = p_k < u_n \left| \psi^{(k)} \times \psi^{(k)} \right| u_n > = p_k \left| c_n^{(k)} \right|^2$$

Now, by Eq. I.1.8, $\left| c_n^{(k)} \right|^2$ is the probability of finding the eigenvalue $a_n^{(k)}$ when performing a measurement bearing on the pure state $|\psi^{(k)}>$. The diagonal element ρ_n^n, which is positive, is thus the statistical mean probability of finding the system in the state $|u_n>$, if it is in a mixed state described by the mean density operator ρ.

ρ_n^n *is the population of the state* $|u_n>$

(b) Nondiagonal elements

These take the form

$$\rho_n^p = p_k \rho_n^{p\,(k)} = p_k < u_n \left| \psi^{(k)} \times \psi^{(k)} \right| u_p > = p_k\, c_n^{(k)} c_p^{(k)\,*}$$

The cross term $c_n^{(k)} c_p^{(k)\,*}$ can vanish even if $c_n^{(k)}$ and $c_p^{(k)\,*}$ are individually nonzero. If ρ_n^p is zero, the statistical mean can be said to have abolished any quantum interference between $|u_n>$ and $|u_p>$. On the contrary, if ρ_n^p is nonzero, there exists a sort of bonding between the evolutions of the probabilities of finding each of the states $| u_n >$ and $| u_p >$. The two states are then said to exhibit *coherence*, and

ρ_n^p *represents the coherence of the states* $| u_n >$ and $| u_p >$

We shall now consider three examples—of increasing complexity—dealing with representations of quantum systems in terms of the density operator. They are intended to give us the necessary mastery of description of matter, albeit in the absence of interaction with a laser beam, and will serve as an approach to the complete treatment of laser-matter interaction.

I.4 Examples of description

I.4.1 Isolated two-level system

The system under consideration is shown in Figure I.4.1(a).

I.4.1.1 Pure case

Here the system is governed by a Hamiltonian operator H_0. For the sake of simplicity, we have restricted the number of levels to 2. The vectors $| a >$ and $| b >$, representing these levels, are eigenvectors of H_0. The associated eigenvalues, denoted by $\hbar\omega_a$ and $\hbar\omega_b$, are the energies of the two levels, respectively. We have

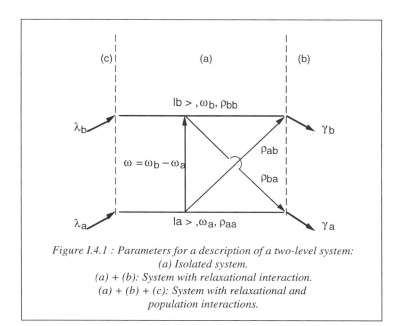

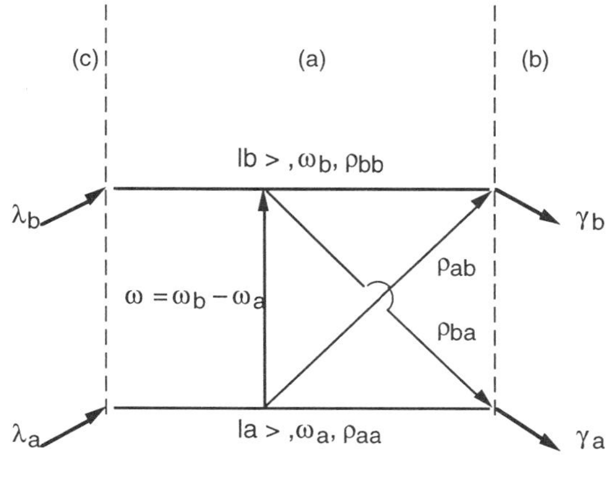

Figure I.4.1 : Parameters for a description of a two-level system:
(a) Isolated system.
(a) + (b): System with relaxational interaction.
(a) + (b) + (c): System with relaxational and population interactions.

$$H_0|a> = \hbar\,\omega_a|a>$$
$$H_0|b> = \hbar\,\omega_b|b> \qquad (I.4.1)$$

The evolution of the populations $\rho_{aa}^{(k)}$ (of level $|a>$) and $\rho_{bb}^{(k)}$ (of level $|b>$), as well as that of the coherences, $\rho_{ab}^{(k)}$ and $\rho_{ba}^{(k)}$, are obtained on solving the Liouville equation (Eq. I.3.9). The calculation has been carried out in Problem I.1, leading to the following results:

(a) The populations, quite obviously, remain constant, since the system is assumed to be isolated.

(b) For the coherences, we get

$$\rho_{ab}^{(k)} = \rho_{ba}^{(k)*} = \rho_{ab}^{(k)}(0)\exp[-i(\omega_a - \omega_b)t] \qquad (I.4.2)$$

The coherences oscillate at a frequency equal to the difference between the eigenfrequencies of the two levels.

I.4.1.2 Mixed case

We need only to replace the density operator $\rho^{(k)}$ by the mean density operator ρ (with deletion of the superscript k in the preceding expressions.) It should be kept in mind that the model of an isolated quantum system is a highly ideal case. There is no way of eliminating completely all coupling between the atoms (molecules) and the electromagnetic field in which they are unavoidably immersed. Nor is it easy to get rid of their coupling to the surrounding medium. At temperatures upward of zero the microsystems are coupled to the ambient thermal radiation, whereas at zero temperature their coupling to the electromagnetic field of vacuum would cause them to evolve toward their ground state of lowest energy. Also, coupling with the medium as well as collisions with surrounding molecules and the walls of the recipient vessels can never be neglected completely.

In the present work, for simplicity, we shall introduce *phenomenological constants of time relaxation* assumed as exponential, in order to take into account the very highly complex coupling with the surrounding medium as well as the spontaneous emission due to interaction with the electromagnetic field of vacuum. The latter cannot be treated strictly in a semiclassical approach in which *only matter is dealt within quantum mechanics*. The correct approach to spontaneous emission *presupposes quantization of the electromagnetic radiation* (see the references in the bibliography, Section I.6). This point, which is of essential importance, will also bear on our description of the laser in Part II of the present work.

I.4.2 Two-level system with relaxational interaction

For its visualization, see Figure I.4.1(a) + (b). The relaxation is introduced in the form of an operator Γ represented by a diagonal matrix in the basis $|a>, |b>$.

For conciseness, we write

$$\Gamma_{aa} = \langle a | \Gamma | a \rangle = \gamma_a \qquad \Gamma_{bb} = \langle b | \Gamma | b \rangle = \gamma_b \tag{I.4.3}$$

It is our aim that γ_a and γ_b shall represent the *populations* of the two levels. To achieve this we have to modify the Liouville equation (Eq.I.3.9). Moreover, we shall no longer distinguish the pure and mixed cases but shall proceed directly to a treatment of the more general mixed case applying the mean density operator. The Liouville equation now becomes

$$\dot{\rho} = (1/i\hbar)[H_0, \rho] - (1/2)[\Gamma\rho + \rho\Gamma] \tag{I.4.4}$$

We now proceed to calculate the populations and coherences.

I.4.2.1 Calculation of the populations

We can write

$$\dot{\rho}_{aa} = (1/i\hbar)[\langle a | H_0 \rho | a \rangle - \langle a | \rho H_0 | a \rangle] - (1/2)[\langle a | \Gamma\rho | a \rangle + \langle a | \rho \Gamma | a \rangle]$$

The first term vanishes. To calculate the second term we apply the completeness relation

$$\mathbb{1} = |a\rangle\langle a| + |b\rangle\langle b| \tag{I.4.5}$$

and insert the identity operator between the operators Γ and ρ. Since the operator Γ is diagonal, we have

$$\dot{\rho}_{aa} = -\gamma_a \rho_{aa} \quad \text{and} \quad \dot{\rho}_{bb} = -\gamma_b \rho_{bb} \tag{I.4.6}$$

As expected, the populations relax exponentially, with the constants γ_a and γ_b, toward other levels which we refrain from defining.

I.4.2.2 Calculation of the coherences

Equation I.4.4 leads to

$$\dot{\rho}_{ab} = (1/i\hbar)[\langle a | H_0 \rho | b \rangle - \langle a | \rho H_0 | b \rangle] - (1/2)[\langle a | \Gamma\rho | b \rangle - \langle a | \rho \Gamma | b \rangle]$$

The first term can also be written as $-i(\omega_a - \omega_b)\rho_{ab}$. With regard to the completeness relation, the second term is $-(1/2)(\gamma_a + \gamma_b)\rho_{ab}$.

Hence, the evolution of the coherences takes the form

$$\dot{\rho}_{ab} = \dot{\rho}^*_{ba} = -[i(\omega_a - \omega_b) + (\gamma_a + \gamma_b)/2]\rho_{ab} \tag{I.4.7}$$

A damping in amplitude of the coherences superimposes itself on the oscillations of their phases.

Note: Within the present model, damping of the coherences has the same origin as relaxation of the populations. This, for example, is the case when the dissipative process under consideration is due to *inelastic collisions* between the microscopic systems. However, processes exist, e.g., *elastic collisions* able to modify the coherences without affecting the populations (an atom conserves its energy in an elastic collision; thus, its state remains unchanged and so does the population of the latter). Such processes are not accessible to a correct description within the above model. They raise the relaxation probability of the coherence without modifying the relaxation of the populations. We shall take them into account by defining a coefficient of relaxation of coherence:

$$\gamma_{ab} \geq (\gamma_a + \gamma_b)/2 \tag{I.4.8}$$

Our model still has to be completed. In fact, in many a case, the medium at resonant interaction with laser radiation shall be dealt with starting from the model of a two-level system with the levels $|a>$ and $|b>$, with the transition $|a> \rightarrow |b>$ characterized by an energy $\hbar(\omega_b - \omega_a)$ close to the energy $\hbar\omega_L$ of the laser photons (two-level approximation). The levels $|a>$ and $|b>$ that relax toward the other levels of the system via coefficients γ_a and γ_b can also gain population from those other levels as the result of various processes (collisions, thermal electromagnetic radiation, external sources of radiation, etc.). It is then necessary to also take into account the population variation of $|a>$ and $|b>$ phenomenologically. This, in particular, is the case for the laser, to be described in Part II, whose amplifying medium is subject to *pumping* indispensable to obtain the amplification which is at the origin of laser oscillation.

In order to take into account the possibility of this kind of population variation, we shall give a last touch to the above specified model.

I.4.3 Two-level system with relaxational interaction and with population rates included: the population rate operator

The system now under consideration is shown in Figure I.4.1(a) + (b) + (c). The population rates are described introducing an operator Λ, with diagonal matrix picture within the basis $|a>$, $|b>$. To make things simpler, we put

$$\begin{aligned} \Lambda_{aa} &= \langle a|\Lambda|a\rangle = \lambda_a \\ \Lambda_{bb} &= \langle b|\Lambda|b\rangle = \lambda_b \end{aligned} \tag{I.4.9}$$

We next define, starting from the mean density operator, a population rate operator ρ (t) as follows:

$$\rho(t) = \sum_j \int_{-\infty}^{t} \lambda_j(t_0)\, \rho(j, t_0, t)\, dt_0 \tag{I.4.10}$$

In Eq. I.4.10, λ_j is the rate of population variation of state j (where j = a, b) at time t_0. Let $\lambda_j(t_0)\, dt_0$ denote the number of microsystems carried over into the state j during the time interval from t_0 to $(t_0 + dt_0)$, whereas $\rho(j,t_0,t)$ is the mean density operator, taken at time t, of the system carried over into the state j at time t_0. The integral bears on the infinite set of times separating the origin, assumed as minus infinity, from the moment of time t now under consideration. Thus, the population rate operator comprises the entire history of the two-level system. For reasons of simplicity, we have chosen the same notation (ρ) for the population rate operator and the density operator.

Let us calculate the time-derivative of the population operator:

$$\dot{\rho}(t) = \lim(dt \to 0)\{[\rho(t + dt) - \rho(t)] / dt\}$$

$$= \lim(dt \to 0)\left\{(1/dt)\left[\sum_j \int_{-\infty}^{t} \lambda_j(t_0)\, \rho(j, t_0, t + dt)\, dt_0 \right.\right.$$

$$\left.\left. + \sum_j \int_{t}^{t+dt} \lambda_j(t_0)\, \rho(j, t_0, t + dt)\, dt_0 - \sum_j \int_{-\infty}^{t} \lambda_j(t_0)\, \rho(j, t_0, t)\, dt_0 \right]\right\}$$

Due to the limits of integration, the second term of the sum reduces to a time such that $t_0 \approx t$.

For $t_0 \approx t$, the operator ρ (j,t,t) represents the density operator of a system in the state j at the time that it is carried over into that state. Thus

$$\lambda_a(t)\, \rho(a, t, t) + \lambda_b(t)\, \rho(b, t, t) = \Lambda$$

since in this case each diagonal element of the mean density operator is successively equal to unity, so that $\dot{\rho}(t)$ becomes

$$\dot{\rho}(t) = \Lambda + \lim(dt \to 0)\left\{(1/dt)\left[\sum_j \int_{-\infty}^{t} \lambda_j(t_0)\, \rho(j, t_0, t + dt)\, dt_0 \right.\right.$$

$$\left.\left. - \sum_j \int_{-\infty}^{t} \lambda_j(t_0)\, \rho(j, t_0, t)\, dt_0 \right]\right\}$$

Here emerges the time derivate of the mean density operator. Thus

$$\dot{\rho}(t) = \sum_j \int_{-\infty}^{t} \lambda_j(t_0)\, \dot{\rho}(j, t_0, t)\, dt_0 + \Lambda$$

And on insertion of the value of $\dot{\rho}(j, t_0, t)$ given by Eq. I.4.4, we arrive at

$$\dot{\rho}(t) = \underbrace{(1/i\hbar)[H_0, \rho(t)]}_{\text{isolated system quantum description}} \underbrace{\underbrace{-(1/2)(\Gamma\rho(t) + \rho(t)\Gamma)}_{\text{relaxation}} + \underbrace{\Lambda}_{\text{population rate}}}_{\text{phenomenological descriptions}} \qquad (I.4.11)$$

Thus we have shown that the two formalisms—that of the vector of state and that of the density operator—enable us to describe the evolution of a microscopic system. In order to take into account coupling between the system and its environment we have introduced *phenomenological parameters of relaxation* easily applicable within the framework of the density operator formalism.

Starting from the density operator and *on introducing phenomenologically the rate of population variation* of the levels, we have finally arrived at the *population rate operator*. This gives us the possibility of dealing comprehensively with *statistical ensembles of microsystems* under measurement, at a given time t. We now have available the necessary tools for the description of resonant laser-matter interaction in the further course of our work.

I.5 Problems and outlined solutions

I.5.1 Problem I.1: Application of the density matrix formalism to calculate the populations and coherences of an isolated two-level system

Consider an atom, assumed isolated and not coupled to the electromagnetic field of vacuum, in a state given by $|\psi\rangle = c_a(t)|a\rangle + c_b(t)|b\rangle$, where $|a\rangle$ and $|b\rangle$ are two eigenstates of the Hamiltonian H_0.

1. Derive the expression for the density matrix.
2. Derive the equations of evolution of its elements.
3. Determine the evolution in time of the populations and coherences.

1. The density matrix is given by Eq. I.3.6: $\rho(t) = |\psi(t)\rangle\langle\psi(t)|$.

2. Neglecting interaction between the atom and thermal radiation, and assuming the absence of spontaneous emission (no coupling to the electromagnetic field of vacuum),

the Liouville equation (see Eq. I.3.9) reads $\dot{\rho} = (-i/\hbar)[H_0, \rho]$. We obtain

$$\dot{\rho}_{aa} = (-i/\hbar) < a | [H_0, \rho] | a > = 0$$
$$\dot{\rho}_{bb} = (-i/\hbar) < b | [H_0, \rho] | b > = 0$$

The diagonal elements of the density are constant.

$$\dot{\rho}_{ab} = (-i/\hbar)[< a | H_0, \rho | b > - < a | \rho, H_0 | b >]$$

However, we have $< a | H_0 = \hbar \omega_a < a |$ and $H_0 | b > = \hbar \omega_b | b >$ (eigenvalue equations stating that $| a >$ and $| b >$ are eigenvectors of H_0, associated respectively with the energies $\hbar\omega_a$ and $\hbar\omega_b$). We get $\dot{\rho}_{ab} = (-i/\hbar)[\hbar \omega_a \rho_{ab} - \hbar \omega_b \rho_{ab}]$. Now putting $\omega_0 = \omega_b - \omega_a$, we get $\dot{\rho}_{ab} = i \omega_0 \rho_{ab}$ and, similarly $\dot{\rho}_{ba} = -i \omega_0 \rho_{ba}$.

Thus, the equations of evolution of the four elements of the density matrix are obtained in the form

$$\dot{\rho}_{aa} = \dot{\rho}_{bb} = 0, \quad \dot{\rho}_{ab} = i \omega_0 \rho_{ab}, \quad \dot{\rho}_{ba} = -i \omega_0 \rho_{ba}$$

3. The populations ρ_{aa} and ρ_{bb} remain constant (their first time derivatives vanish). This result accounts for the isolated system under consideration. On integration of the differential equations for the nondiagonal elements we obtain $\rho_{ab} = \rho_{ab}(0) \exp(i \omega_0 t)$ and $\rho_{ba} = \rho_{ba}(0) \exp(-i \omega_0 t)$. The coherences exhibit constant amplitude and a phase oscillating at the frequency $\omega_0 = \omega_b - \omega_a$, the difference between the eigenfrequencies of the two states $| a >$ and $| b >$.

I.5.2 Problem I.2: Application of the vector of state formalism to the description of a system at relaxational interaction

Consider, in vacuum, a system of N-like, noninteracting atoms. Let $| a >$ and $| b >$ be eigenstates of the atomic Hamiltonian H_0.

We consider a state represented as $| \psi(t) > = c_a(t) | a > + c_b(t) | b >$ and neglect spontaneous emission when tackling the first three questions.

1. Derive the equations of evolution for $c_a(t)$ and $c_b(t)$ and give the complete form of $| \psi(t) >$.

2. $N | c_a(t) |^2$ and $N | c_b(t) |^2$ are meant to denote the populations of the levels $| a >$ and $| b >$. Show that this is well-justified. Derive the variation in time of these populations.

3. Prove that the coherence $c_a(t) c_b^*(t)$ evolves with frequency $(\omega_b - \omega_a)$.

4. In order to provide for coupling between the levels a and b and the electromagnetic field of vacuum (spontaneous emission leading from $| a >$ and $| b >$ to other states), one introduces relaxation constants γ_a and γ_b. Calculate the evolution of the populations and coherences.

5. Does the present model take into account the possibility of spontaneous emission between $|a>$ and $|b>$?

1. We have recourse to the postulate P_6 (see Eq. I.1.10). We obtain

$$i\hbar\left[\dot{c}_a(t)|a> + \dot{c}_b(t)|b>\right] = H_0\left[c_a(t)|a> + c_b(t)|b>\right]$$

On left multiplication by bra $<a|$ and bra $<b|$ we obtain, successively, $\dot{c}_a(t) = -i\,\omega_a\,c_a(t)$ and $c_b(t) = -i\,\omega_b\,c_b(t)$ (similarly as in Problem I.1, $\hbar\omega_a$ and $\hbar\omega_b$ are the eigenvalues corresponding respectively to the eigenvectors $|a>$ and $|b>$ of the nonperturbed Hamiltonian H_0). On integration of the two preceding differential equations, $|\psi(t)>$ takes the form

$$|\psi(t)> = c_a(0)\exp(-i\,\omega_a\,t)|a> + c_b(0)\exp(-i\,\omega_b\,t)|b>$$

2. The postulate P_4 (see Eq. I.1.8) states that the probabilities $P(\hbar\omega_a)$ and $P(\hbar\omega_b)$ of obtaining—in a measurement—values $\hbar\omega_a$ and $\hbar\omega_b$ of the energy per atom are

$$P(\hbar\,\omega_a) = <\psi|a\times a|\psi> = |c_a(0)|^2 = |c_a(t)|^2$$
$$P(\hbar\,\omega_b) = <\psi|b\times b|\psi> = |c_b(0)|^2 = |c_b(t)|^2$$

Now, if measurement gives $\hbar\omega_a$ (or respectively $\hbar\omega_b$), one of the atoms is in state $|a>$ (respectively $|b>$). Thus, $N|c_a(t)|^2$ and $N|c_b(t)|^2$ are, respectively, the populations of states $|a>$ and $|b>$. The preceding calculations show that they are constant.

3. The coherence is expressed by $c_a(t)\,c_b^*(t) = c_a(0)\,c_b(0)\exp\left[i\,(\omega_b - \omega_a)\,t\right]$. Its phase oscillates with the frequency $(\omega_b - \omega_a)$.

4. We now introduce the relaxations writing the time-dependence of the coefficients $c_a(t)$ and $c_b(t)$ in the form

$$c_a(t) = c_a(0)\exp(-i\,\omega_a t - \gamma_a t/2) \quad \text{and} \quad c_b(t) = c_b(0)\exp(-i\,\omega_b t - \gamma_b t/2)$$

We obtain the above result on insertion of $-\gamma_i c_i(t)/2$ into the equations of evolution of c_a and c_b:

$$\dot{c}_a(t) = -(i\,\omega_a + \gamma_a/2)\;c_a(t);\; \dot{c}_b(t) = -(i\,\omega_b + \gamma_b/2)\,c_b(t)$$

As expected, the evolution of the population proceeds according to the following expressions:

$$|c_a(t)|^2 = c_a^2(0)\exp(-\gamma_a t) \quad \text{and} \quad |c_b(t)|^2 = c_b^2(0)\exp(-\gamma_b t)$$

(In density matrix formalism we again arrive at the corresponding expression I.4.6).

The coherence obeys the expression

$$c_a(t)\, c_b^*(t) = c_a(0)\, c_b(0) \exp\left\{ \left[i\,(\omega_b - \omega_a) - (\gamma_a + \gamma_b)/2 \right] t \right\}$$

(see Eq. I.4.7).

5. The above model fails to provide for spontaneous emission between the states $|\, a >$ and $|\, b >$.

I.5.3 Problem I.3: Phenomenon of relaxation in quantum physics

Consider a molecule in vacuum with Hamiltonien H_0. Let $|\, n >$ be the set of eigenstates of H_0, with energies E_n.

1. Have recourse to the definition of the stationary state. Prove that if a molecule is in a nonstationary state $|\, \psi(0) > = \sum_n c_n(0)\, |\, n >$, at the initial moment of time, it will conserve the same probability of being in the state $|\, n >$ as long as it is not subject to a perturbation. This description remains in agreement with experimental observation only with regard to the stable ground state. A molecule initially in an excited unstable state n, described as $|\, \psi_n^{(0)}(t) >$ in the absence of perturbations, is always subject to some perturbation that makes it return to its ground state. In order to take this relaxation process into account phenomenologically, one assumes that the probability of presence in the state n decreases with time exponentially.

We put $|\, \psi_n(t) > = \exp\left[-\gamma_n t/2 \right] |\, \psi_n^{(0)}(t) >$. Prove that this result can be reached on the assumption that the molecule is subjected to a non-Hermitian diagonal perturbation V. Determine the perturbation.

2. Define the lifetime of the excited level. Excited states with high (low) values of γ_n are referred to as unstable (metastable).

3. What is the physical origin of this relaxation?

4. The above description deliberately disregards one of the postulates of quantum mechanics. Which of them? And how?

1. A state is said to be stationary if its energy is time-independent. Let us calculate the population of state $|\, n >$, given by $|\, c_n(t)|^2$ (see Problem I.2). Since the energy of a conservative system is not a function of time, the Hamiltonian H_0 does not involve t. The postulate P_6 reads

$$i\hbar\, |\, \dot{\psi}(t) > \; = i\hbar \left(\sum_m \dot{c}_m(t)\, |\, m > \right) = H_0 \left(\sum_m c_m(t)\, |\, m > \right)$$

$$= \sum_m c_m(t)\, E_m\, |\, m >$$

We now perform left multiplication by $< n\, |$. We get $i\hbar\, \dot{c}_n(t) = c_n(t)\, E_n$. And after integration we obtain $c_n(t) = c_n(0) \exp\left(-i\, E_n t / \hbar\right)$. The population $|\, c_n(t)|^2$ is constant and equal to $c_n^2(0)$.

If we suppose that the system is submitted to a diagonal and non-Hermitian perturbation V, we write

$$\left(H_0+V\right)|\psi_n(t)> = \left(H_0+V\right)\sum_m c_m(t)|m>$$
$$=\sum_m c_m(t)E_m^{(0)}|m> + \sum_m c_m(t)V|m> = i\hbar\dot{c}_m(t)|m>$$

We now perform left multiplication by the bra of $<n|$:

$$i\hbar\dot{c}_n(t) = c_n(t)E_n^{(0)} + \sum_m c_m(t)<n|V|m>$$

Since V is diagonal, we get $i\hbar\dot{c}_n(t) = c_n(t)\left[E_n^{(0)} + V_{nn}\right]$

Integration gives $c_n(t) = c_n(0)\exp\left[-i\left(E_n^{(0)} + V_{nn}\right)t/\hbar\right]$.

Hence $|\psi_n(t)> = c_n(0)\exp\left[-i\left(E_n^{(0)} + V_{nn}\right)t/\hbar\right]|n>$

since at the moment of time t = 0, only the state | n > is populated [all the $c_m(0)$ with $m \neq n$ vanish].

We put $|\psi_n^{(0)}(t)> = c_n^{(0)}\exp\left(-i\ E_n^{(0)}t/\hbar\right)$ and obtain $|\psi_n(t)> = \exp\left(-i\ V_{nn}t/\hbar\right)|\psi_n^{(0)}(t)>$. Thus $V_{nn} = -i\gamma_n\hbar/2$, and we obtain that *V* is not Hermitian, since $\left(V_{nn} \neq V_{nn}^*\right)$.

2. The lifetime of the excited level | n > is given by γ_n^{-1}.

3. The above relaxation is due to spontaneous emission (coupling of the molecule to the electromagnetic field of vacuum).

4. The postulate P_2 provides for hermiticity of the operators of quantum physics. Now *V* is not Hermitian. However, the introduction of the relaxation described by the operator *V*—a purely phenomenological matter—is a step beyond the limits of quantum physics.

I.5.4 Problem I.4: Expression of the electric dipole moment within the density matrix formalism

The electric dipole moment of a system is defined as $\mathbf{p} = \sum_k q_k \mathbf{r}_k$, where q_k is the electric charge of a particle k situated in a point described by the vector $\mathbf{r}_k$. We shall consider the case of a one-electron system, presenting two levels | a > and | b > and possessing the moment $\mathbf{p} = -e\mathbf{r}$.

1. Derive the expression for the dipole moment operator in the basis | a >, | b >.

2. Applying the density operator, calculate the mean value of the dipole moment.

3. Determine the relation between the dipole moment and coherence.

1. The dipole moment operator is $\boldsymbol{p} = -e\boldsymbol{r}$. The coefficients of its matrix representation within the basis $|a>$, $|b>$ are

$$<a|\boldsymbol{p}|a> = -e<a|\boldsymbol{r}|a> \text{ and } <b|\boldsymbol{p}|b> = -e<b|\boldsymbol{r}|b>$$

Now $\boldsymbol{r}$ is an odd function of space and, for reasons of symmetry, the diagonal elements vanish if the states $|a>$ and $|b>$ possess a well-defined parity, which is the case if H_0 commutes with the parity operator π ($p_{aa} = p_{bb} = 0$). The nondiagonal elements take the form $p_{ab} = p^*_{ba} = <a|\boldsymbol{p}|b>$.

2. With regard to Eq. I.3.8 we have

$$<|\mathbf{p}|> = \mathrm{Tr}(\rho\,\boldsymbol{p}) = -e\,\mathrm{Tr}(\rho\,\boldsymbol{r}) = -e\,\mathrm{Tr}\begin{pmatrix} \rho_{aa} & \rho_{ab} \\ \rho_{ba} & \rho_{bb} \end{pmatrix}\begin{pmatrix} 0 & r_{ab} \\ r_{ba} & 0 \end{pmatrix}$$

$$= -e\,\mathrm{Tr}\begin{pmatrix} \rho_{ab}\,r_{ba} & \rho_{aa}\,r_{ab} \\ \rho_{bb}\,r_{ba} & \rho_{ba}\,r_{ab} \end{pmatrix} = -e\,(\rho_{ab}\,r_{ba} + c\,c)$$

3. ρ_{ab} represents the coherence existing between the states $|a>$ and $|b>$. Thus, proportionality holds between the mean value of the dipole moment on the one hand and the real part of the coherence on the other if r_{ab} and, consequently, ρ_{ab} are real (the last condition can always be fulfilled on choosing the phase of $|a>$ or $|b>$ judiciously).

I.5.5 Problem I.5: Evolution of a quantum system submitted to a constant nondiagonal perturbation

We consider an atom, represented by an isolated two-level $(|a>, |b>)$ system noncoupled to the electromagnetic field of vacuum and acted on by a constant perturbation, expressible by a nondiagonal operator V. Let us assume $H = H_0 + V$.

1. Applying completeness relation, prove that $<a|V\rho|a> = V_{aa}\,\rho_{aa} + V_{ab}\,\rho_{ab}$.

2. Prove that the time derivatives of the population terms $\dot{\rho}_{aa}$ and $\dot{\rho}_{bb}$ are simply functions of the expression $V_{ab}\,\rho_{ab}$.

3. Prove that the time derivative of the coherence term $\dot{\rho}_{at}$ is simply a function of the difference $\rho_{aa} - \rho_{bb}$.

1. The completeness relation (see Eq. I.4.5) gives

$$<a|V\rho|a> = <a|V\,\big[\,|a\rangle\langle a| + |b\rangle\langle b|\,\big]\rho|a>$$

$$= \rho_{aa}<a|V|a> + \rho_{ba}<a|V|b> = V_{aa}\,\rho_{aa} + V_{ab}\,\rho_{ba}$$

2. Having recourse to the Liouville equation (see Eq. I.3.9) $\dot{\rho} = (-i/\hbar)[H, \rho]$, we write

$$\dot{\rho}_{aa} = (-i/\hbar)\left[< a \mid (H_0 + V)\rho \mid a > - < a \mid \rho (H_0 + V) \mid a >\right]$$
$$= (-i/\hbar)\left[< a \mid V\rho \mid a > - < a \mid \rho V \mid a >\right].$$

Again making use of the completeness relations to fractionate the products $V\rho$ and ρV, we obtain

$$\begin{cases} \dot{\rho}_{aa} = (-i/\hbar)\left(V_{ab}\,\rho_{ba} - V_{ba}\,\rho_{ab}\right) \\ \dot{\rho}_{bb} = (-i/\hbar)\left(V_{ba}\,\rho_{ab} - V_{ab}\,\rho_{ba}\right) \end{cases}$$

3. For the coherences, we similarly get

$$\begin{cases} \dot{\rho}_{ba} = -i\omega_0\,\rho_{ba} - (i/\hbar)\left(\rho_{aa} - \rho_{bb}\right) V_{ba} \\ \dot{\rho}_{ab} = i\omega_0\,\rho_{ab} + (i/\hbar)\left(\rho_{aa} - \rho_{bb}\right) V_{ab} \end{cases}$$

These equations will prove highly important. They will serve as an introduction to our description of laser-matter interaction in Part III of the present book.

I.6 Bibliography

COHEN-TANNOUDJI, C., DIU, B., AND LALOË, F. *Quantum Mechanics*, John Wiley & Sons, New York, 1976.

MESSIAH, A. *Quantum Mechanics*, North Holland Publishing Company, John Wiley & Sons, New York, 1965.

CHAPTER II

Description of the Optical Properties of Media: Tensors and their Applications

In Chapter I we have shown that quantum physics provides a satisfactory description of microscopic entities, such as atoms and molecules, and that the mean value < b >—we have in mind *the quantum mean value*—of measurements performed on an observable B is given by Eq. I.3.8: < b > = Tr (ρB.) However, Chapter I leaves us in the dark as to *the physical nature* of the observable involved; i.e., in our case, the microscopic optical property which governs the behavior of matter illuminated by a laser wave. Within the classical treatment of electromagnetic waves to be applied throughout this work (see Section IV.3), a laser wave conveys two fields—an electric field **E** and a magnetic field **H**—both of which are *vectorial* in nature. In Chapter VII these fields will be shown to induce optical properties which, too, are *vectorial.* Thus the quantities connecting the cause and the effect are, mathematically, of the form of equations between vectors. In three-dimensional space, which we shall be using in our description, the two vectors have three components each. Hence, their relationship involves at least nine components, jointly constituting *a quantity endowed with physical dimensions* (in a conveniently chosen system of units like cgs or SI) to which we shall be referring as a *tensor*, denoted by square brackets [ ]. In Chapter I we have dealt with *matrices.* They closely resemble tensors but differ from them fundamentally in that a matrix is an *operator* (a "transformer") with components that are *dimensionless* whereas tensors are, in general, *quantities endowed with a physical dimension.*

We shall thus have to become familiar with tensors. We shall refrain from adducing mathematically rigorous definitions, for which we refer the reader to the bibliography at the end of this chapter. What we shall need most is a *criterion of tensoriality* enabling us to distinguish *tensors* from *pseudotensors*, a distinction of great importance in magneto-optics (for example, when dealing with the Faraday effect and the inverse Faraday effect).

Moreover, we shall draw attention to the convenience of *tensor notation. Tensor algebra* will be reviewed briefly, and two specific tensors—we have in mind the *Kronecker* and the *Levi-Civita tensor*—will be shown to facilitate our further calculations. Tensor formalism will be proved to be a potent tool when applied on electric multipoles and the potentials and fields they give rise to within the bulk of material systems. These permanent internal fields play a very considerable role in optics; for example, they are responsible for first-harmonic scattering by a disordered assemblage of centrosymmetric microsystems (see Chapter VIII). Finally, in the Appendices at the end of this chapter we tabulate the independent and nonzero components of the tensors representing the electric multipoles and optical susceptibilities of material entities belonging to various classes of symmetry.

II.1 A general review

II.1.1 Two-coordinate systems of general use

II.1.1.1 Cartesian coordinates

The elementary displacement from a point P to a point P' is given by

$$\mathbf{dL} = dx\, \mathbf{e}_x + dy\, \mathbf{e}_y + dz\, \mathbf{e}_z$$

Obviously, the element of volume is

$$dv = dx\, dy\, dz$$

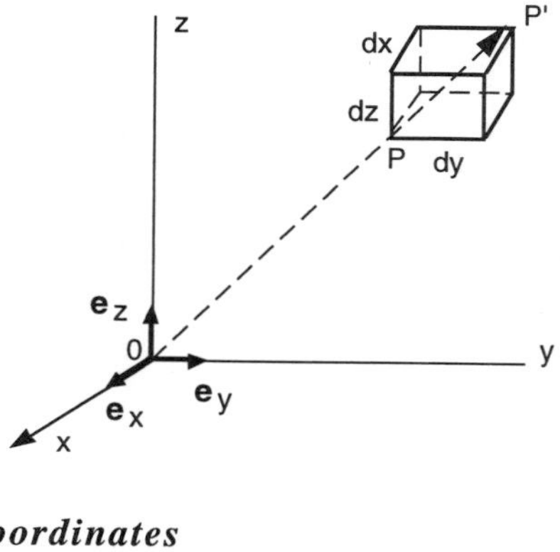

II.1.1.2 Spherical coordinates

(a) Element of displacement

The elementary displacement takes the form

$$\mathbf{dL} = r\, d\theta\, \mathbf{e}_\theta + r \sin\theta\, d\varphi\, \mathbf{e}_\varphi + dr\, \mathbf{e}_r$$

The element of volume is

$$dv = r^2 \sin\theta\, d\theta\, d\varphi\, dr$$

The elementary solid angle is given by

$$d\Omega = \left(ds / r^2\right) = r^2 \sin\theta \, d\varphi \, d\theta / r^2$$
$$= \sin\theta \, d\theta \, d\varphi$$

The integral over all space is

$$\int_0^{2\pi} d\varphi \int_0^{\pi} \sin\theta \, d\theta = 4\pi$$

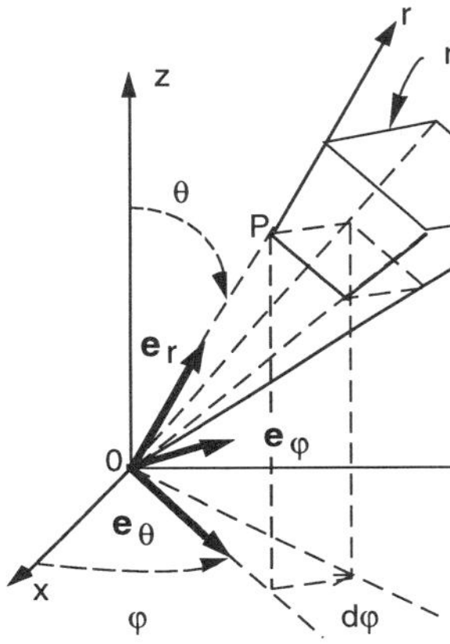

(b) The fundamental formula of spherical trigonometry reads

$$\cos\theta_{12} = \cos\theta_1 \cos\theta_2 + \sin\theta_1 \sin\theta_2 \cos(\varphi_2 - \varphi_1) \qquad \text{(II.1.1)}$$

We writte successively

$$\mathbf{r}_1 = \mathbf{r} \cos\theta_1 + \mathbf{d}_1 \sin\theta_1$$

$$\mathbf{r}_2 = \mathbf{r} \cos\theta_2 + \mathbf{d}_2 \sin\theta_2$$

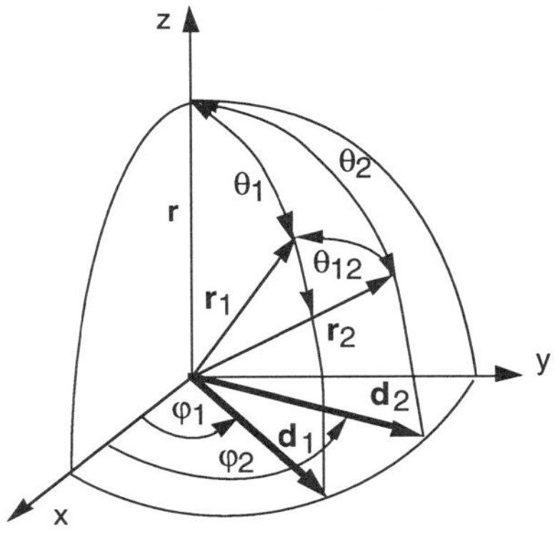

The vectors $\mathbf{r}$, $\mathbf{r}_i$, and $\mathbf{d}_i$ are normalized to unity. We now calculate the scalar products:

$$\mathbf{r}_1 \mathbf{r}_2 = \cos\theta_{12} = \cos\theta_1 \cos\theta_2 + \sin\theta_1 \sin\theta_2 \, \mathbf{d}_1 \mathbf{d}_2$$

Now, $\mathbf{d}_1 \mathbf{d}_2 = \cos(\varphi_2 - \varphi_1)$ and we arrive at Eq. II.1.1.

II.1.2 Linear transformations

We have introduced two bases in three-dimensional space ($\mathbf{e}_x$, $\mathbf{e}_y$, $\mathbf{e}_z$ and $\mathbf{e}_\theta$, $\mathbf{e}_r$, $\mathbf{e}_\varphi$). Let us write, more generally, for two bases in the same three-dimensional space, ($\mathbf{e}_\alpha$, $\mathbf{e}_\beta$, $\mathbf{e}_\gamma$) and ($\mathbf{E}_i$, $\mathbf{E}_j$, $\mathbf{E}_k$). Within these bases, a vector $\mathbf{V}$ takes the form

$$\mathbf{V} = v_\alpha \mathbf{e}_\alpha = V_i \mathbf{E}_i \qquad \text{(II.1.2)}$$

where we have applied the convention of summation over recurrent (dummy) indices. We shall be applying this convention everywhere, except for some particular cases which we shall distinguish carefully.

Now the vector $\mathbf{E}_i$ decomposes in the basis e_α as follows

$$\mathbf{E}_i = c_{i\alpha} \mathbf{e}_\alpha \qquad \text{(II.1.3)}$$

On insertion of Eq. II.1.3 into Eq. II.1.2 we obtain

$$v_\alpha \mathbf{e}_\alpha = c_{i\alpha} V_i \mathbf{e}_\alpha$$

Thus, the components correspond to one another as follows:

$$v_\alpha = c_{i\alpha} V_i$$

The coefficients $c_{i\alpha}$ are referred to as the *directional cosines* of the transformation $(\alpha, \beta, \gamma) \to (i, j, k)$. They are seen to form a square matrix with $3 \times 3 = 9$ components.

The reader will readily check the following relationships between the directional cosines $c_{i\alpha}$ and the directional cosines $c'_{i\alpha}$ of the inverse transformation: $V_i = c'_{\alpha i} v_\alpha$ (see Inset II.1).

$$\begin{aligned}
&— \; c_{i\alpha} = \text{minor}\left(c'_{\alpha i}\right) / \left| c'_{\alpha i} \right| \\
&— \; c'_{\alpha i} = \text{minor}\left(c_{i\alpha}\right) / \left| c_{i\alpha} \right| \\
&— \; \left| c_{i\alpha} \right| \left| c'_{\alpha i} \right| = 1 \\
&— \; c_{i\alpha} \, c'_{\alpha j} = \delta_{ij}
\end{aligned} \qquad \text{(II.1.4)}$$

These relations hold quite generally. In the simpler case of orthonormal bases (to which we shall have frequent recourse) they become

— $|c_{i\alpha}| = |c'_{\alpha i}| = \pm 1$ $\begin{pmatrix} +\text{: dextrogyric transformation} \\ -\text{: levogyric transformation} \end{pmatrix}$

— $c_{i\alpha} = c'_{\alpha i}$ (II.1.5)

— $c_{i\alpha}\ c_{j\alpha} = \delta_{ij}$

Orthogonal transformations are more conveniently described in terms of *Euler angles* (see Inset II.1).

II.2 Tensors

II.2.1 How to identify a tensor; the criterion of tensoriality

Consider a vectorial space E_3, constructed on the set R of real numbers, and related to two orthogonal bases $(e_\alpha, e_\beta, e_\gamma)$ and (E_i, E_j, E_k) in E_3. Assume, for example, three vectors **U**, **V**, **W**, belonging to E_3 and expressed in the basis $(e_\alpha, e_\beta, e_\gamma)$. Now consider a trilinear functional f(**U**, **V**, **W**), arisen by applying E_3 to R. This functional, which reads

$$f(\mathbf{U}, \mathbf{V}, \mathbf{W}) = t_{\alpha\beta\gamma}\, U_\alpha\, V_\beta\, W_\gamma \qquad \text{(II.2.1)}$$

contains the components $(U_\alpha, V_\beta, W_\gamma)$ of the three vectors within the basis $(e_\alpha, e_\beta, e_\gamma)$.

We now go over from the basis $(e_\alpha, e_\beta, e_\gamma)$ to the basis (E_i, E_j, E_k). The functional f is a scalar, unaffected by the transformation $(e_\alpha, e_\beta, e_\gamma) \to (E_i, E_j, E_k)$. Thus, we are justified in writing

$$f(\mathbf{U}, \mathbf{V}, \mathbf{W}) = T_{ijk}\, U_i\, V_j\, W_k$$

where U_i, V_j, and W_k are, respectively, the components **U**, **V,** and **W** in the basis (E_i, E_j, E_k). By Eq. II.1.3, we get

$$U_\alpha = c_{i\alpha}\, U_i$$

$$V_\beta = c_{j\beta}\, V_j \qquad \text{(II.2.2)}$$

$$W_\gamma = c_{k\gamma}\, W_k$$

On insertion of Eq.II.2.2 into Eq. II.2.1 we have

$$f(\mathbf{U}, \mathbf{V}, \mathbf{W}) = t_{\alpha\beta\gamma} c_{i\alpha} c_{j\beta} c_{k\gamma} U_i V_j W_k$$

whence

$$T_{ijk} = c_{i\alpha} c_{j\beta} c_{k\gamma} t_{\alpha\beta\gamma} \qquad \text{(II.2.3)}$$

The functional f(**U**,**V**,**W**) defines a *tensor of rank 3*—$t_{\alpha\beta\gamma}$ in the basis $(e_\alpha, e_\beta, e_\gamma)$ or T_{ijk} in $(E_i, E_j, E_k.)$ The transformation of the tensor under consideration involves the product of *three* directional cosines. More generally, a tensor rank n transforms in accordance with an expression of the type II.2.3, albeit

Inset II.1 ***Euler angles and the transformation matrices associated with them***

The three Euler angles are plotted to the right. They express the transformations between the systems of coordinates (1, 2, 3) and (x, y, z). The transformation matrices are given below.

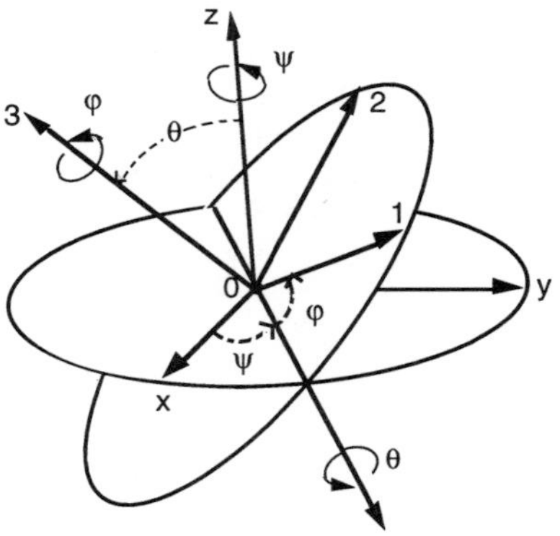

$$\begin{Bmatrix} 1 \\ 2 \\ 3 \end{Bmatrix} = \begin{pmatrix} \cos\varphi\cos\psi - \sin\varphi\sin\psi\cos\theta & \cos\varphi\sin\psi + \sin\varphi\cos\psi\cos\theta & \sin\varphi\sin\theta \\ -\sin\varphi\cos\psi - \cos\varphi\sin\psi\cos\theta & -\sin\varphi\sin\psi + \cos\varphi\cos\psi\cos\theta & \cos\psi\sin\theta \\ \sin\varphi\sin\theta & \cos\psi\sin\theta & \cos\theta \end{pmatrix} \begin{Bmatrix} x \\ y \\ z \end{Bmatrix}$$

(II.1.6)

$$\begin{Bmatrix} x \\ y \\ z \end{Bmatrix} = \begin{pmatrix} \cos\varphi\cos\psi - \sin\varphi\sin\psi\cos\theta & -\sin\varphi\cos\psi - \cos\varphi\sin\psi\cos\theta & \sin\psi\sin\theta \\ \cos\varphi\sin\psi + \sin\varphi\cos\psi\cos\theta & -\sin\varphi\sin\psi + \cos\varphi\cos\psi\cos\theta & -\cos\psi\sin\theta \\ \sin\varphi\sin\theta & \cos\psi\sin\theta & \cos\theta \end{pmatrix} \begin{Bmatrix} 1 \\ 2 \\ 3 \end{Bmatrix}$$

The formula for the element of rotation reads

$$d\Omega = \sin\theta \, d\theta \, d\varphi \, d\psi \qquad \text{(II.1.7)}$$

On integration it leads to: $\Omega = \int_0^{2\pi} d\varphi \int_0^{2\pi} d\psi \int_0^{\pi} \sin\theta \, d\theta = 8\,\pi^2$

tensor of rank n transforms in accordance with an expression of the type II.2.3, albeit involving products of n directional cosines describing the change of coordinates. The result derived above provides us with a *criterion of tensoriality* enabling us to recognize a tensor and moreover to determine its rank immediately. We shall denote tensors by square brackets [ ]; however, more often, we shall identify a tensor by its indices, the number of which betrays its rank. For details and some references to physics, consult the following table.

Note 1: A scalar, independently of the system of reference, should be considered as being a tensor of rank zero.

Note 2: Equation II.1.2 proves that a vector is a tensor of rank 1.

The following two notations are equivalent:

tensorial notation: $[T_i] = c_{i\alpha}[t_\alpha]$

$$\text{vectorial notation: } \begin{Bmatrix} T_i \\ T_j \\ T_k \end{Bmatrix} = \begin{pmatrix} 3 \times 3 \text{ matrix M of} \\ \text{the directional} \\ \text{cosines} \\ \text{of change in basis} \end{pmatrix} \begin{Bmatrix} t_\alpha \\ t_\beta \\ t_\gamma \end{Bmatrix} \qquad \text{(II.2.4)}$$

Note 3: Similarly, for rank 2, we have the following two tensorial notations, which are mutually equivalent:

TABLE II.2.1 — **Some optical (and derived) properties and their tensorial characteristics.**

Optical and derived physical properties	Rank of the tensor	Its full description	Its notation	The number of its components in three-dimensional space
Refractive index of an isotropic medium	0	Tensor of rank 0 or *scalar*	[T]	$3^0 = 1$
Electric field; electric polarization ...	1	Tensor of rank 1 or *vector*	$[T_i]$	$3^1 = 3$
Optical susceptibility of *order* 1; electric permittivity ...	2	Tensor of rank 2	$[T_{i_1 i_2}]$	$3^2 = 9$
Optical susceptibility of *order* 2; piezoelectricity ...	3	Tensor of rank 3	$[T_{i_1 i_2 i_3}]$	$3^3 = 27$
Optical susceptibility of *order* 3; électrostriction ...	4 . . .	Tensor of rank 4 . . .	$[T_{i_1 i_2 i_3 i_4}]$ . . .	$3^4 = 81$ . . .
	n	Tensor of rank n	$[T_{i_1 \ldots i_n}]$	3^n

$$[T_{ij}] = c_{i\alpha}\, c_{j\beta}\, [t_{\alpha\beta}]$$
$$[T_{ij}] = (M)[t_{\alpha\beta}](M^{-1}) \qquad \text{(II.2.5)}$$

For ranks upward of 2, it is no longer feasible to use a matrix (M.) One has to use Eq. II.2.3.

II.2.2 Pseudotensors

Consider the particular case of a transformation involving inversion in space, when a vector of the initial basis is replaced by its inverse (symmetry with respect to the origin of coordinates). The matrix (M) operating this transformation is simply

$$\begin{pmatrix} -1 & 0 & 0 \\ 0 & -1 & 0 \\ 0 & 0 & -1 \end{pmatrix}$$

It makes a vector $\mathbf{V}$ go over into its spatial inverse: $\mathbf{V}' = -\mathbf{V}$. The transformation proceeds as follows:

$$\left\{ \mathbf{V}' \right\} = \begin{pmatrix} -1 & 0 & 0 \\ 0 & -1 & 0 \\ 0 & 0 & -1 \end{pmatrix} \left\{ \mathbf{V} \right\} \qquad \text{(II.2.6)}$$

Here, all the components of V undergo a change in sign.

We now shall concentrate on a particular kind of "vector" V, defined as the result of the vectorial product operation: $\mathbf{V} = \mathbf{V}_1 \times \mathbf{V}_2$. If $\mathbf{V}_1$ and $\mathbf{V}_2$ have respectively the components (X_1; Y_1; Z_1) and (X_2; Y_2; Z_2), this operation leads to $\mathbf{V}$ with components

$$(Y_1 Z_2 - Y_2 Z_1\ ;\ Z_1 X_2 - Z_2 X_1\ ;\ X_1 Y_2 - X_2 Y_1)$$

One is readily convinced that the "vector" thus defined is no true vector because its components remain unchanged under space inversion. Its transformation is governed by the following equation:

$$\left\{ \mathbf{V}' \right\} = -1 \begin{pmatrix} -1 & 0 & 0 \\ 0 & -1 & 0 \\ 0 & 0 & -1 \end{pmatrix} \left\{ \mathbf{V} \right\}$$

Now (– 1) is equal to the determinant of the matrix (M) of directional cosines. This determinant is referred to as a Jacobian and is denoted by J. Thus, a vectorial product transforms according to

$$\{\mathbf{V}'\} = J\,(M)(M)\{\mathbf{V}\} \qquad \text{(II.2.7)}$$

Vectorial physical quantities transforming in accordance with an expression of the type II.2.7 involving the Jacobian of the transformation matrice (M) are called *pseudovectors*. Pseudotensors are defined by analogy. The reader may have noted that a true vector of rank N behaves like $(-1)^N$ under space inversion, whereas the respective pseudotensor behaves like $(-1)^{N+1}$. An example of a pseudovector is provided by the magnetic field **H**. It can be obtained from the vectorial potential **A** according to the formula

$$\mathbf{H} = \mathbf{curl\ A} = \nabla \times \mathbf{A} \qquad \text{(II.2.8)}$$

where ∇ is the vector *nabla*, with the components ($\partial/\partial x$; $\partial/\partial y$; $\partial/\partial z$) in cartesian coordinates. **H** is a pseudovector—a fact of great importance, e.g. in magneto-optics (the Faraday and inverse Faraday effects, which we shall consider later on).

II.2.3 Elementary tensorial algebra

II.2.3.1 Addition of tensors

Tensors can be added only if they are of the same rank and are defined in the same point of the same system of coordinates. Thus, for two tensors of rank 2 we have

$$T_{ij} = T'_{ij} + T''_{ij} \qquad \text{(II.2.9)}$$

Much more important in practice is the fact that *a tensor* T_{ij} *of rank* 2 *can be expressed univocally as the sum of a symmetric tensor* $T^{(s)}_{ij}$ *and an antisymmetric tensor* $T^{(a)}_{ij}$. (A tensor of rank 2 is said to be symmetric if its components on either side of the principal diagonal are correspondingly equal, but is antisymmetric if they are inverse.)

Obviously we always have

$$T_{ij} = T^{(s)}_{ij} + T^{(a)}_{ij}$$

with

$$T^{(s)}_{ij} = [T_{ij} + T_{ji}]/2 \qquad \text{(II.2.10)}$$

(the components of $T^{(s)}_{ij}$ remain unchanged on permutation of i and j), and

$$T^{(a)}_{ij} = [T_{ij} - T_{ji}]/2 \qquad \text{(II.2.11)}$$

(those of $T^{(a)}_{ij}$ undergo a change in sign on permutation of i and j).

There is no overstatement in saying that decomposition of tensors into symmetric and antisymmetric parts is of the highest importance. We shall have to deal with this procedure throughout the present book. For example, in Chapter IV we shall show that the electric field **E** of a laser wave, when applied to a molecule, gives rise to a polarization **p** of electronic origin; in a first approximation and provided that **E** is not excessively intense (the linear case), we can write

$$\mathbf{p} \approx [\alpha]\,\mathbf{E} \tag{II.2.12}$$

where $[\alpha]$ is the polarization of order 1 of the molecule. $[\alpha]$ is a tensor of rank 2, which can be written

$$\alpha_{ij} = \alpha_{ij}^{(s)} + \alpha_{ij}^{(a)}$$

It will be shown that the antisymmetric part $\alpha_{ij}^{(a)}$ can differ from zero for *chiral* molecules, thus giving rise not only to *rotatory optical activity* but moreover to *light scattering of antisymmetric nature.*

II.2.3.2 Contraction of tensors

Contraction of a tensor resides in carrying out summation over a pair of its indices—two of its free indices are made equal and thus become dummy indices (summation indices). As a result of this operation the initial rank n of the tensor is lowered to n – 2. If n is sufficiently high, contraction can be repeated according to the following scheme:

$$\left[T^{(n)}\right] \xrightarrow[\text{contractions}]{p} \left[T^{(n-2p)}\right]$$

To illustrate this immensely important technique, consider a tensor of rank n = 3, thus T_{ijk}.

On using Eq. II.2.3 to express T_{ijk} after *one* contraction (we have put k = j and carried out summation over j) we have $T_{ijj} = c_{i\alpha}\, c_{j\beta}\, c_{j\gamma}\, T_{\alpha\beta\gamma}$. Equation II.1.5 shows that $c_{j\beta}\, c_{j\gamma} = \delta_{\beta\gamma}$., whence $T_{ijj} = c_{i\alpha}\, \delta_{\beta\gamma}\, T_{\alpha\beta\gamma}$. On performing summation in $\delta_{\beta\gamma}\, T_{\alpha\beta\gamma}$, the only nonzero components are found to be those with $\beta = \gamma$, so that we get $T_{ijj} = c_{i\alpha}\, T_{\alpha\beta\beta}$.

This transformation formula is that of a tensor of rank 1. In this way, starting from a tensor of rank 3, we have lowered its rank by 2 and, what is more, we have shown that *contraction is equivalent to multiplication by the Kronecker symbol* performing summation over dummy indices. Contraction can be repeated many times, leading to following results:

$$T_{ij}\,\delta_{ij} = T_{ii} \qquad \text{(a scalar)}$$

$$T_{ijk}\,\delta_{jk} = T_{ijj} \qquad \text{(a vector)}$$

$$T_{ijkl}\,\delta_{ij}\,\delta_{kl} = T_{iijj} \qquad \text{(a scalar)}$$

$$T_i\,T_{jk}\,\delta_{jk} = T_i\,T_{jj} \qquad \text{(a vector)}$$

The Kronecker delta, as it were, exchanges its indices in the (tensorial) entity to which it is applied. Below, we give an example of this behavior.

In the case of an isotropic medium we can define a mean optical refractive index n by $n^2 = \varepsilon^{(r)}$. With the mean values of the tensors of rank 2 occurring in Eq. II.2.15 we obtain

$$n^2 - 1 = \chi / \varepsilon_0 \tag{II.2.16}$$

Inset II.2 ***The relation between the electric displacement vector* (D) *and the electric field vector* (E)**

According to the fundamental formulae of Maxwell (see Section IV.5.5 in the bibliography of Chapter IV) the displacement vector **D** is defined as

$$\mathbf{D} = \varepsilon_0 \left[\varepsilon_{ij}^{(r)}\right] \mathbf{E} = \varepsilon_0 [\mathbb{1}] \mathbf{E} + \mathbf{P}$$

where $\left[\varepsilon_{ij}^{(r)}\right]$ is the tensor of rank 2 of relative electric permittivity. ε_0 is the permittivity of vacuum, and $[\mathbb{1}]$ stands for the unit tensor of rank 2.

The preceding formula can be rewritten as

$$D_i = \varepsilon_0 \, \varepsilon_{ij}^{(r)} \, E_j = \varepsilon_0 \, E_i + P_i \qquad \text{(II.2.13)}$$

However, $E_i = E_j \, \delta_{ij}$, whence we get

$$\varepsilon_{ij}^{(r)} - \delta_{ij} = P_i / \varepsilon_0 \, E_j \qquad \text{(II.2.14)}$$

and, moreover,

$$\mathbf{P} = [\chi_{ij}] \, \mathbf{E} \qquad \text{or} \qquad P_i = \chi_{ij} \, E_j$$

enabling us to write

$$\varepsilon_{ij}^{(r)} - \delta_{ij} = \chi_{ij} / \varepsilon_0 \qquad \text{(II.2.15)}$$

To define the mean values, we have to introduce the notion of *trace*.

II.2.3.3 Trace of a tensor

The trace (denoted by Tr) is defined as resulting from contraction with δ_{ij} (carried out once or repeatedly). For tensors of ranks 2 and 4, the trace Tr is

$$\begin{aligned} \mathrm{Tr}[T_{ij}] &= T_{ij} \, \delta_{ij} = T_{ii} \\ \mathrm{Tr}[T_{ijkl}] &= T_{ijkl} \, \delta_{ij} \, \delta_{kl} = T_{iijj} \end{aligned} \qquad \text{(II.2.17)}$$

The trace is a scalar quantity. For rank 2, T_{ii} is *the sum of the diagonal components of the tensor*. In this case the mean value of the tensor is defined as follows:

$$\text{The mean value of } [T_{ij}] = T_{ii} / 3 = \mathrm{Tr}[T_{ij}] / 3 \qquad \text{(II.2.18)}$$

that is, to one-third of the sum of its diagonal components. Above, we have also used the unit tensor of rank 2, denoted by $[\mathbb{1}]$; its properties will be considered shortly, along with those of a particular tensor of rank 3, very often used in tensor calculus.

The concept of the trace of a tensor enables us to prove that *a symmetric tensor* $T_{ij}^{(s)}$ *of rank 2 can be decomposed univocally into the sum of a symmetric tensor*

Inset II.3 ***The Kronecker tensor [δ_{ij}] and the Levi-Civita tensor [δ_{ijk}]***

The Kronecker tensor is a unit tensor of rank 2. Above, we have denoted it by [𝟙], but shall now be denoting it by [δ_{ij}]. Each of its three diagonal components equals 1, whereas its nondiagonal components are zero irrespective of the coordinates in which the tensor is expressed. Is it a true tensor or a pseudotensor? The reader will easily verify that it transforms as $\delta_{ij} = c_{i\alpha}\, c_{j\beta}\, \delta_{\alpha\beta}$, i.e., as $(-1)^N$, with N = 2. It is even. *It is thus a true symmetric tensor.*

The Levi-Civita tensor is a tensor of rank 3, denoted by [ε_{ijk}]. It, too, is defined independently of the system of coordinates, and solely by the following values of its components:

$$\varepsilon_{ijk} \begin{cases} = \pm 1 & \text{if } i \neq j \neq k \text{ and if, starting from the initial ordering i, j, k, an even number of permutations is carried out (then + 1) or if the number of permutations is odd (then } -1). \\ = 0 & \text{if at least two of the indices are equal.} \end{cases}$$

How does it transform, say, under space inversion, leading from the system of reference (e_α, e_β, e_γ) to (E_i, E_j, E_k)? Let us calculate the component ε_{123} in (E_i, E_j, E_k.) We obtain $c_{11}\, c_{22}\, c_{33} = -1$. Thus, ε_{123}, by definition, has to equal +1 in any system of reference. Consequently, using the Jacobian (–1) of the transformation, we have to write

$$\varepsilon_{ijk} = (-1)\, c_{i\alpha}\, c_{j\beta}\, c_{k\gamma}\, \varepsilon_{\alpha\beta\gamma} \qquad \text{(II.2.19)}$$

The Levi-Civita tensor transforms under space invertion as $(-1)^{N+1}$, with N = 3. It is odd. *Thus, it is an antisymmetric pseudotensor.*

The vectorial product of two Levi-Civita tensors (obtained by multiplying, one by one, each component of the one by each component of the other) is a true tensor of rank 6:

$$\varepsilon_{ijk} \cdot \varepsilon_{\alpha\beta\gamma} = \begin{vmatrix} \delta_{i\alpha} & \delta_{j\alpha} & \delta_{k\alpha} \\ \delta_{i\beta} & \delta_{j\beta} & \delta_{k\beta} \\ \delta_{i\gamma} & \delta_{j\gamma} & \delta_{k\gamma} \end{vmatrix} \qquad \text{(II.2.20)}$$

Contractions carried out on Eq. II.2.20 lead to the following simple, widely used results:

A first contraction: $\left(\varepsilon_{ijk} \cdot \varepsilon_{\alpha\beta\gamma}\right) \delta_{i\alpha} = \delta_{j\beta}\, \delta_{k\gamma} - \delta_{j\gamma}\, \delta_{k\beta}$

A second contraction: $\left(\varepsilon_{ijk} \cdot \varepsilon_{\alpha\beta\gamma}\right) \delta_{i\alpha}\, \delta_{j\beta} = 2\delta_{k\gamma}$ (II.2.21)

A third contraction: $\left(\varepsilon_{ijk} \cdot \varepsilon_{\alpha\beta\gamma}\right) \delta_{i\alpha}\, \delta_{j\beta}\, \delta_{k\gamma} = +6$

(To gain experience and become better acquainted with the procedure, the reader is invited to show that Eqs. II.2.21 can be obtained simply starting from Eq. II.2.20.)

$T_{ij}^{(s,\ iso)}$*, proportional to the tensor* δ_{ij} *and referred to as isotropic (three mutually equal diagonal components) and a symmetric tensor* $T_{ij}^{(s,\ aniso)}$ *with zero trace referred to as anisotropic.*

It suffices to write

$$T_{ij}^{(s)} = T_{ij}^{(s,\ iso)} + T_{ij}^{(s,\ aniso)} = \left[(1/3)\, T_{kk}\, \delta_{ij}\right] + \left[T_{ij}^{(s)} - (1/3)\, T_{kk}\, \delta_{ij}\right] \qquad \text{(II.2.22)}$$

In fact, we have

$$\mathrm{Tr}\left[T_{ij}^{(s,\ aniso)}\right] = T_{ij}^{(s,\ aniso)}\, \delta_{ij} = T_{ij}^{(s)}\, \delta_{ij} - T_{kk} = T_{ii} - T_{kk} = 0$$

An important application of the above decomposition concerns molecular scattering of light by molecules (Rayleigh scattering).

The symmetric part of the polarizability, $\alpha_{ij}^{(s)}$, decomposes into an *isotropic* part (giving rise to "isotropic" scattering related to the mean molecular polarizability) and an *anisotropic* part (giving rise to "anisotropic" scattering representing the anisotropy of the polarizability) of the particle. This point will be considered in more detail in Inset III.1. In the course of the past 20 years the latter effect has been the subject of intensive fundamental research, essentially carried out on dense molecular fluids composed of strongly anisotropic species.

The following table reviews briefly the application of tensorial notation in the description of mathematical and physical quantities to be considered later on.

TABLE II.2.2 — **Vectorial and tensorial notations.**

	Vectorial notation	Tensorial notation
Vectors	$\mathbf{A} = A_x\, \mathbf{i} + A_y\, \mathbf{j} + A_z\, \mathbf{k}$	A_i
Scalar product	$\mathbf{A}\ \mathbf{B}$	$A_i\, B_i$
Vectorial product	$\mathbf{A} \times \mathbf{B}$	$\varepsilon_{ijk}\, A_j\, B_k$
Mixed product	$(\mathbf{A} \times \mathbf{B})\ \mathbf{C}$	$\varepsilon_{ijk}\, A_j\, B_k\, C_i$
Gradient	$\nabla\, \phi$	$\nabla_i\, \phi$
Divergence	$\nabla \mathbf{A}$	$\nabla_i\, A_i$
Curl	$\nabla \times \mathbf{A}$	$\varepsilon_{ijk}\, \nabla_j\, A_k$
Laplacian	$\Delta = \nabla\ \nabla$	$\nabla_i\, \nabla_i$
Maxwell's equations	$\nabla\ \mathbf{E} = -\dfrac{\rho}{\varepsilon_0}$ $\nabla \times \mathbf{E} = -\dfrac{\partial \mathbf{B}}{\partial t}$ $\nabla\ \mathbf{B} = 0$ $\nabla \times \mathbf{B} = \dfrac{\mathbf{J}}{\varepsilon_0\, c^2} + \dfrac{1}{c^2}\dfrac{\partial \mathbf{E}}{\partial t}$	$\nabla_i\ E_i = -\dfrac{\rho}{\varepsilon_0}$ $\varepsilon_{ijk}\, \nabla_j\, E_k = -\dfrac{\partial B_i}{\partial t}$ $\nabla_i\, B_i = 0$ $\varepsilon_{ijk}\, \nabla_j\, B_k = \dfrac{J_i}{\varepsilon_0\, c^2} + \dfrac{1}{c^2}\dfrac{\partial Ei}{\partial t}$
The fields as functions of the scalar and vectorial potentials	$\mathbf{E} = -\nabla\, \phi - \dfrac{\partial}{\partial t}\mathbf{A}$ $\mathbf{B} = \nabla \times \mathbf{A}$	$E_i = -\nabla_i\, \phi - \dfrac{\partial}{\partial t} A_i$ $B_i = \varepsilon_{ijk}\, \nabla_j \times A_k$

Above, we have defined the tensor. However, its application is beset with difficulties often caused by the great number of its components. Luckily, for reasons we shall discuss and classify, the number of tensor components can undergo a reduction (as it is the case for the nondiagonal components of an isotropic tensor of rank 2, which are *zero* and thus vanish from the scene). In very many cases the number of components undergoes a reduction if relationships exist between them; we then have to deal with a smaller number of *independent* tensor components (for instance, the three diagonal components of an anisotropic tensor of rank 2 are interrelated in a manner to cause their sum to vanish). Very often, the problem facing us consists in determining the number of *nonzero components* and *independent* components of a tensor. To make the reader familiar with the different ways whereby a reduction of the number of tensor components can be achieved, we shall have recourse to examples from the domain of linear and nonlinear laser optics, thus preparing him or her for Part III of this book.

II.3 Reduction in the number of the components of a tensor: its nonzero and independent components

II.3.1 Reduction arising from the symmetry of the physical property under consideration

Generally, a tensor describes a physical property, *intrinsically* or *conditionally* endowed with some type of *symmetry*. Its symmetry is decisive for the vanishing of tensor components and their interrelations.

II.3.1.1 Natural reduction

As mentioned above (see Section II.2.3.1), when a laser wave acts on an atom or molecule, the electric field of the wave induces polarization **P**, given in a first approximation by Eq. II.2.12. In the case of a macroscopic body we can write more explicitly, in tensorial form,

$$P_i(\omega_L) \approx \chi_{ij}^{(1)}(-\omega_L, \omega_L)\, E_j(\omega_L) \qquad \text{(II.3.1)}$$

where ω_L is the circular frequency of the laser wave, and $\chi_{ij}^{(1)}(-\omega_L, \omega_L)$ is the optical susceptibility, a tensor of rank 2.

The circular frequencies $(-\omega_L)$ and (ω_L), put in parentheses, bear on the components i and j of the tensor: $(-\omega_L)$ states that the i-component of the polarization vector **P** oscillates at the frequency ω_L, whereas (ω_L) states that the j-component of the electric field oscillates at ω_L. The "minus" sign is introduced to obey the convention that the sum of the frequencies inside parentheses of a tensor (here, accompanying χ) shall be zero.

Strictly speaking, Eq. II.3.1 gives only an approximate description of the polarization. We shall now supplement the picture by adducing three examples.

(a) Optical activity by intervention of the spatial gradient of the electric field

Equation II.3.1 can be rendered formally in a more complete form as follows:

$$P_i(\omega_L) \approx \chi_{ij}^{(1)}(-\omega_L, \omega_L) E_j(\omega_L) + \eta_{ijk}(-\omega_L, \omega_L, 0) \nabla_k E_j(\omega_L) \qquad \text{(II.3.2)}$$

The three symbols $(-\omega_L)$, (ω_L), (0) now concern the i-, j-, and k-components, respectively. We assume a plane wave (see Chapter IV), given by

$$E_j(\omega_L) = E_j(0) \exp i(\mathbf{k}\,\mathbf{r} - \omega_L t)$$

interacting with the material. Then

$$\nabla_k E_j(\omega_L) = i\,|\mathbf{k}|\, S_k\, E_j(\omega_L)$$

where S_k is the k-component of the propagation vector **S**. We obtain

$$P_i(\omega_L) / E_j(\omega_L) = \chi_{ij}^{(1)}(-\omega_L, \omega_L) + i\,|\mathbf{k}|\,\eta_{ijk}(-\omega_L, \omega_L, 0)\, S_k$$

which, on insertion into Eq. II.2.14 and dropping the frequencies, and assuming the material to be isotropic, leads to

$$\varepsilon_{ij}^{(r)} - \delta_{ij} = \chi^{(1)}\delta ij / \varepsilon_0 + 2\pi\, i\, \eta\, \varepsilon_{ijk}\, S_k / \varepsilon_0\, \lambda \qquad \text{(II.3.3)}$$

It should be noted that, in place of the tensor η_{ijk}, we have written $\eta\varepsilon_{ijk}$. We refer the reader to Section III.1.2.1.e, for the proof that

$$\eta = \varepsilon_{\alpha\beta\gamma}\, \eta_{\alpha\beta\gamma} / 6 \qquad \text{(II.3.4)}$$

Equation II.3.3 permits us to structure the relative electric permittivity tensor. In an orthonormal system of coordinates Oxyz and a field with components Ex and Ey propagating along z, it takes the following form:

$$\left[\varepsilon_{ij}^{(r)}\right] = \begin{pmatrix} \varepsilon^{(r)} & i\varepsilon'^{(r)} & 0 \\ -i\varepsilon'^{(r)} & \varepsilon^{(r)} & 0 \\ 0 & 0 & \varepsilon^{(r)} \end{pmatrix} \qquad \text{(II.3.5)}$$

The term $\nabla_k E_j(\omega_L)$, accounting for *spatial dispersion* of the wave along Oz within the material, has thus led to the emergence of two imaginary nondiagonal terms in tensor $\left[\varepsilon_{ij}^{(r)}\right]$:

$$\varepsilon'^{(r)} = |\mathbf{k}|\, S_z\, \eta \qquad \text{(II.3.6)}$$

Applying Eq. II.2.13 we can write

$$(\varepsilon^{(r)} - n^2)\, E_x + i\varepsilon'^{(r)}\, E_y = 0$$
$$- i\varepsilon'^{(r)}\, E_x + (\varepsilon^{(r)} - n^2)\, E_y = 0$$

and consequently equate the two solutions:

$$\begin{cases} E_y / E_x = \exp(i\pi / 2) & \text{associated with} \quad n_-^2 = \varepsilon^{(r)} - \varepsilon'^{(r)} \\ E_y / E_x = \exp(-i\pi / 2) & \text{associated with} \quad n_+^2 = \varepsilon^{(r)} + \varepsilon'^{(r)} \end{cases} \qquad \text{(II.3.7)}$$

We thus deal with circular waves (the phases of E_y and E_x are at right angles), propagating with different phase velocities endowing the material with *optical rotation*. Here, we recognize Born's formalism (1933): The dephasing induced is given by $\varphi = 2\pi d\Delta n/\lambda = 2\pi d\varepsilon'^{(r)}/n\lambda$, where d is the thickness of the sample, leading to a rotation by $\theta = \varphi/2$ of a plane-polarized wave traversing the sample.

Born had introduced a gyration tensor G_{kl} given by

$$\varepsilon_{ijl}\, G_{kl} = 2\pi\, \eta\, \varepsilon_{ijk} / \lambda \qquad \text{(II.3.8)}$$

Note that the term $\eta_{ijk}\nabla_k E_j$ has to be odd for spatial inversion (**P** being a true vector.) Now, since both ∇_k and E_j are odd, η_{ijk} has to be odd and has to behave like $(-1)^3$. Hence η_{ijk} is a true tensor. This leads to the following results:

η_{ijk}	is a true tensor of rank 3
η	is a pseudoscalar
G_{kl}	is a pseudotensor of rank 2
θ	is a pseudoscalar

Rotation due to intrinsic optical activity is a pseudorotation.

The sense of the rotation undergoes inversion under spatial inversion—that is, on inversion of the sense of propagation of the wave.

Let us now go back to the general structure of the optical activity tensor η_{ijk}. It possesses 27 components. We can prove quite generally (see the bibliography at 7.5) that it is antisymmetric in its indices i and j:

$$\eta_{ijk} = -\,\eta_{jik} \qquad \text{(II.3.9)}$$

Thus, the tensor η_{ijk} possesses only 18 nonzero components (the three components with identical indices have to vanish since the field cannot be collinear with respect to the propagation vector) interrelated by nine equations of the type II.3.9. Hence we have nine independent components as imposed by the very nature of the property under consideration. These components will be dealt with one by one and identified in Appendix II.8.1.4.

(b) Faraday effect and inverse Faraday effect

The Faraday effect involves a *static* magnetic field (a field of zero frequency.) The effect is described by an equation similar to Eq. II.3.2:

$$P_i(\omega_L) \approx \chi^{(1)}_{ij}(-\omega_L, \omega_L) E_j(\omega_L) + \eta'_{ijk}(-\omega_L, \omega_L, 0) H_k(0) E_j(\omega_L) \quad (II.3.10)$$

On repeating the preceding analysis, we calculate the rotation induced on traversal of the thickness d in an isotropic material with refractive index n, obtaining

$$\theta \approx i \pi d \eta' H_z(0) / \varepsilon_0 \lambda n \quad (II.3.11)$$

The quantity $i \pi \eta' / \varepsilon_0 \lambda n = V$ defines the *Verdet constant V.* Although their treatment follows similar lines, the Faraday effect differs strongly from intrinsic optical activity. Since **P** is a true vector, $\eta'_{ijk}H_kE_j$ is odd. Now the pseudovector **H** behaves like $(-1)^2$ and is even. E_j being odd, η'_{ijk} is even and behaves like $(-1)^4$, thus being a *pseudo-tensor*. We arrive at the following characteristics:

η'_{ijk}	is a pseudo-tensor of rank 3
η'	is a true scalar
V	is a true scalar
θ	is a true scalar

Faraday rotation is a true rotation.

The sense of Faraday rotation is unaffected by spatial inversion along the direction of propagation of the wave.

Faraday's effect is widely applied in laser engineering (see Part II.) It provides the practical basis for the construction of *optical diodes*, forbidding the development of one of the two running laser waves in ring cavities. This point is explained in Inset II.4.

The inverse effect exists as well. Its discovery is due to P. Pershan (1966): an intense circularly polarized laser wave induces magnetization at zero frequency with i component given by

Inset II.4 ***How to make an optical diode***

Applied jointly, the two effects discussed above (intrinsic optical activity and Faraday' effect) permit the combination shown in Figure II.3.1:

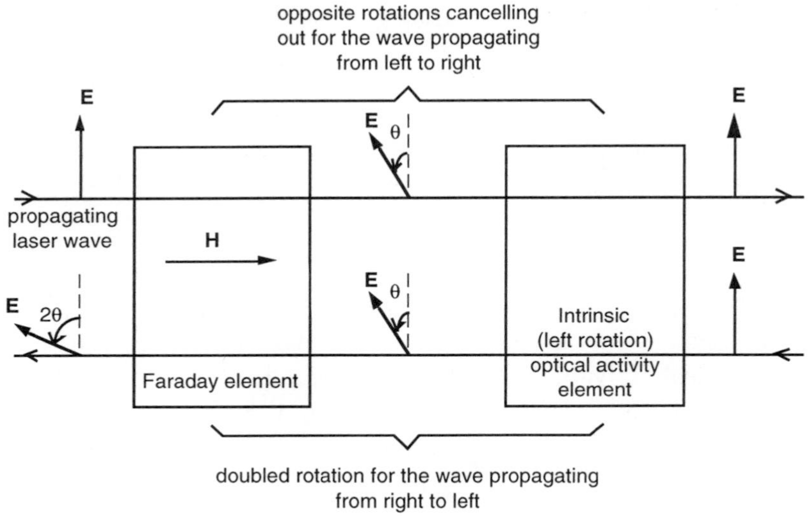

Figure II.3.1: Joint application of intrinsic optical activity and Faraday's effect to differentiate the polarizations of two laser waves propagating in two opposite directions.

The change by 2θ in the direction of polarization of the wave propagating from right to left is incompatible with the other elements of the cavity (not shown above); it ceases to propagate when the optical diode is switched on. The wave propagating from left to right undergoes no rotation of its polarization; it is not affected by the action of the diode.

$$M_i(0) \approx \eta'_{ijk}(0, \omega_L, -\omega_L)\, E_j(\omega_L)\, E_k^*(-\omega_L) \qquad \text{(II.3.12)}$$

For an isotropic medium we again can write (dropping the frequencies)

$$M_i(0) \approx \eta' \varepsilon_{ijk}\, E_j\, E_k^* \qquad \text{(II.3.13)}$$

where $M_i(0)$ is a pseudoscalar.

Now, considering a dextrogyric wave of wavelength λ propagating along Oz and having the intensity

$$I^+ = E^+ E^{+*}/2 = |E^+|^2/2$$

where

$$E^{+} = (E_x + i\,E_y)/\sqrt{2}$$

we obtain

$$M_z(0) \approx \left(2\,\varepsilon_0\,\lambda\,n/\pi\right) V\,I^{+} \qquad \text{(II.3.14)}$$

This magnetization is proportional to the Verdet constant and to the intensity of the laser wave. A procedure identical to that followed when dealing with the optical activity tensor η_{ijk} enables us to show that the tensor η'_{ijk} describing the Faraday and inverse Faraday effects is, naturally, antisymmetric in its indices i and j; it possesses, at the most, 18 nonzero components, nine of which are independent (see Appendix II.8.1.5.)

(c) Optical susceptibility of order 2

We have said that the expression II.3.1 derived for the i-component of the polarization **P** was only a first approximation. We have already discussed the ways in which **P** is sensitive to the spatial gradient of an electric field as well as the strength of a magnetic field; moreover, **P** is a quantity fundamentally nonproportionnal to the electric field strength **E**, and the traditional approach consists in expanding it in growing powers of **E**. Thus, $\chi^{(1)}_{ij}(-\omega_L, \omega_L)\,E_j(\omega_L)$ appears to us as the term of order 1—i.e., $P^{(1)}_i(\omega_L)$—of the expansion of **P**. The second term defines second -order polarization:

$$P^{(2)}_i(2\omega_L) = \chi^{(2)}_{ijk}(-2\,\omega_L, \omega_L, \omega_L)\,E_j(\omega_L)\,E_k(\omega_L) \qquad \text{(II.3.15)}$$

The tensor $\chi^{(2)}_{ijk}$ is naturally symmetric in its indices j and k since the field components in Eq. II.3.15 play completely symmetric roles. It is easily shown that their partial symmetry introduces nine relations between the 27 components of $\chi^{(2)}_{ijk}$, which, thus, has only 18 independent components.

II.3.1.2 Conditional reduction

Our use of the word "conditional" appears justified by considerations concerning $\chi^{(2)}_{ijk}$. If the nonlinearity dealt with above becomes apparent in a situation when the circular frequencies ω_L and $2\omega_L$ are remote from the eigenfrequencies ω_0 of the system (see Chapter IV), the tensor $\chi^{(2)}_{ijk}$ can be considered as completely symmetric. Then, five relations of equality hold between the six components having all three indices different, and 12 relations hold between those having two indices identical. The tensor, in this case, has only 10 independent components. This kind of reduction, of conditional nature, is referred to as *Kleinman's rule*.

II.3.2 Reduction by symmetry elements of the system

In addition to simplifications in the numbers of zero and independent components due to the symmetry exhibited by the *property* under consideration, a considerable reduction originates in the elements of symmetry of the *system* under consideration. This is due to the postulate that *each tensor component must remain invariant under the action of each element* of symmetry characterizing the object studied, be it microscopic or macroscopic. Thus, the physical property described has, at least, the symmetry of the respective object. Group theory provides the key to the classification of the elements of symmetry of systems (see the bibliography in II.7.2.) A method for deciding on the invariance of a component $T_{ab...x}$ under the action of the operator associated with a given element of symmetry has been proposed by Fumi. It makes use of the fact that the component under consideration, expressed in coordinates Oxyz (where a, b ... x are any one of the three coordinates x, y, z), transforms like the product ab ... x of the corresponding coordinates. If the product is invariant under the action of the operator of symmetry, the component $T_{a, b, \ldots x}$ is nonzero. Otherwise, it vanishes. To get an idea of Fumi's procedure, let us consider the problem of existence of the tensor $\chi^{(2)}_{ijk}$ in a system endowed with a *center of symmetry* (all the symmetry groups with a center of symmetry, in particular gases and disordered liquids.) The product ijk (where i, j, k are any of the three coordinates x, y, z) transforms into (– ijk) under the action of the operator (*the center of symmetry.*) Thus, the product is not invariant, and *all* the components of $\chi^{(2)}_{ijk}$ vanish so that *no* polarization $P^{(2)}_i$ (Eq. II.3.15) can take place.

Later on, we shall see that the foregoing result is of great importance in the field of nonlinear optics. It should be noted, however, that Fumi's method, though rapid and handy, suffers from certain inadequacies, particularly regarding rhombohedral and hexagonal systems. For certain classes of these systems it is not possible to choose cartesian axes transforming into themselves under the action of the elements of symmetry. In these cases the direct method proposed by Fumi is not applicable, and more general analytical procedures have to be used (see the bibliography in II.7.6.) Problems 6.1 and 6.2 will deal with the reduction resulting from the symmetry of the system under consideration. Moreover, Table II.3.1 visualizes the elements of symmetry of almost all the classes belonging to the seven crystallographic systems and occurring in axially symmetric compounds, and in most cases it enables us to find the tensorial invariances rapidly. Finally, Appendix II.8.1 lists the nonzero and independent components of the susceptibility tensors of order 1, 2, and 3 as well as those of the tensors of intrinsic and induced (Faraday) optical activity.

The various elements of symmetry are visualized in what follows.

TABLE II.3.1 — **Geometric illustration of the main symmetry classes of the seven crystallographic systems and of axially symmetric bodies. Except for the three figures which are drawn in perspective, all figures are seen along the 0z axis; the page represents the x0y plane. Dark gray parts point out of the front of the page, while light gray parts must be seen as pointing behind the page; the parts left white are in the plane of the page.**

Triclinic
(arbitrary prism)

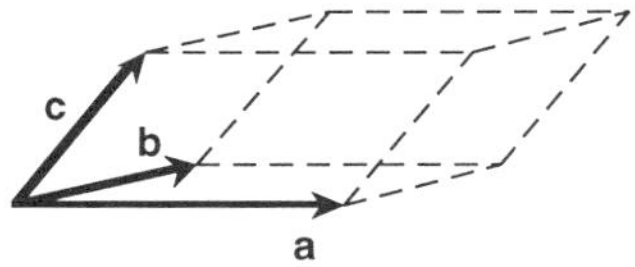

1 (C_1)

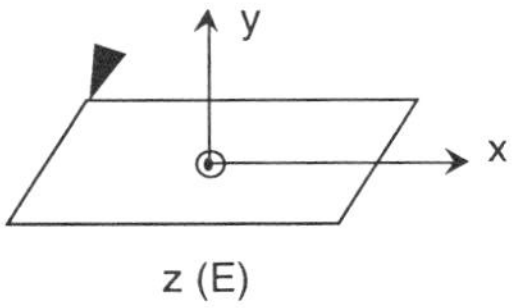

$\bar{1}$ (C_i)

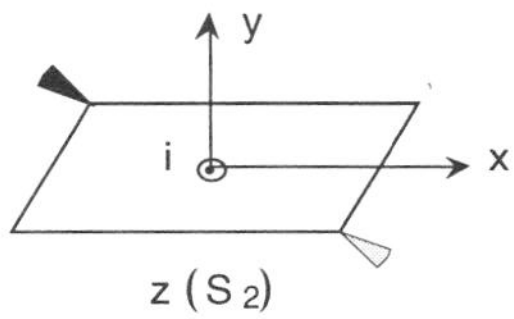

Monoclinic
(orthogonal prism with parallelogram basis)

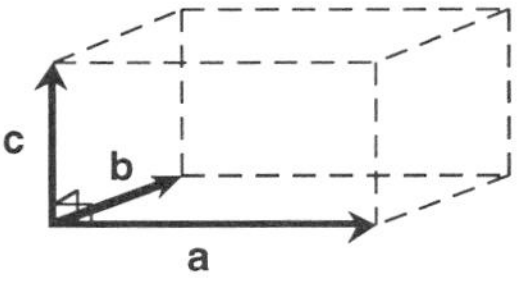

2 (C_2)

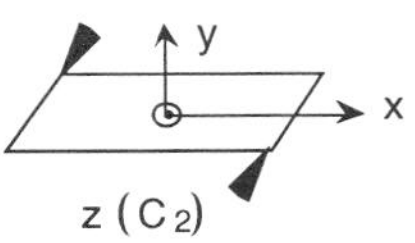

m (C_s)

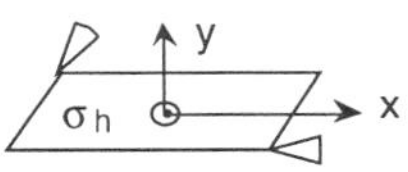

2/m (C_{2h})

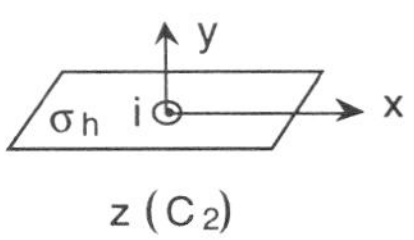

Orthorhombic
(orthogonal prism with rectangular basis)

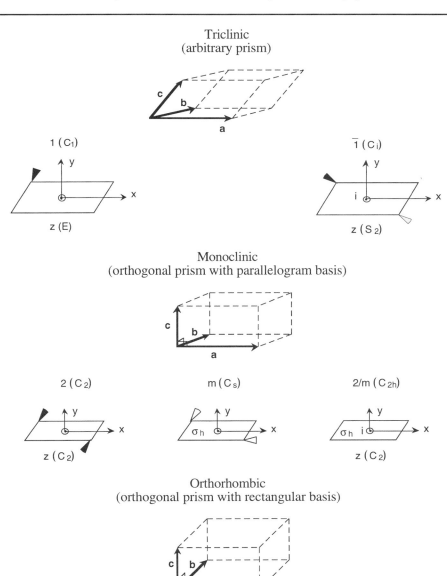

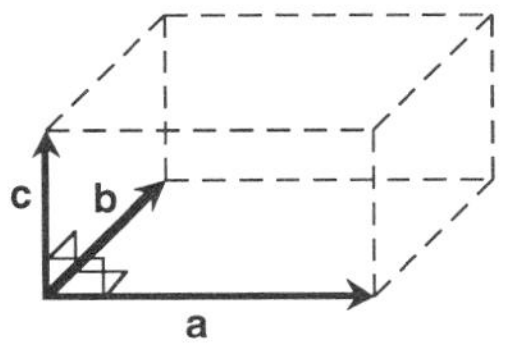

TABLE II.3.1 (*Continued*)

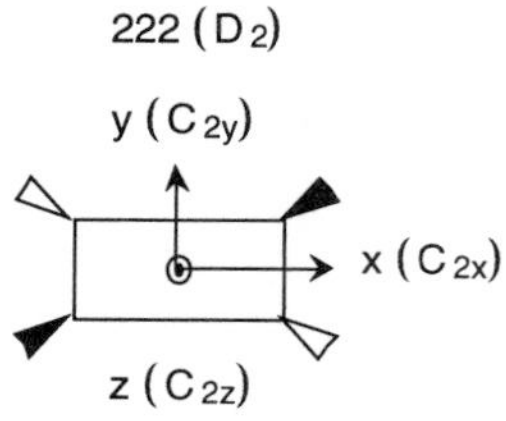

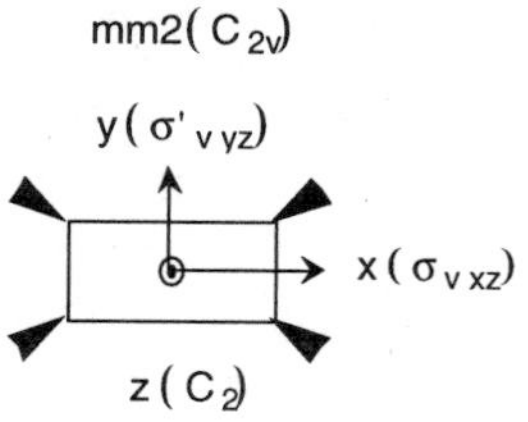

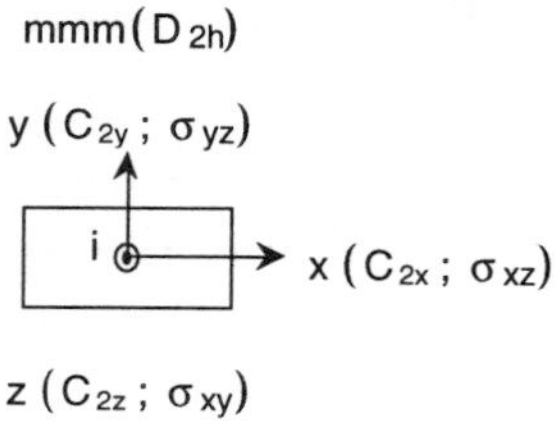

Quadratic
(orthogonal prism with square basis)

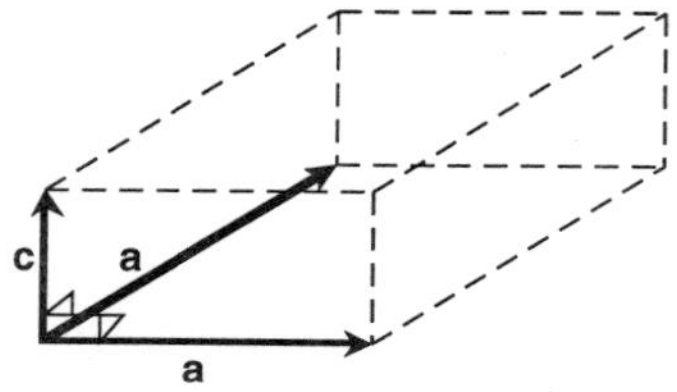

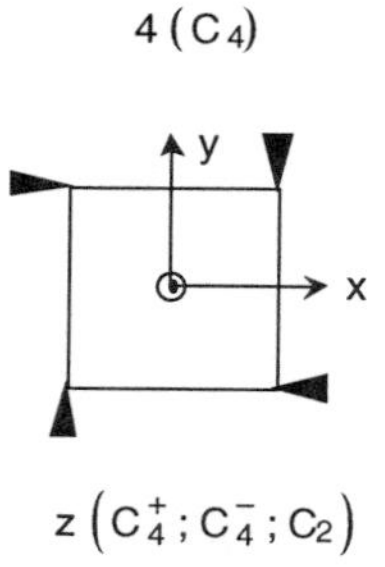

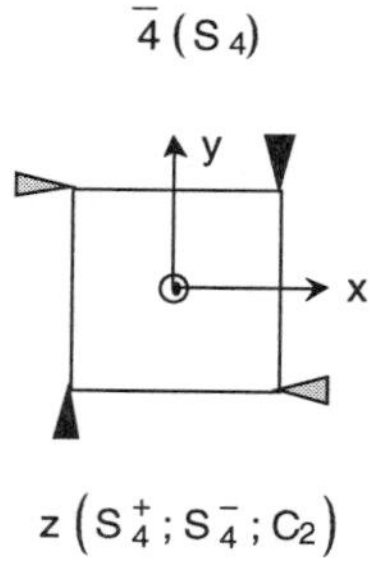

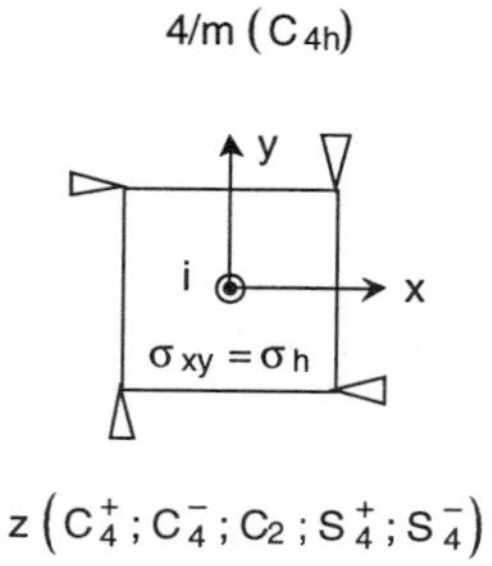

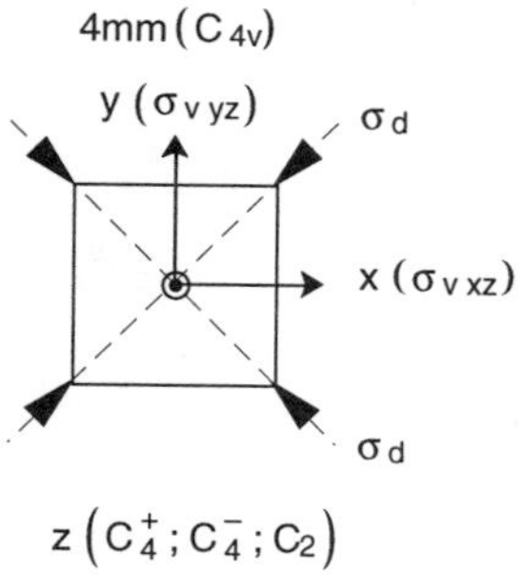

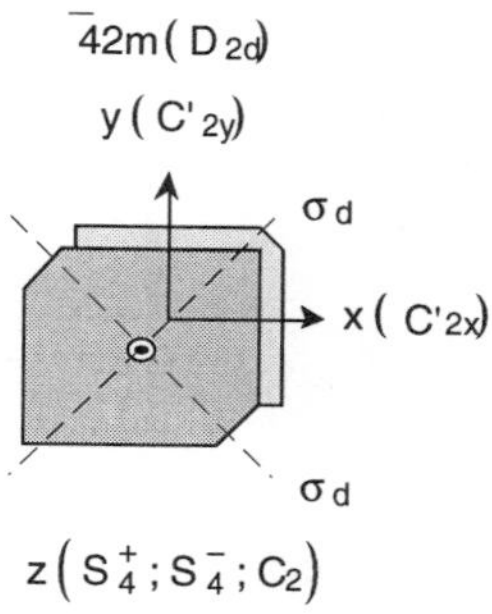

TABLE II.3.1 (*Continued*)

4/mmm(D_{4h})
y (C'_{2y} ; $\sigma_{v\,yz}$)

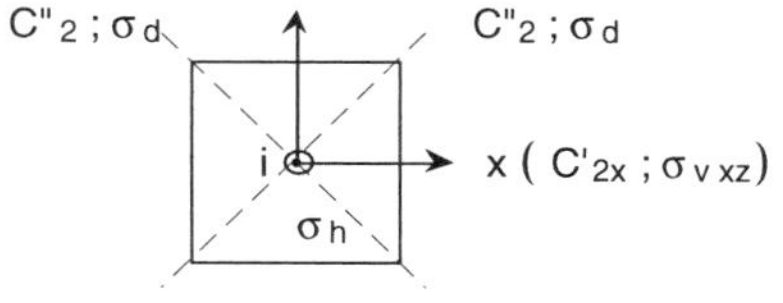

z (C_4^+ ; C_4^- ; C_{2z} ; S_4^+ ; S_4^-)

422 (D_4)
y (C'_{2y})

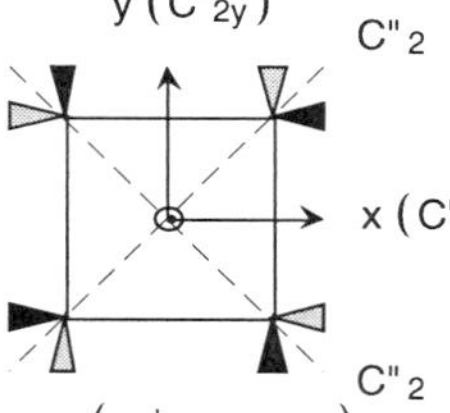

z (C_4^+ ; C_4^- ; C_2)

Rhombohedral
(inclined prism with rhombus faces)

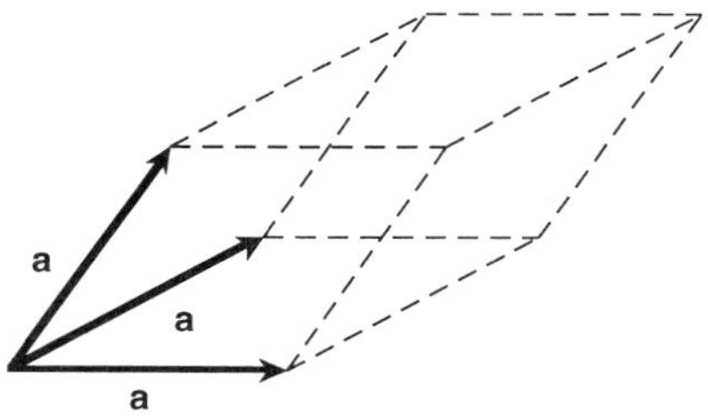

3 (C_3)

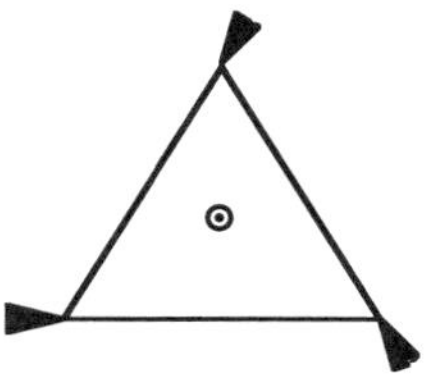

z (C_3^+ ; C_3^-)

$\bar{3}$ (S_6)

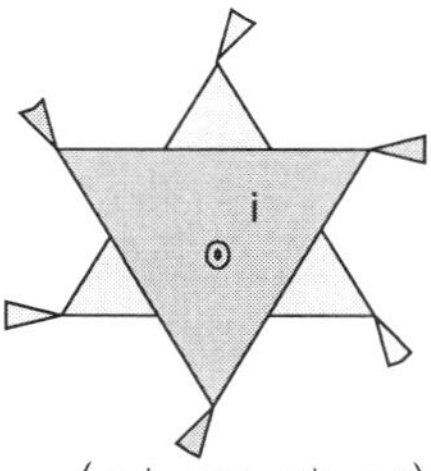

z (S_6^+ ; S_6^- ; C_3^+ ; C_3^-)

3m (C_{3v})

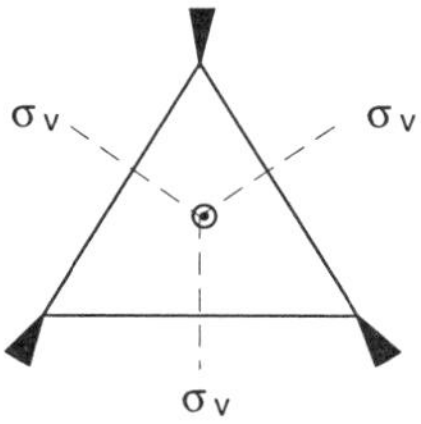

z (C_3^+ ; C_3^-)

32 (D_3)

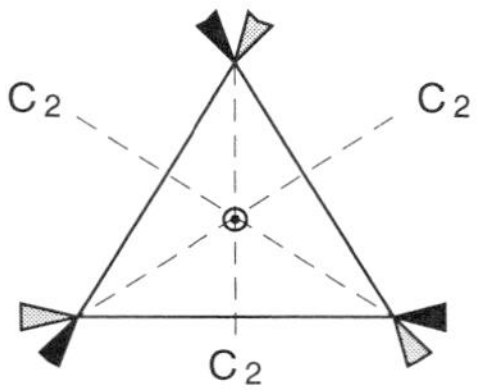

z (C_3^+ ; C_3^-)

$\bar{3}$m (D_{3d})

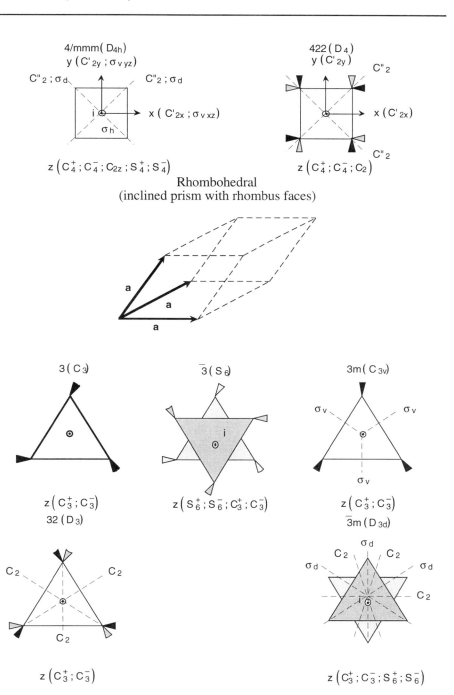

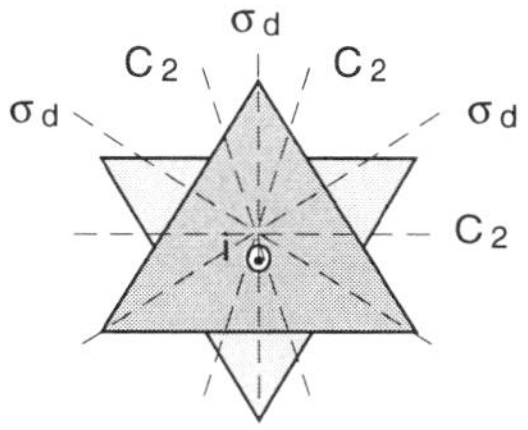

z (C_3^+ ; C_3^- ; S_6^+ ; S_6^-)

TABLE II.3.1 (*Continued*)

Hexagonal
(orthogonal prism with equilateral rhombus basis)

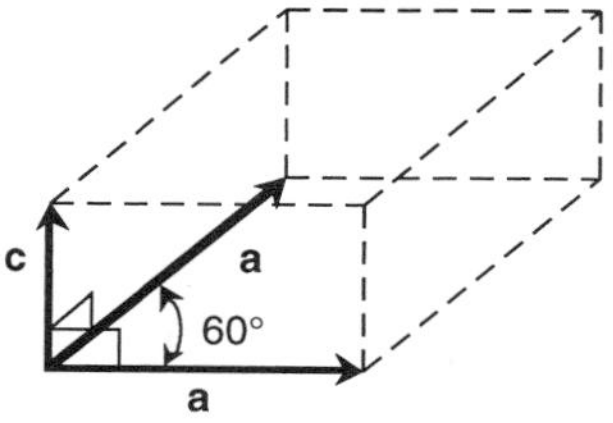

$6\ (C_6)$

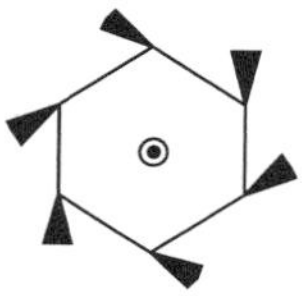

$z\,(C_6^+; C_6^-; C_3^+; C_3^-; C_2)$

$\bar{6}\ (C_{3h})$

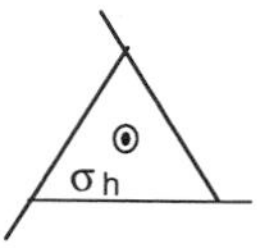

$z\,(C_3^+; C_3^-; S_3^+; S_3^-)$

$6/m\ (C_{6h})$

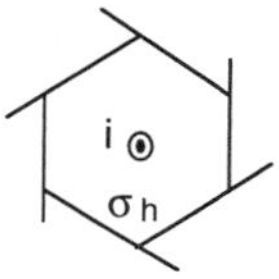

$z\,(C_6^+; C_6^-; C_3^+; C_3^-;$
$C_2; S_3^+; S_3^-; S_6^+; S_6^-)$

$\bar{6}m2\ (D_{3h})$

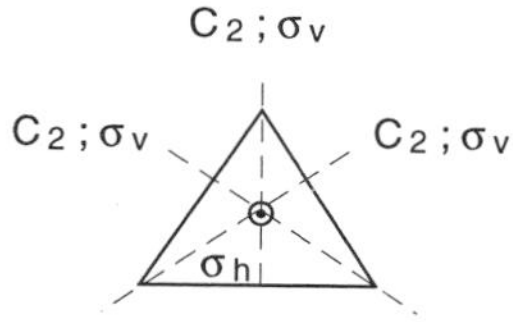

$z\,(C_3^+; C_3^-; S_3^+; S_3^-)$

$6mm\ (C_{6v})$

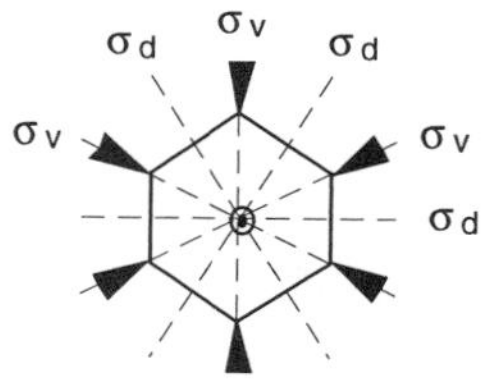

$z\,(C_6^+; C_6^-; C_3^+; C_3^-; C_2)$

$622\ (D_6)$

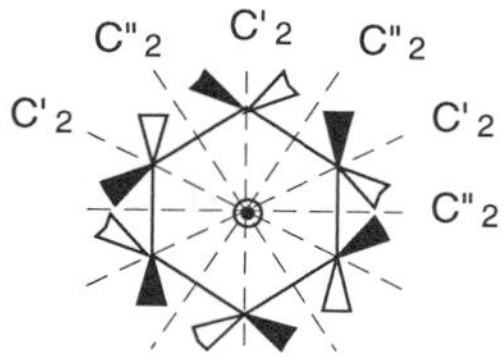

$z\,(C_6^+; C_6^-; C_3^+; C_3^-; C_2)$

$6/mmm\ (D_{6h})$

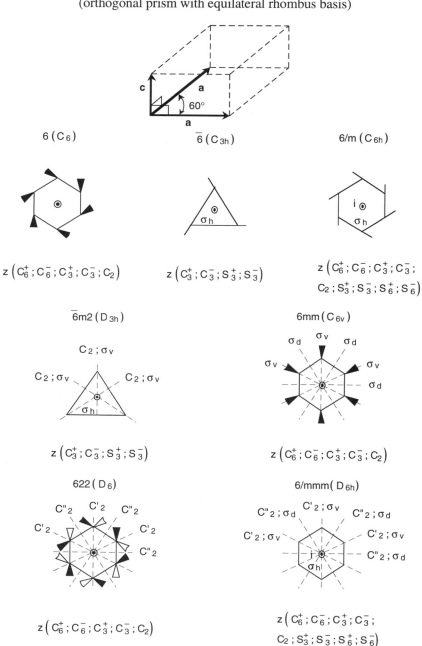

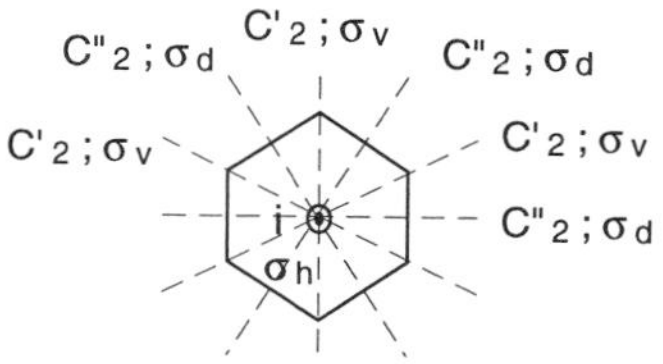

$z\,(C_6^+; C_6^-; C_3^+; C_3^-;$
$C_2; S_3^+; S_3^-; S_6^+; S_6^-)$

TABLE II.3.1 (*Continued*)

Cubic (view in perspective of two classes)
(rectangular prism with square faces)

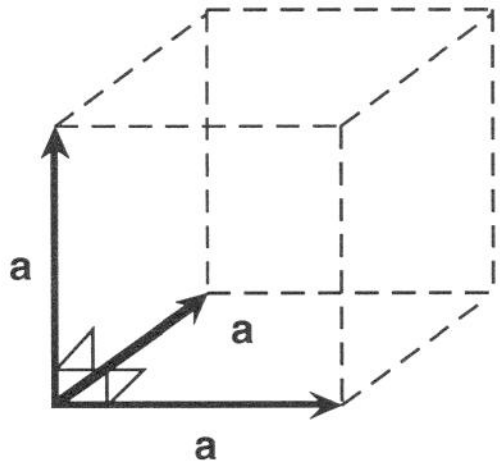

$\bar{4}3m$ (T_d)

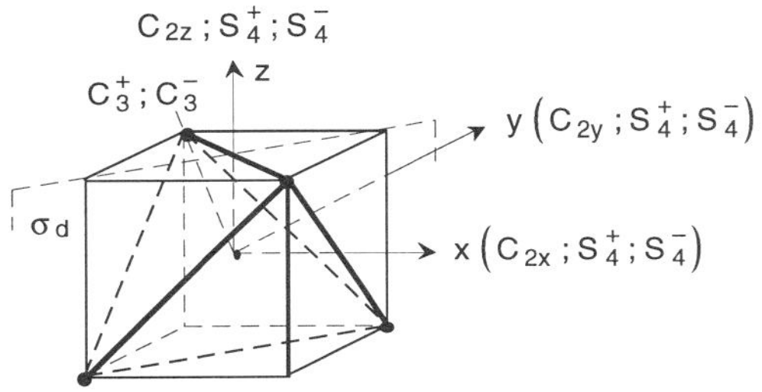

$m3m$ (O_h)

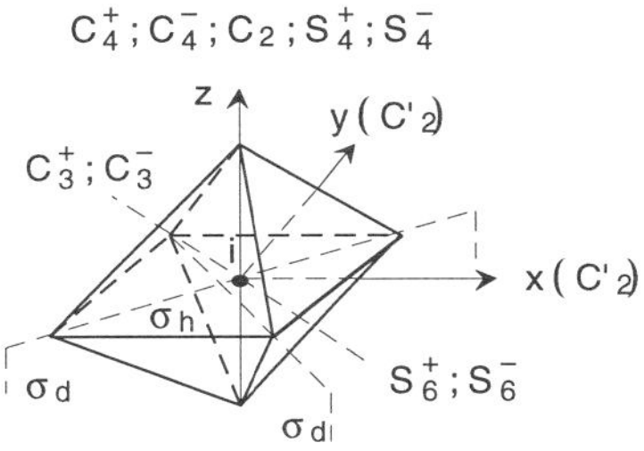

$\bar{4}3m$ (T_d) — not drawn:

- 3 axes of order 3: the 3 other diagonals of the cube
- 5 bisector planes σ_d through 2 opposite edges

$m3m$ (O_h) — not drawn:

- 6 axes of order 3: the 3 other straight lines connecting 2 opposite faces.
- 6 inverse axes of order 6: idem.
- 2 planes σ_h: the 2 others squares.
- 4 axes C'_2: the diagonals of the 2 square planes.
- 4 axes C_4: the two straight lines connecting the opposite corners of the octahedron.
- 4 axes S_4: idem.
- 2 axes C_2: idem.
- 4 planes σ_d: the bisecting planes through 2 opposite corners.

TABLE II.3.1 (*Continued*)

Axially symmetric body

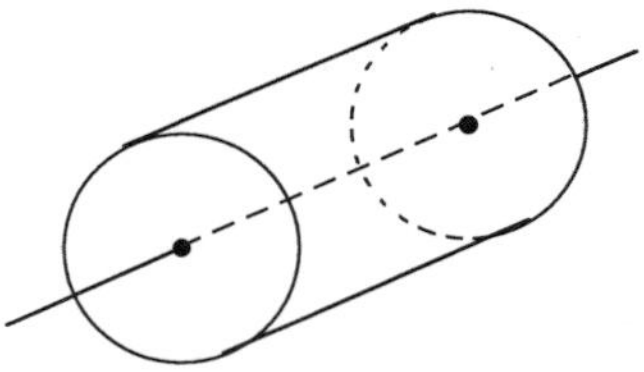

∞m ($C_{\infty v}$)

(view in perspective)

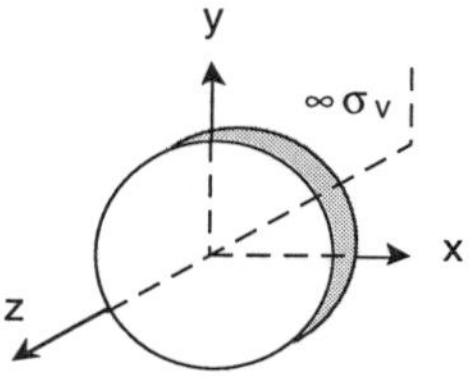

z (C_∞^+ ; C_∞^-)

∞/mm ($D_{\infty h}$)

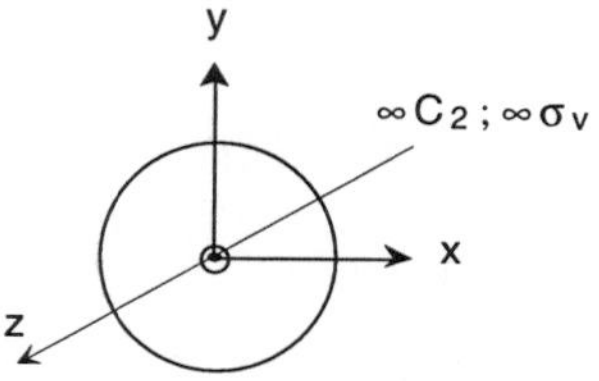

z (C_∞^+ ; C_∞^- ; S_∞^+ ; S_∞^-)

Let us illustrate these reduction properties in a more detailed way by writing explicitly the molecular electric multipoles and the internal fields they induce in the medium. Indeed, the internal fields play an essential part in laser-matter interaction. It will be shown that they are even the cause of certain physical effects of nonlinear optics.

II.4 Electromagnetic tensorial properties of microsystems

II.4.1 Tensorial expression of permanent electric multipoles

Let us consider a static distribution of electric charges e_s, positive or negative, located at points P_{e_s} (see diagram.)

We can now define the physical quantities usually associated with charges e_s and their corresponding position vectors $\mathbf{r}_s$, namely the electric multipoles.

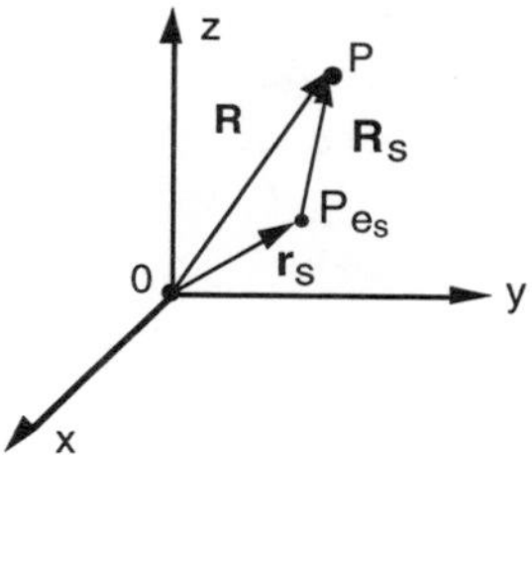

II.4.1.1 Monopole

By definition, the monopole is the scalar

$$e = \sum_s e_s \qquad \text{(II.4.1)}$$

II.4.1.2 Dipole

The dipole is the vector which is defined tensorially as

$$P_i = e_s\, r_{si} \qquad \text{(II.4.2)}$$

II.4.1.3 Quadrupole

Traditionally, the quadrupole is written as $q_{ij} = e_s r_{si} r_{sj}$. In practice, however, a second definition is very often used:

$$Q_{ij} = \left(3\, q_{ij} - q_{kk}\, \delta_{ij}\right) / 2 = (e_s / 2)\left(3\, r_{si}\, r_{sj} - r_s^2\, \delta_{ij}\right) \qquad \text{(II.4.3)}$$

Multiplying this expression by δ_{ij} gives the trace of Q_{ij}, which vanishes. By construction, this anisotropic tensor of rank 2 is symmetrical. Therefore, it has *five independent components.*

II.4.1.4 Octopole

Like the preceding multipoles, the octopole is defined by the formula $e_s r_{si} r_{sj} r_{sk}$ which leads to the following completely symmetrical, anisotropic tensor:

$$\Omega_{ijk} = (e_s / 2)\left[5\, r_{si}\, r_{sj}\, r_{sk} - r_s^2 \left(r_{si}\, \delta_{jk} + r_{sj}\, \delta_{ki} + r_{sk}\, \delta_{ij}\right)\right] \qquad \text{(II.4.4)}$$

Multiplying by δ_{ij}, gives

$$\Omega_{iik} = (e_s / 2)\left[5\, r_s^2\, r_{sk} - r_s^2\, (r_{sk} + r_{sk} + 3\, r_{sk})\right] = 0 \qquad \text{(II.4.5)}$$

so that we find

$$\Omega_{iij} = \Omega_{ijj} = \Omega_{iji} = 0$$

This equation states that the three traces of the tensor, Ω_{iij}, Ω_{ijj} and Ω_{iji}, vanish. On the other hand, we showed earlier (see Section II.3.1.2) that a totally symmetric tensor of rank 3 has 10 independent components. Equation II.4.5 imposes three extra conditions, linking together component having two identical indices. Therefore, tensor Ω_{ijk} has at most *seven independent components*.

II.4.1.5 Hexadecapole

Starting from the general expression $e_s r_{si} r_{sj} r_{sk} r_{sl}$, the hexadecapole is defined by the following totally symmetric and anisotropic tensor of rank 4:

$$\phi_{ijkl} = (e_s / 8)\Big[35\, r_{si}\, r_{sj}\, r_{sk}\, r_{sl} - 5\, r_s^2 \sum_{\mathcal{P}_{ik}} r_{si}\, r_{sk}\, \delta_{jl} + r_s^4 \left(\delta_{ij}\, \delta_{kl} + \delta_{ik}\, \delta_{jl} + \delta_{il}\, \delta_{jk}\right)\Big] \qquad \text{(II.4.6)}$$

$\sum_{\mathcal{P}_{ik}}$ represents the sum of the six terms obtained by circular permutation of i, j, k, l on the product $r_{si}\, r_{sk}$.

By multiplying by the appropriate Kronecker delta, one is led to discover six equations which reduce the traces to zero:

$$\phi_{iijk} = \phi_{ijik} = \phi_{ijki} = \phi_{iijj} = \phi_{ijji} = \phi_{ijij} = 0 \qquad \text{(II.4.7)}$$

However, a tensor of rank 4 has 81 components. If it is totally symmetric, only 15 of these are independent (see Appendix II.8.3.4). The six equations just mentioned reduce the total number of *independent components to nine in the case of the tensor representing the hexadecapole.*

II.4.1.6 Generalization to n-poles

Let us define a 2^n pole moment in a very general way:

$$m^{(2n)}_{i_1 i_2 \ldots i_n} = e_s\, r_{si_1}\, r_{si_2} \ldots r_{si_n} \qquad \text{(II.4.8)}$$

From this expression, we shall construct totally symmetric and anisotropic tensors for values of n equal to 0, 1, 2, ... It is interesting to determine the order of magnitude of these various moments. Let us therefore take $e_s \sim 10^{-19}$ C and $r_s \sim 10^{-10}$ m. Table II.4.1 shows the results, as well as some values for a few very simple molecules. Moreover, Appendix II.8.3 gives the independent and nonvanishing components of the previously defined multipole moments for all symmetry classes.

TABLE II.4.1 — **Main characteristics of the electric 2^n pole moments.**

Rank n of the tensor	2^n	Name of the moment	Order of magnitude and examples	
			SI units	CGS units
0	1	Monopole	10^{-19}	10^{-10}
1	2	Dipole	10^{-29}	10^{-18} = 1 Debye
2	4	Quadrupole	10^{-39}	10^{-26} (CO_2: 5×10^{-26})
3	8	Octopole	10^{-49}	10^{-34} (CCl_4: 15×10^{-34})
4	16	Hexadecapole	10^{-59}	10^{-42} (SF_6: 5×10^{-42})
⋮	⋮	⋮	⋮	⋮
n	2^n	2^n Pole	$10^{-(19+10n)}$	$10^{-(10+8n)}$

We are now left to determine the electric field radiated by the multipole moments due to spatial charge distributions.

II.5 Electric field radiated by permanent multipole moments

The electric potential created by a multipole at point P (see diagram of Section II.4.1) can be written as

$$\phi(R) = (1/4\pi\varepsilon_0)\, e_s / |\mathbf{R}_s| = (1/4\pi\varepsilon_0)\, e_s\, G(R_s) \tag{II.5.1}$$

when the function $G(R_s)$ is defined as

$$G(R_s) = 1/|\mathbf{R} - \mathbf{r}_s| = 1/|\mathbf{R}_s| = 1/R_s \tag{II.5.2}$$

For a point P far away from all the charges (i.e., such that $|\mathbf{r}_s| << |\mathbf{R}|$), $G(R_s)$ can be expanded as a series in powers of r_s:

$$G(R_s) \approx [G(R_s)]_{r_s=0} + r_{si}\left[\partial G(R_s)/\partial r_{si}\right]_{r_s=0} + \frac{1}{2} r_{si} r_{sj}\left[\partial^2 G(R_s)/\partial r_{si}\,\partial r_{sj}\right]_{r_s=0} + \cdots \tag{II.5.3}$$

In order to calculate the derivatives (i.e., the terms in brackets of the expression above), one is led to define the following symmetric and anisotropic tensors:

$$\begin{aligned} T_i &= \left[\partial G(R_s)/\partial r_{si}\right]_{r_s=0} = -\nabla_i G(R) \\ T_{ij} &= \left[\partial^2 G(R_s)/\partial r_{si}\,\partial r_{sj}\right]_{r_s=0} = -\nabla_i \nabla_j G(R) \end{aligned} \tag{II.5.4}$$

and so on.

Using these tensors to write $\phi(R)$—which is proportional to $e_s\, G(R_s)$—reveals the contracted products of the multipole moments. Thus, the potential $\phi(R)$ takes on the following condensed form:

$$\phi(R) = (1/4\,\pi\varepsilon_0)\left[(e/R) + T_i P_i - T_{ij} Q_{ij}/3 + T_{ijk}\,\Omega_{ijk}/15 - T_{ijkl}\,\phi_{ijkl}/105 + \cdots\right] \tag{II.5.5}$$

The internal field radiated at point P by the permanent electric multipoles is expressed by $\mathbf{F}(R) = -\nabla\phi(R)$, so we have

$$F_i(R) = (1/4\,\pi\varepsilon_0)\left[e\, T_i - T_{ij} P_j + T_{ijk} Q_{jk}/3 - T_{ijkl}\,\Omega_{jkl}/15 + \cdots\right] \tag{II.5.6}$$

The formal simplicity of this expression should convince the reader how powerful the tensorial notation is, and how great the advantage of using it systematically whenever dealing with physical properties of microsystems.

It is interesting to note that for electrically neutral entities (like molecules for instance) which are also *centrosymmetrical,* the first two terms of the above mentioned expansion vanish, since e = 0 and **P** = 0. In this case—and if allowed by the molecule's symmetry—the dominant term of the internal field is due to quadrupole radiation and is given by

$$F_i\,(R) \approx T_{ijk}\,Q_{jk} \,/\, 12\,\pi\,\varepsilon_0$$

where

$$T_{ijk} = -\,\nabla_k\,\nabla_j\,\nabla_i\,G\,(R) = -\,\nabla_k\,\nabla_j\,\nabla_i\,(1/R)$$

The successive derivations are easy to calculate. For example, for the first one we find

$$-\,\nabla_i\,(1/R) = R_i \,/\, R^3$$

so that

$$T_{ijk} = -\,3\,R^{-7}\left[5\,R_i\,R_j\,R_k - R^2\left(R_i\,\delta_{jk} + R_j\,\delta_{ki} + R_k\,\delta_{ij}\right)\right] \qquad \text{(II.5.7)}$$

At zero frequency, the i-component of the internal field F can be written in the following way-which is the form we shall be using subsequently:

$$F_i\,(R) \approx -\left(R^{-7} / 4\pi\varepsilon_0\right) Q_{jk}\left[5\,R_i\,R_j\,R_k - R^2\left(R_i\,\delta_{jk} + R_j\,\delta_{ki} + R_k\,\delta_{ij}\right)\right] \qquad \text{(II.5.8)}$$

II.6 Problems and outlined solutions

II.6.1 Problem II.1: Determination of the nonzero and mutually independent components of the susceptibility tensor of order 2 for a system belonging to the symmetry class $C_{\infty v}$

Use the Fumi method to determine the nonzero and mutually independent components, as well as the relations between dependent components, of the susceptibility tensor of order 2 for a system belonging to symmetry class $C_{\infty v}$.

The diagram shown below indicates the positions of the two symmetry elements (C_2 and C_4) of class $C_{\infty v}$ used in the solution.

Axis 0z is the symmetry axis of infinite order. The tensor has 27 components. Using the natural (N) and the conditional (C) properties of symmetry, we can write the following:

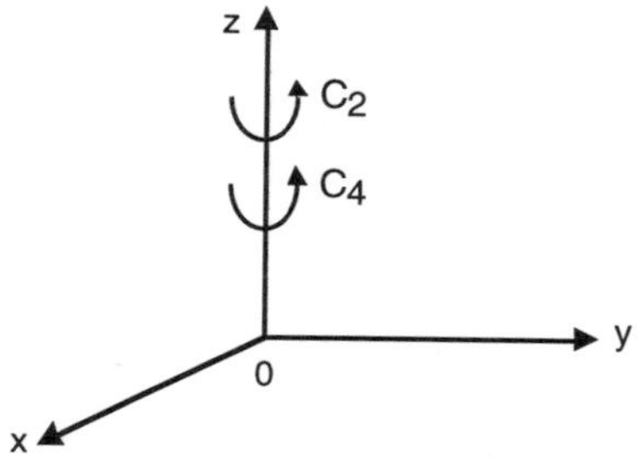

— The relations between components having three distinct indices:

$$\chi_{xyz} \overset{(N)}{=} \chi_{xzy} \overset{(C)}{=} \chi_{yzx} \overset{(N)}{=} \chi_{yxz} \overset{(C)}{=} \chi_{zxy} \overset{(N)}{=} \chi_{zyx}$$

(5 identities)

— The relations between components having two identical indices:

Let us take the case of indices x and y:

$$\chi_{xxy} \overset{(N)}{=} \chi_{xyx} \overset{(C)}{=} \chi_{yxx} \quad ; \quad \chi_{yyx} \overset{(N)}{=} \chi_{yxy} \overset{(C)}{=} \chi_{xyy}$$

(4 identities)

This must be repeated for the other two pairs of indices (x, z) and (y, z), so that we end up with 12 equations.

Accordingly, we find that tensor χ_{ijk} has 10 independent components:

$$\chi_{xxx};\ \chi_{yyy};\ \chi_{zzz};\ \chi_{xyz};\ \chi_{xxy};\ \chi_{yyx};\ \chi_{yyz};\ \chi_{zzy};\ \chi_{xxz};\ \chi_{zzx}$$

Fumi's method allows us to reduce some of these components to zero. Operators C_2 and C_4 transform reference frame 0xyz as follows:

$$C_2 \begin{pmatrix} x \\ y \\ z \end{pmatrix} = \begin{pmatrix} -x \\ -y \\ z \end{pmatrix} ; \quad C_4 \begin{pmatrix} x \\ y \\ z \end{pmatrix} = \begin{pmatrix} y \\ -x \\ z \end{pmatrix}$$

The results obtained with these two operators are listed in the table below:

Operator	Coordinate product which changes sign upon transformation	Correspondent component which reduces to zero
C_2	xxx	χ_{xxx}
C_2	yyy	χ_{yyy}
C_2	xxy	χ_{xxy}
C_2	yyx	χ_{yyx}
C_2	zzy	χ_{zzy}
C_2	xxz	χ_{xxz}
C_2	zzx	χ_{zzx}
C_4	xyz	χ_{xyz}

Thus, eight of the 10 independent components of χ_{ijk} vanish in the case of a system belonging to class $C_{\infty v}$. Only χ_{zzz} and χ_{xxz} are left, so that the tensor has seven nonzero components, two of which are independent:

$$\chi_{zzz}\ ;\ \chi_{xxz} \overset{(N)}{=} \chi_{xzx} \overset{(C)}{=} \chi_{zxx} \overset{(C_4)}{=} \chi_{zyy} \overset{(C)}{=} \chi_{yzy} \overset{(N)}{=} \chi_{yyz}$$

(The reader may verify that no element of symmetry belonging to $C_{\infty v}$ will reduce any of these seven components to zero.)

II.6.2 Problem II.2: Determination of the nonzero and mutually independent components of the susceptibility tensor of order 3 for a system belonging to the symmetry class Td

The same question as in Problem II.1 but in the case of the susceptibility tensor of order 3, considered to be totally symmetric, for a system belonging to symmetry class Td.

The diagram shows the position of the three elements of symmetry (C_2; S_{4z} and S_{4x}) used to solve the problem. (S_{4z} is an inverse rotational axis of order 4: It induces a rotation of $\pi/4$ about the 0z axis followed by an inversion with respect to the 0xy plane.) The operators act on reference frame 0xyz in the following way:

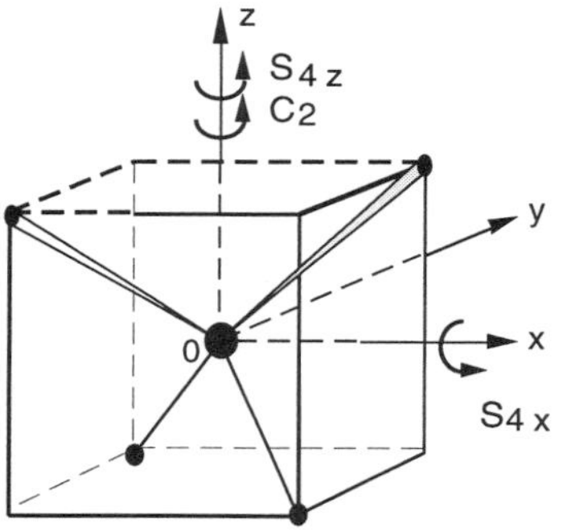

$$C_2\begin{Bmatrix} x \\ y \\ z \end{Bmatrix} = \begin{Bmatrix} -x \\ -y \\ z \end{Bmatrix};\quad S_{4z}\begin{Bmatrix} x \\ y \\ z \end{Bmatrix} = \begin{Bmatrix} y \\ -x \\ -z \end{Bmatrix};\quad S_{4x}\begin{Bmatrix} x \\ y \\ z \end{Bmatrix} = \begin{Bmatrix} -x \\ z \\ -y \end{Bmatrix}$$

In fact, there are three C_2 axes and six S_4 axes in class Td, pointing pairwise through the six sides of the cube.

The reader may verify that all products of four coordinates of which two or three are identical are inverted by the C_2 operators. Therefore, the 24 components having three identical indices, and the 36 components having two identical indices, reduce to zero. Twenty-one nonzero components remain in tensor χ_{ijkl}. Let us show now that only two of these are independent. Operator S_{4z} shows that xxxx = yyyy. In the same way, operator S_{4x} shows that yyyy = zzzz. It follows that $\chi_{xxxx} = \chi_{yyyy} = \chi_{zzzz}$.

Moreover, by using the natural (N) and conditional (C) properties of symmetry of the tensor, we can conclude that

$$\chi_{xxyy} \overset{(N)}{=} \chi_{xyxy} \overset{(N)}{=} \chi_{xyyx} \overset{(C)}{=} \chi_{yxyx} \overset{(N)}{=} \chi_{yyxx} \overset{(N)}{=} \chi_{yxxy}$$

$$\overset{(S_{4x})}{=} \chi_{xxzz} \overset{(N)}{=} \chi_{xzxz} \overset{(N)}{=} \chi_{xzzx} \overset{(C)}{=} \chi_{zxzx} \overset{(N)}{=} \chi_{zzxx} \overset{(N)}{=} \chi_{zxxz}$$

$$\overset{(S_{4x})}{=} \chi_{yyzz} \overset{(N)}{=} \chi_{yzyz} \overset{(N)}{=} \chi_{yzzy} \overset{(C)}{=} \chi_{zyzy} \overset{(N)}{=} \chi_{zzyy} \overset{(N)}{=} \chi_{zyyz}$$

Clearly, calculations with a tensor of this kind will be relatively easy.

II.6.3 Problem II.3: Expression of the susceptibility tensor of order 1 of an axially symmetric molecule

Show that α_{ij}, the optical susceptibility tensor of order 1 of an axially symmetric molecule (as expressed in the laboratory reference frame 0xyz), can be expressed simply as a function of the mean molecular polarizability $\overline{\alpha} = (\alpha_{//} + 2\,\alpha_{\perp})/3$ and of its anisotropy $\gamma = \alpha_{//} - \alpha_{\perp}$ ($\alpha_{\alpha\beta}$ are the components of the tensor as expressed in the principal reference frame 1, 2, 3 of the molecule; we use the notation $\alpha_3 = \alpha_{//}$; $\alpha_1 = \alpha_2 = \alpha_{\perp}$).

We shall use relation II.2.5 (upper part), which translates the change of base from one reference frame to another $0_{123} \rightarrow 0_{xyz}$:

$$\alpha_{ij} = c_{i\alpha}\, c_{j\beta}\, \alpha_{\alpha\beta}$$

$c_{i\alpha}$ and $c_{j\beta}$ are two directional cosines of the change of base. In the principal molecular reference frame, the only nonzero components are $\alpha_{11} = \alpha_{22} = \alpha_{\perp}$ and $\alpha_{33} = \alpha_{//}$, so that $\alpha_{ij} = (c_{i1}\,c_{j1} + c_{i2}\,c_{j2})\,\alpha_{\perp} + c_{i3}\,c_{j3}\,\alpha_{//}$.

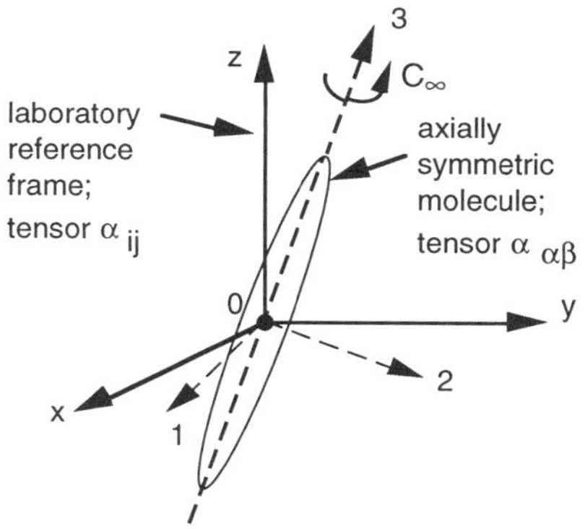

But Eq. II.1.5 (lower part) states that $c_{i1}\,c_{j1} + c_{i2}\,c_{j2} = \delta_{ij} - c_{i3}\,c_{j3}$. We therefore have $\alpha_{ij} = \alpha_{\perp}\,\delta_{ij} + \gamma\, c_{i3}\,c_{j3}$,.

This can be rewritten as $\alpha_{ij} = \left[\alpha_{\perp} + (\gamma/3)\right]\delta_{ij} + \gamma\left(c_{i3}\,c_{j3} - \delta_{ij}/3\right)$, which is again equal to $\alpha_{ij} = \overline{\alpha}\,\delta_{ij} + \gamma\left(c_{i3}\,c_{j3} - \delta_{ij}/3\right)$.

II.6.4 Problem II.4: Calculation of the quadruple electric field radiated by a molecule belonging to the symmetry class D_{6h}

Let us consider a molecule belonging to symmetry class D_{6h}. Determine the quadrupole moment of this molecule as expressed in its own principal reference frame 0_{123}. Give its expression in the laboratory reference frame. Calculate the static electric field radiated at a point placed at a distance R from the origin 0.

In the molecule's principal reference frame, the molecular quadrupole has only three nonzero components Q_{11}, Q_{22}, and Q_{33} (see Appendix II.8.3.2). With the same reasoning as in Problem II.3, the symmetry of the molecule imposes that $Q_{11} = Q_{22} = Q_\perp$; $Q_{33} = Q_{//}$. But since the tensor has a vanishing trace, we have $Q_\perp = -Q_{//}/2$. Thus, the only nonzero components of the molecular quadrupole tensor $[Q_{\alpha\beta}]$ are $Q_{11} = Q_{22} = -Q_{//}/2$; $Q_{33} = Q_{//}$.

As we did above, let us proceed to a change of reference frame from the molecular to the laboratory coordinates: $Q_{ij} = c_{i\alpha} c_{j\beta} Q_{\alpha\beta} = (Q_{//}/2)(c_{i1} c_{j1} - c_{i2} c_{j2}) + Q_{//} c_{i3} c_{j3}$. This immediately gives $Q_{ij} = -(Q_{//}/2)(\delta_{ij} - c_{i3} c_{j3}) + Q_{//} c_{i3} c_{j3}$; in other words, $Q_{ij} = (Q_{//}/2)(3\, c_{i3} c_{j3} - \delta_{ij})$.

Equation II.5.8 gives the radiated static electric field:

$$F_i(R) \approx -(Q_{//} R^{-7}/8\pi\varepsilon_0)\, c_{j3} c_{k3}\left[15\, R_i R_j R_k - 3\, R^2 (R_i \delta_{jk} + R_j \delta_{ki} + R_k \delta_{ij})\right]$$

II.7 Bibliography

II.7.1 Tensors and tensorial computing can be studied in:

NYE, J. F. *Physical Properties of Crystals,* Clarendon Press, Oxford, 1985.

II.7.2 Group theory can be studied in:

COTTON, F. A. *Chemical Applications of Group Theory*, John Wiley & Sons, New York, 1963.

II.7.3 Classical electromagnetism can be studied in:

JACQSON, J. D. *Classical Electrodynamics*, John Wiley & Sons, New York, 1962.

II.7.4 Statistical physics can be studied in the works listed in III.5.1.

II.7.5 A demonstration of the antisymmetry with respect to indices j and k of the optical activity tensor can be found in:

LANDAU, L., AND LIFCHITZ, E. *Statistical Physics*, Pergamon Press, London, 1963.

II.7.6 Fumi's method can be studied in, for instance:

BHAGAVANTAM, S. *Crystal Symmetry and Physical Properties*, Academic Press, New York, 1966.

II.8 Appendices

II.8.1 Independent and nonzero components of susceptibility and (natural or induced) optical activity tensors for all classes of symmetry. Only the indices of the tensorial components will be given (the 0z axis is always the axis of highest symmetry).

II.8.1.1 Susceptibility of order 1 (the number of independent components is indicated in parentheses)

1. **Triclinic** (6)

$$\begin{bmatrix} xx & xy & zx \\ xy & yy & yz \\ zx & yz & zz \end{bmatrix}$$

2. **Monoclinic** (4)

$$\begin{bmatrix} xx & 0 & zx \\ 0 & yy & 0 \\ zx & 0 & zz \end{bmatrix}$$

3. **Orthorhombic** (3)

$$\begin{bmatrix} xx & 0 & 0 \\ 0 & yy & 0 \\ 0 & 0 & zz \end{bmatrix}$$

4. **Quadratic Rhombohedral Hexagonal Axially symmetric bodies** (2)

$$\begin{bmatrix} xx & 0 & 0 \\ 0 & xx & 0 \\ 0 & 0 & zz \end{bmatrix}$$

5. **Cubic Isotropic medium** (1)

$$\begin{bmatrix} xx & 0 & 0 \\ 0 & xx & 0 \\ 0 & 0 & xx \end{bmatrix}$$

II.8.1.2 Susceptibility of order 2

$$\begin{bmatrix} [(xxx)] & [(xyy)] & [(xzz)] & [(xyz)] & xzy & xzx & [(xxz)] & [(xxy)] & xyx \\ [yxx] & [(yyy)] & [(yzz)] & [(yyz)] & yzy & yzx & [yxz] & [yxy] & yyx \\ [zxx] & [zyy] & [(zzz)] & [zyz] & zzy & zzx & [zxz] & [zxy] & zyx \end{bmatrix}$$

A tensor which is symmetric with respect to a pair of indices j, k, has only 18 independent components (those written in square brackets). For a totally symmetric tensor, this number is reduced to 10 (those written in parentheses). We shall list them individually hereafter. For those classes endowed with a center of symmetry, all components vanish.

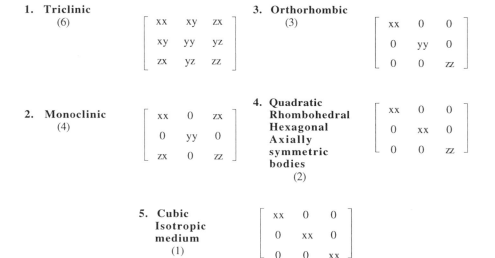

1. **Triclinic**
Class 1 (C_1): 27 nonzero components, 10 of which are independent (those in parentheses in the above table).

2. **Monoclinic**
Class 2 (C_2): 13 nonzero components, 4 of which are independent.

zzz,
zxx = xzx = xxz
zyy = yzy = yyz
xyz = xzy = yxz = yzx = zxy = zyx

Class m (C_s): 14 nonzero components, 6 of which are independent.

yyy, xxx
xyy = yxy = yyx,
yxx = xyx = xxy
xzz = zxz = zzx
yzz = zyz = zzy

3. **Orthorhombic**
Class 222 (D_2): 6 nonzero components, 1 of which is independent.

xyz = xzy = yxz = yzx = zxy = zyx

Class 2mm (C_{2v}): 7 nonzero components, 3 of which are independent.

zzz
zxx = xzx = xxz
zyy = yzy = yyz

4. **Quadratic**
Class 4 (C_4): 7 nonzero components, 2 of which are independent.

zzz
zxx = xzx = xxz = zyy = yzy = yyz

Class $\bar{4}$ (S_4): 12 nonzero components, 2 of which are independent.

zxx = xzx = xxz = – zyy = – yzy = – yyz
xyz = xzy = yxz = yzx = zxy = zyx

Class 422 (D_4): all components vanish.

Class 4mm (C_{4v}): 7 nonzero components, 2 of which are independent.

zzz
zxx = xzx = xxz = zyy = yzy = yyz

Class $\bar{4}$2m (D_{2d}): 6 nonzero components, 1 of which is independent.

xyz = xzy = yxz = yzx = zxy = zyx

5. **Rhombohedral**
Class 3 (C_3): 15 nonzero components, 4 of which are independent.

zzz
xxx = – xyy = – yxy = – yyx
yyy = – yxx = – xyx = – xxy
zxx = xzx = xxz = zyy = yzy = yyz

Class 32 (D_3): 4 nonzero components, 1 of which is independent.

xxx = – xyy = – yxy = – yyx

Class 3m (C_{3v}): 11 nonzero components, 3 of which are independent.

zzz
yyy = – yxx = – xyx = – xxy
zxx = xzx = xxz = zyy = yzy = yyz

6. **Hexagonal**
Class 6 (C_6): 7 nonzero components, 2 of which are independent.

zzz,
zxx = xzx = xxz = zyy = yzy = yyz

Class $\bar{6}$ (C_{3h}): 8 nonzero components, 2 of which are independent.

xxx = – xyy = – yxy = – yyx
yyy = – yxx = – xyx = – xxy

Class 622 (D_6): all components vanish.

Class 6mm (C_{6v}): 7 nonzero components, 2 of which are independent.

zzz
zxx = xzx = xxz = zyy = yzy = yyz

Class $\bar{6}2m$ (D_{3h}): 4 nonzero components, 1 of which is independent.

yyy = – yxx = – xyx = – xxy

7. **Cubic**
Class 432 (0): all components vanish.

Classes $\bar{4}3m$ (T_d) and 23 (T): 6 nonzero components, 1 of which is independent.

xyz = xzy = yxz = yzx = zxy = zyx

8. **Isotropic medium**
Classes I, I_h, K, K_h: all components vanish.

9. **Axially symmetric body**
Classes ∞ (C_∞) and ∞m ($C_{\infty v}$): 7 nonzero components, 2 of which are independent.

zzz,
zxx = xzx = xxz = zyy = yzy = yyz

Classes ∞/m ($C_{\infty h}$) and ∞/mm ($D_{\infty h}$): all components vanish.

II.8.1.3 Susceptibility of order 3 (the tensor is assumed as completely symmetric)

1. **Triclinic**
Classes 1 (C_1) and $\bar{1}$ (C_i): 81 nonzero components, 15 of which are independent.

xxxx, yyyy, zzzz
xxyy = xyxy = xyyx = yyxx = yxyx = yxxy
xxzz = xzxz = xzzx = zzxx = zxzx = zxxz
yyzz = yzyz = yzzy = zzyy = zyzy = zyyz
xxxy = xxyx = xyxx = yxxx
yyyx = yyxy = yxyy = xyyy
xzzz = zxzz = zzxz = zzzx
yzzz = zyzz = zzyz = zzzy
xxxz = xxzx = xzxx = zxxx
yyyz = yyzy = yzyy = zyyy
xxyz = xyxz = xyzx = xxzy = xzxy = xzyx = yxxz = yxzx = yzxx = zxxy = zxyx = zyxx
yyxz = yxyz = yxzy = yyzx = yzyx = yzxy = xyyz = xyzy = xzyy = zyyx = zyxy = zxyy
xzzy = xzyz = xyzz = yxzz = yzxz = yzzx = zxyz = zyxz = zyzx = zxzy = zzyx = zzxy

2. **Monoclinic**
Classes 2 (C_2), m (C_s), and 2/m (C_{2h}): 41 nonzero components, 9 of which are independent.

xxxx, yyyy, zzzz
xxyy = xyxy = xyyx = yyxx = yxyx = yxxy
xxzz = xzxz = xzzx = zzxx = zxzx = zxxz
yyzz = yzyz = yzzy = zzyy = zyzy = zyyz
xxxy = xxyx = xyxx = yxxx
yyyx = yyxy = yxyy = xyyy
xzzy = xzyz = xyzz = yxzz = yzxz = yzzx = zxyz = zyxz = zyzx = zxzy = zzyx = zzxy

3. **Orthorhombic**
Classes 222 (D_2), 2mm (C_{2v}), mmm (D_{2h}): 21 nonzero components, 6 of which are independent.

xxxx, yyyy, zzzz
xxyy = xyxy = xyyx = yyxx = yxyx = yxxy
xxzz = xzxz = xzzx = zzxx = zxzx = zxxz
yyzz = yzyz = yzzy = zzyy = zyzy = zyyz

4. Quadratic

Classes 4 (C_4), $\bar{4}$ (S_4), 4/m (C_{4h}): 29 nonzero components, 5 of which are independent.

xxxx = yyyy, zzzz
xxyy = xyxy = xyyx = yyxx = yxyx
= yxxy
xxzz = xzxz = xzzx = zzxx = zxzx
= zxxz = yyzz = yzyz = yzzy
= zzyy = zyzy = zyyz
xxxy = xxyx = xyxx = yxxx
= – yyyx – yyxy = – yxyy = – xyyy

Classes 422 (D_4), 4mm (C_{4v}), $\bar{4}$2m (D_{2d}): 21 nonzero components, 4 of which are independent.

xxxx = yyyy, zzzz
xxyy = xyxy = xyyx = yyxx = yxyx
= yxxy
xxzz = xzxz = xzzx = zzxx = zxzx
= zxxz = yyzz = yzyz = yzzy
= zzyy = zyzy = zyyz

5. Rhombohedral

Classes 3 (C_3) and $\bar{3}$ (S_6): 53 nonzero components, 5 of which are independent.

zzzz
xxxx = yyyy = xxyy + xyxy + xyyx
xxyy = xyxy = xyyx = yyxx = yxyx
= yxxy
xxzz = xzxz = xzzx = zzxx = zxzx
= zxxz = yyzz = yzyz = yzzy
= zzyy = zyzy = zyyz
xxxz = xxxx = xzxx = zxxx
= – yxyz = – yzyx = – yyxy
= – yxzy = – yyzx = – yzxy
= – xyyz = – xyzy = – xzyy
= – zxyy = – zyxy = – xyyx
yyyz = yyzy = yzyy = zyyy = – xyxz
= – xzxy = – xxyz = – xyzx
= – xxzy = – xzyx = – yxxz
= – yxzx = – yzxx = – zyxx
= – zxyx = – zxxy

Classes 3m (C_{3v}), $\bar{3}$m (D_{3d}), 32 (D_3): 37 nonzero components, 4 of which are independent.

zzzz
xxxx = yyyy = xxyy + xyxy + xyyx
xxyy = xyxy = xyyx = yyxx = yxyx
= yxxy
xxzz = xzxz = xzzx = zzxx = zxzx
= zxxz = yyzz = yzyz = yzzy
= zzyy = zyzy = zyyz
xxxz = xxzx = xzxx = zxxx = – yxyz
= – yzyx = – yyxz = – yxzy = – yyzx
= – yzxy = – xyyz = – xyzy
= – xzyy = – zxyy = – zyxy = – zyyx

6. Hexagonal

Classes 6 (C_6), $\bar{6}$ (C_{3h}), 6/m (C_{6h}), 622 (D_6), 6mm (C_{6v}), $\bar{6}$2m (D_{3h}), 6/mmm (D_{6h}): 21 nonzero components, 3 of which are independent.

zzzz
xxxx = yyyy = xxyy + xyxy + xyyx,
xxyy = xyxy = xyyx = yyxx = yxyx
= yxxy
xxzz = xzxz = xzzx = zzxx = zxzx
= zxxz = yyzz = yzyz = yzzy
= zzyy = zyzy = zyyz

7. Cubic

21 nonzero components, 2 of which are independent.

xxxx = yyyy = zzzz
xxyy = xyxy = xyyx = yyxx = yxyx
= yxxy = xxzz = xzxz = xzzx
= zzxx = zxzx = zxxz = yyzz
= yzyz = yzzy = zzyy = zyzy
= zyyz

8. Isotropic medium

21 nonzero components, 1 of which is independent. (one more relation than in the cubic case: xxxx = xxyy + xyxy + xyyx.)

9. Axially symmetric body

Classes C_∞, $C_{\infty v}$, $C_{\infty h}$, $D_{\infty h}$: identically as in the hexagonal case.

II.8.1.4 Optical activity (the tensor is antisymmetric in its first two indices)

1. Triclinic

Class 1 (C_1): 18 nonzero components, 9 of which are independent.

xyy = − yxy, yzz = − zyz, xyz = − yxz
yxx = − xyx, zxx = − xzx, xzy = − zxy
xzz = − zxz, zyy = − yzy, yzx = − zyx

Class $\bar{1}$ (C_i): all components vanish.

2. Monoclinic

Class 2 (C_2): 10 nonzero components, 5 of which are independent.

zxx = − xzx, zyy = − yzy
xyz = − yxz, xzy = − zxy, yzx = − zyx

Class m (C_s): 8 nonzero components, 4 of which are independent.

xyy = − yxy, yxx = − xyx
xzz = − zxz, yzz = − zyz

Class 2/m (C_{2h}): all components vanish.

3. Orthorhombic

Class 222 (D_2): 6 nonzero components, 3 of which are independent.

xyz = − yxz, xzy = − zxy, yzx = − zyx

Class 2mm (C_{2v}): 4 nonzero components, 2 of which are independent.

zxx = − xzx, zyy = − yzy

Class mmm (D_{2h}): all components vanish.

4. Quadratic

Class 4 (C_4): 10 nonzero components, 3 of which are independent.

zxx = − xzx = zyy = − yzy
xyz = − yxz
xzy = − zxy = zyx = − yzx

Class $\bar{4}$ (S_4): 8 nonzero components, 2 of which are independent.

zxx = − xzx = yzy = − zyy
xzy = − zxy = yzx = − zyx

Class 4/m (C_{4h}): all components vanish.

Class 422 (D_4): 6 nonzero components, 2 of which are independent.

xyz = − yxz
xzy = − zxy = zyx = − yzx

Class 4mm (C_{4v}): 4 nonzero components, 1 of which is independent.

zxx = − xzx = zyy = − yzy

Class $\bar{4}$2m (D_{2d}): 4 nonzero components, 1 of which is independent.

xzy = − zxy = yzx = − zyx

Class 4/mmm (D_{4h}): all components vanish.

5. Rhombohedral

Class 3 (C_3): 10 nonzero components, 3 of which are independent.

zxx = − xzx = zyy = − yzy
xyz = − yxz
xzy = − zxy = zyx = − yzx

Class $\bar{3}$ (S_6): all components vanish.

Class 32 (D_3): 6 nonzero components, 2 of which are independent.

xyz = − yxz
xzy = − zxy = zyx = − yzx

Class 3m (C_{3v}): 4 nonzero components, 1 of which is independent.

zxx = − xzx = zyy = − yzy

Class $\bar{3}$m (D_{3d}): all components vanish.

6. **Hexagonal**

Class 6 (C_6): 10 nonzero components, 3 of which are independent.

zxx = − xzx = zyy = − yzy
xyz = − yxz
xzy = − zxy = zyx = − yzx

Class $\bar{6}$ (C_{3h}): all components vanish.

Class 6/m (C_{6h}): all components vanish.

Class 622 (D_6): 6 nonzero components, 2 of which are independent.

xyz = − yxz
xzy = − zxy = zyx = − yzx

Class 6mm (C_{6v}): 4 nonzero components, 1 of which is independent.

zxx = − xzx = zyy = − yzy

Class $\bar{6}$2m (D_{3h}): all components vanish.

Class 6/mmm (D_{6h}): all components vanish.

7. **Cubic**

Classes 23 (T) and 432 (0): 6 nonzero components, 1 of which is independent.

xyz = − yxz = yzx = − zyx = zxy = − xzy

Classes m3 (T_h), m3m (0_h) and $\bar{4}$3m (T_d): all components vanish.

8. **Isotropic medium**

Classes I, K: 6 nonzero components, 1 of which is independent.

xyz = − yxz = yzx = − zyx = zxy = − xzy

Classes I_h, K_h: all components vanish.

9. **Axially symmetric body**

Class ∞ (C_∞): 10 nonzero components, 3 of which are independent.

zxx = − xzx = zyy = − yzy
xyz = − yxz
xzy = − zxy = zyx = − yzx

Class ∞m ($C_{\infty v}$): 4 nonzero components, 1 of which is independent.

zxx = − xzx = zyy = −yzy

Classes ∞/m ($C_{\infty h}$) and ∞/mm ($D_{\infty h}$): all components vanish.

II.8.1.5 Faraday effect and inverse Faraday effect (the tensor is antisymmetric in its first two indices; we have assumed that the two tensors are identical—a hypothesis that is found to hold if electronic absorption in the system is neglected)

1. **Triclinic**

Classes 1 (C_1) and $\bar{1}$ (C_i):
18 nonzero components, 9 of which are independent.

xyy = − yxy, yxx = − xyx, xzz = − zxz
yxx = − xyx, zxx = − xzx, zyy = − yxy
xyz = − yxz, xzy = − zxy, yzx = − zyx

2. **Monoclinic**

Classes 2 (C_2), m (C_s), and 2/m (C_{2h}):
10 nonzero components, 5 of which are independent.

zxx = − xzx, zyy = − yzy
xzy = − zxy, xyz = − yxz, yzx = − zyx

3. Orthorhombic
Classes 222 (D_2), 2mm (C_{2v}), and mmm (D_{2h}): 6 nonzero components, 3 of which are independent.

xyz = – yxz, xzy = – zxy, yzx = – zyx

4. Quadratic
Classes 4 (C_4), $\bar{4}$ (S_4), and 4/m (C_{4h}): 10 nonzero components, 3 of which are independent.

zxx = – xzx = zyy = – yzy
xyz = – yxz
xzy = – zxy = zyx = – yzx

Classes 422 (D_4), 4mm (C_{4v}), $\bar{4}$2m (D_{2d}), and 4mmm (D_{4h}): 6 nonzero components, 2 of which are independent.

xyz = – yxz
xzy = – zxy = zyx = – yzx

5. Rhombohedral
Classes 3 (C_3) and $\bar{3}$ (S_6): 10 nonzero components, 3 of which are independent.

zxx = – xzx = zyy = – yzy
xyz = – yxz
xzy = – zxy = zyx = – yzx

Classes 3m (C_{3v}), 32 (D_3) and $\bar{3}$m (D_{3d}): 6 nonzero components, 2 of which are independent..

xyz = – yxz
xzy = – zxy = zyx = – yzx

6. Hexagonal
Classes 6 (C_6), $\bar{6}$ (C_{3h}) and 6/m (C_{6h}): 10 nonzero components, 3 of which are independent.

zxx = – xzx = zyy = – yzy
xyz = – yxz
xzy = – zxy = zyx = – yzx

Classes 622 (D_6), 6mm (C_{6v}), $\bar{6}$2m (D_{3h}), and 6/mmm (D_{6h}): 6 nonzero components, 2 of which are independent.

xyz = – yxz
xzy = – zxy = zyx = – yzx

7. Cubic
Classes 23 (T), m3 (T_h), 432 (0), m3m (0_h), and $\bar{4}$3m (T_d): 6 nonzero components, 1 of which is independent.

xyz = – yxz = yzx = – zyx = zxy
= – xzy

8. Isotropic medium
Classes I, I_h, K, K_h: 6 nonzero components, 1 of which is independent.

xyz = – yxz = yzx = – zyx = zxy
= – xzy

9. Axially symmetric body
Classes ∞ (C_∞) and ∞/m ($C_{\infty h}$):
10 nonzero components, 3 of which are independent.

zxx = – xzx = zyy = – yzy
xyz = – yxz
xzy = – zxy = zyx = – yzx

Classes ∞m ($C_{\infty v}$) and ∞/mm ($D_{\infty h}$):
6 nonzero components, 2 of which are independent.
xyz = – yxz
xzy = – zxy = zyx = – yzx

II.8.2 Numbers of independent (I) and nonzero (N) components of the electric dipolar (P), quadrupolar (Q), octopolar (Ω), and hexadecapolar (ϕ) moments for all the classes of symmetry

System	Symbol of the group		Numbers of independent and nonzero components							
	International	Shoenflies	P_α		$Q_{\alpha\beta}$		$\Omega_{\alpha\beta\gamma}$		$\phi_{\alpha\beta\gamma\delta}$	
			I	N	I	N	I	N	I	N
Triclinic	1	C_1	3	3	5	9	7	27	9	81
	$\bar{1}$	C_i	0	0	5	9	0	0	9	81
Monoclinic	m	C_s	2	2	3	5	4	14	5	41
	2	C_2	1	1	3	5	3	13	5	41
	2/m	C_{2h}	0	0	3	5	0	0	5	41
Orthorhombic	2mm	C_{2v}	1	1	2	3	2	7	3	21
	222	D_2	0	0	2	3	1	6	3	21
	mmm	D_{2h}	0	0	2	3	0	0	3	21
Quadratic	4	C_4	1	1	1	3	1	7	3	29
	$\bar{4}$	S_4	0	0	1	3	2	12	3	29
	4m	C_{4h}	0	0	1	3	0	0	3	29
	4/mm	C_{4v}	1	1	1	3	1	7	2	21
	$\bar{4}2m$	D_{2d}	0	0	1	3	1	6	2	21
	4/mmm	D_{4h}	0	0	1	3	0	0	2	21
	422	D_4	0	0	1	3	0	0	2	21
Rhombohedral	3	C_3	1	1	1	3	3	15	3	53
	$\bar{3}$	S_6	0	0	1	3	0	0	3	53
	3m	C_{3v}	1	1	1	3	2	11	2	37
	32	D_3	0	0	1	3	1	4	2	37
	$\bar{3}m$	D_{3d}	0	0	1	3	0	0	2	37
Hexagonal	6	C_6	0	0	1	3	2	8	1	21
	$\bar{6}$	C_{3h}	1	1	1	3	1	7	1	21
	6/m	C_{6h}	0	0	1	3	0	0	1	21
	$\bar{6}2m$	D_{3h}	0	0	1	3	1	4	1	21
	6mm	C_{6v}	1	1	1	3	1	7	1	21
	622	D_6	0	0	1	3	0	0	1	21
	6/mmm	D_{6h}	0	0	1	3	0	0	1	21
Cubic	23	T	0	0	0	0	1	6	1	21
	m3	T_h	0	0	0	0	0	0	1	21
	$\bar{4}3m$	T_d	0	0	0	0	1	6	1	21
	432	0	0	0	0	0	0	0	1	21
	m3m	0_h	0	0	0	0	0	0	1	21
Isotropic medium	Sphere	K	0	0	0	0	0	0	0	0
		K_h	0	0	0	0	0	0	0	0
Axially symmetric bodies	∞	C_∞	1	1	1	3	1	7	1	21
	∞m	$C_{\infty v}$	1	1	1	3	1	7	1	21
	∞/m	$C_{\infty h}$	0	0	1	3	0	0	1	21
	∞/mm	$D_{\infty h}$	0	0	1	3	0	0	1	21

II.8.3 Independent and nonzero components of the multipole moments for all the classes of symmetry

II.8.3.1 Dipole moment

System	Symbol of the group: International	Symbol of the group: Schoenflies	Number of nonzero components	Nonzero components
Triclinic	1	C_1	3	p_x, p_y, p_z
	$\bar{1}$	C_i	0	
Monoclinic	m	C_s	2	p_x, p_y
	2	C_2	1	p_z
	2/m	C_{2h}	0	
Orthorhombic	2mm	C_{2v}	1	p_z
	222	D_2	0	
	mmm	D_{2h}	0	
Quadratic	4	C_4	1	p_z
	$\bar{4}$	S_4	0	
	4/m	C_{4h}	0	
	4mm	C_{4v}	1	p_z
	$\bar{4}2m$	D_{2d}	0	
	4/mmm	D_{4h}	0	
	422	D_4	0	
Rhombohedral	3	C_3	1	p_z
	$\bar{3}$	S_6	0	
	3m	C_{3v}	1	p_z
	32	D_3	0	
	$\bar{3}m$	D_{3d}	0	
Hexagonal	$\bar{6}$	C_{3h}	0	
	6	C_6	1	p_z
	6/m	C_{6h}	0	
	$\bar{6}2m$	D_{3h}	0	
	6mm	C_{6v}	1	p_z
	622	D_6	0	
	6/mmm	D_{6h}	0	
Cubic	23	T	0	
	m3	T_h	0	
	$\bar{4}3m$	T_d	0	
	432	0	0	
	m3m	0_h	0	
Isotropic medium	sphere	K	0	
		K_h	0	
Axially symmetric bodies	∞	C_∞	1	p_z
	∞ m	$C_{\infty v}$	1	p_z
	∞/m	$C_{\infty h}$	0	
	∞/mm	$D_{\infty h}$	0	

II.8.3.2 Quadrupole moment

1. **Triclinic**
   For classes 1 (C_1) and $\bar{1}$ (C_i), the tensor has 9 nonzero components, 5 of which are independent.

   xx, yy, xy = yx, xz = zx, yz = zy,

   zz = – (xx + yy)

2. **Monoclinic**
   For classes m (C_s), 2 (C_2), and 2/m (C_{2h}), the tensor has 5 nonzero components, 3 of which are independent.

   xx, yy, xy = yx, zz = – (xx + yy)

3. **Orthorhombic**
   For classes 2mm (C_{2v}), 222 (D_2), and mmm (D_{2h}), the tensor has 3 nonzero components, 2 of which are independent.

   xx, yy, zz = – (xx + yy)

4. **Quadratic**
   For classes 4 (C_4), $\bar{4}$ (S_4), $\bar{4}$/m (C_{4h}), 4mm (C_{4v}), $\bar{4}$2m (D_{2d}), 4/mmm (D_{4h}), and 422 (D_4), the tensor has 3 nonzero components, 1 of which is independent.

   xx = yy, zz = – (xx + yy)

5. **Rhombohedral**
   For classes 3 (C_3), $\bar{3}$ (S_6), 3m (C_{3v}), 32 (D_3), and $\bar{3}$m (D_{3d}), the tensor has 3 nonzero components, 1 of which is independent.

   xx = yy, zz = – (xx + yy)

6. **Hexagonal**
   For classes $\bar{6}$ (C_{3h}), 6 (C_6), 6/m (C_{6h}), $\bar{6}$2m (D_{3h}), 6mm (C_{6v}), 622 (D_6), and 6/mmm (D_{6h}), the tensor has 3 nonzero components, 1 of which is independent.

   xx = yy, zz = – (xx + yy)

7. **Cubic**
   For classes 23 (T), m3 (T_h), $\bar{4}$3m (T_d), 432 (0), and m3m (0_h), all tensor components vanish.

8. **Isotropic medium**
   All components vanish.

9. **Axially symmetric body**
   For classes ∞ (C_∞), ∞m ($C_{\infty v}$), ∞/m ($C_{\infty h}$), and ∞/mm ($D_{\infty h}$), the tensor has 3 nonzero components, 1 of which is independent.

   xx = yy, zz = – (xx + yy)

II.8.3.3 Octopole moment

1. **Triclinic**
   For class 1 (C_1): 27 nonzero components, 7 of which are independent.

   xyy = yyx = yxy
   xxy = yxx = xyx
   xyz = xzy = yxz = yzx = zxy = zyx
   xxz = zxx = xzx
   yyz = zyy = yzy
   xzz = zzx = zxz
   yzz = zzy = zyz

   xxx = – (xyy + xzz), yyy = – (xxy + yzz)
   zzz = – (xxz + yyz)

   For class $\bar{1}$ (C_i): all components vanish.

2. **Monoclinic**
   Class m (C_s): 14 nonzero components, 4 of which are independent.

   xyy = yyx = yxy
   xxy = yxx = xyx
   xzz = zzx = zxz
   yzz = zzy = zyz

   xxx = – (xyy + xzz), yyy = – (xxy + yzz)

Class 2 (C_2): 13 nonzero components, 3 of which are independent.

xxz = zxx = xzx
yyz = zyy = yzy
xyz = xzy = yxz = yzx = zxy = zyx

zzz = – (xxz + yyz)

Class 2/m (C_{2h}): all components vanish.

3. **Orthorhombic**
Class 2mm (C_{2v}): 7 nonzero components, 2 of which are independent.

xxz = zxx = xzx
yyz = zyy = yzy

zzz = – (xxz + yyz)

Class 222 (D_2): 6 nonzero components, 1 of which is independent.

xyz = xzy = yxz = yzx = zxy = zyx

Class mmm (D_{2h}): all components vanish.

4. **Quadratic**
Class 4 (C_4): 7 nonzero components, 1 of which is independent.

xxz = zxx = xzx = yyz = zyy = yzy

zzz = – (xxz + yyz)

Class $\bar{4}$ (S_4): 12 nonzero components, 2 of which are independent.

xxz = zxx = xzx = – yyz = – zyy = – yzy
xyz = xzy = yxz = yzx = zxy = zyx

Class 4/m (C_{4h}): all components vanish.

Class 4 mm (C_{4v}): 7 nonzero components, 1 of which is independent.

xxz = zxx = xzx = yyz = zyy = yzy

zzz = – (xxz + yyz)

Class $\bar{4}$2m (D_{2d}): 6 nonzero components, 1 of which is independent.

xyz = xzy = yxz = yzx = zxy = zyx

Class 422 (D_4): all components vanish.

Class 4/mmm (D_{4h}): all components vanish.

5. **Rhombohedral**
Class 3 (C_3): 15 nonzero components, 3 of which are independent.

xxx = – xyy = – yxy = – yyx
xxz = xzx = zxx = yyz = zyy = yzy
yyy = – yxx = – xyx = – xxy

zzz = – (xxz + yyz)

Class $\bar{3}$ (S_6): all components vanish.

Class 3m (C_{3v}): 11 nonzero components, 2 of which are independent.

xxz = xzx = zxx = yyz = yzy = zyy
yyy = – yzz = – xyx = – xxy

zzz = – (xxz + yyz)

Class 32 (D_3): 4 nonzero components, 1 of which is independent.

xxx = – xyy = – yxy = – yyx

Class $\bar{3}$m (D_{3d}): all components vanish.

6. **Hexagonal**
Classes 6 (C_6) and 6mm (C_{6v}): 7 nonzero components, 1 of which is independent.

xxz = zxx = xzx = yyz = zyy = yzy

zzz = – (xxz + yyz)

Class $\bar{6}$ (C_{3h}): 8 nonzero components, 2 of which are independent.

xxx = – xyy = – yxy = – yyx
yyy = – yxx = – xyx = – xxy

Class $\bar{6}$2m (D_{3h}): 4 nonzero components, 1 of which is independent.

yyy = – yxx = – xyx = – xxy

Classes 6/m (C_{6h}), 622 (D_6), and 6/mmm (D_{6h}): all components vanish.

7. **Cubic**
Classes 23 (T) and $\bar{4}$3m (T_d): 6 nonzero components, 1 of which is independent.

xyz = xzy = yxz = yzx = zxy = zyx

Classes m3 (T_h), 432 (0), and m3 (0_h): all components vanish.

8. **Isotropic medium:** all components vanish.

9. **Axially symmetric body**
Class ∞ (C_∞) and ∞m ($C_{\infty v}$): 7 nonzero components, 1 of which is independent.

xxz = zxx = xzx = yyz = yzy = zyy
zzz = – (xxz + yyz)

Classes ∞/m ($C_{\infty h}$) and ∞/mm ($D_{\infty h}$): all components vanish.

II.8.3.4 Hexadecapole moment

1. **Triclinic**
Classes 1 (C_1) and $\bar{1}$ (C_i): 81 nonzero components, 9 of which are independent.

xxyy = xyxy = xyyx = yyxx = yxyx
= yxxy
xxzz = xzxz = xzzx = zzxx = zxzx
= zxxz
yyzz = yzyz = yzzy = zzyy = zyzy
= zyyz

xxxx = – (yyxx + zzxx)
yyyy = – (xxyy + zzyy)
zzzz = – (xxzz + yyzz)

yyyx = yyxy = yxyy = xyyy
zzyx = xzzy = xzyz = xyzz = yxzz
= yzxz = yzzx = zxyz = zyxz
= zyzx = zxzy = zzxy
yyzx = yyxz = yxyz = yxzy = yzyx
= yzxy = xyyz = xyzy = xzyy
= zyyx = zxyy = zyxy

zzzx = zzxz = zxzz = xzzz
xxyz = xyxz = xyzx = xxzy = xzxy
= xzyx = yxxz = yxzx = yzxx
= zxxy = zxyx = zyxx

zzzy = zzyz = zyzz = yzzz

xxxy = xxyx = xyxx = yxxx
= – (yyyx + zzyx)
xxxz = xxzx = xzxx = zxxx
= – (yyzx + zzzx)
yyyz = yyzy = yzyy = zyyy
= – (xxyz + zzzy)

2. **Monoclinic**
Classes m (C_s), 2 (C_2), and 2/m (C_{2h}): 41 nonzero components, 5 of which are independent.

xxyy = xyxy = xyyx = yyxx = yxyx
= yxxy
xxzz = xzxz = xzzx = zzxx = zxzx
= zxxz
yyzz = yzyz = yzzy = zzyy = zyzy
= zyyz

xxxx = – (yyxx + zzxx)
yyyy = – (xxyy + zzyy)
zzzz = – (xxzz + yyzz)

yyyx = yyxy = yxyy = xyyy
zzyx = xzzy = xzyz = xyzz = yxzz
= yzxz = yzzx = zxyz = zyxz
= zxzy = zzxy = zyzx

xxxy = xxyx = xyxx = yxxx
= – (yyyx + zzyx)

3. **Orthorhombic**
Classes 2mm (C_{2v}), 222 (D_2), and mmm (D_{2h}): 21 nonzero components, 3 of which are independent.

xxyy = xyxy = xyyx = yyxx = yxyx
= yxxy
xxzz = xzxz = xzzx = zzxx = zxzx
= zxxz
yyzz = yzyz = yzzy = zzyy = zyzy
= zyyz

xxxx = − (yyxx + zzxx)
yyyy = − (xxyy + zzyy)
zzzz = − (xxzz + yyzz)

4. Quadratic

Classes 4 (C_4), $\bar{4}$ (S_4), and 4/m (C_{4h}): 29 nonzero components, 2 of which are independent.

xxyy = xyxy = xyyx = yyxx = yxyx
= yxxy
xxzz = xzxz = xzzx = zzxx = zxzx
= zxxz = yyzz = yzyz = yzzy
= zzyy = zyzy = zyyz
= xxxy = yxyx = xyxx = yxxx
= − yyyx = − yyzy = − yxyy
= − xyyy

xxxx = yyyy = − (xxyy + zzyy)
zzzz = − (xxzz + yyzz)

Classes 422 (D_4), 4mm (C_{4v}), 4/mmm (D_{4h}), and $\bar{4}$2m (D_{2d}): 21 nonzero components, 2 of which are independent.

xxyy = xyxy = xyyx = yyxx = yxyx
= yxxy
xxzz = xzxz = xzzx = zzxx = zxzx
= zxxz = yyzz = yzyz = yzzy
= zzyy = zyzy = zyyz

xxxx = yyyy = − (xxyy + zzyy)
zzzz = − (xxzz + yyzz)

5. Rhombohedral

Classes 3 (C_3), and $\bar{3}$ (S_6): 53 nonzero components, 3 of which are independent.

xxyy = xyxy = xyyx = yyxx = yxyx
= yxxy
= − (1/4) xxzz = − (1/4) xzxz
= − (1/4) xzzx = − (1/4) zzxx
= − (1/4) zxzx = − (1/4) zxxz
= − (1/4) yyzz = − (1/4) yzyz
= − (1/4) yzzy = − (1/4) zzyy
= − (1/4) zyzy = − (1/4) zyyz

xxxz = xxzx = xzxx = zxxx
= − yxyz = − yyzx = − yzyx
= − yyxz = − yxzy = − yzxy
= − xyyz = − xyzy = − xzyy
= − zxyy = − zyxy = − zyyx

yyyz = yyzy = yzyy = − zyyy
= − xyxz = − xxzy = − xzxy
= − xxzy = − xyzx = − xzyx
= − yxxz = − yxzx = − yzxx
= − zyxx = − zxyx = − zxxy

xxxx = yyyy = 3xxyy = − (3/4) xxzz

zzzz = − 2xxzz = 8xxyy

Classes 32 (D_3), 3m (C_{3v}), and $\bar{3}$m (D_{3d}): 37 nonzero components, 2 of which are independent.

xxyy = xyxy = xyyx = yyxx = yxyx
= yxxy = − (1/4) xxzz
= − (1/4) zxzx = − (1/4) xzzx
= − (1/4) zzxx = − (1/4) xzxz
= − (1/4) zxxz = − (1/4) yyzz
= − (1/4) yzyz = − (1/4) yzzy
= − (1/4) zzyy = − (1/4) zyzy
= − (1/4) zyyz

yyyz = yyzy = yzyy = zyyy
= − xyxz = − xxzy = − xzxy
= − xxyz = − xyzx = − xzyx
= − yxxz = − yxzx = − yzxx
= − zyxx = − zxyx = − zxxy

xxxx = yyyy = 3xxyy = − (1/4) xxzz

zzzz = − 2xxzz = 8xxyy

6. Hexagonal

Classes 6 (C_6), $\bar{6}$ (C_{3h}), 6/m (C_{6h}), 622 (D_6), 6mm (C_{6v}), $\bar{6}$2m (D_{3h}), and 6/mmm (D_{6h}): 21 nonzero components, 1 of which is independent.

xxyy = xyxy = xyyx = yyxx = yxyx
= yxxy = − (1/4) xxzz
= − (1/4) xzxz = − (1/4) xzzx
= − (1/4) zzxx = − (1/4) zxzx
= − (1/4) zxxz = − (1/4) yyzz
= − (1/4) yzyz = − (1/4) yzzy
= − (1/4) zzyy = − (1/4) zyzy
= − (1/4) zyyz

xxxx = yyyy = 3xxyy = − (1/4) xxzz

zzzz = − 2xxzz = 8xxyy

7. Cubic
Classes 23 (T), m3 (T_h), 432 (0), $\bar{4}$3m (T_d), and m3m (0_h): 21 nonzero components, 1 of which is independent.

xxyy = xyxy = xyyx = yyxx = yxyx
= yxxy = xxzz = xzxz = xzzx
= zzxx = zxzx = zxxz = yyzz
= zyzy = zyyz = zzyy = yzyz
= yzzy

xxxx = yyyy = zzzz = – 2xxyy

8. Isotropic medium: all components vanish.

9. Axially symmetric body
Classes ∞ (C_∞), ∞m ($C_{\infty v}$), ∞/m ($C_{\infty h}$), and ∞/mm ($D_{\infty h}$): The same as for the hexagonal system.

CHAPTER III

Passage from the Microscopic to the Macroscopic: Statistical Physics

In the first two chapters we discussed the representation of a microscopic system, the results of measurements performed thereon, and its tensorial optical properties.

The results hitherto obtained would suffice for the treatment of experiments carried out on a *single* atom or an *isolated* molecule; such experiments may prove feasible in the near future by laser manipulation and trapping of quantal objects, procedures we shall touch on in Chapter VII. However, when introducing the density operator in Chapter I, we already opened the way for the description of collections of microscopic systems in different states, i.e., for the transition from microscopic to macroscopic.

In fact, the vast majority of our experimentation bears on macroscopic systems consisting of great numbers of microscopic entities, of unceasingly growing chemical and structural complexity. The molecules of these macroscopic systems can exhibit various kinds of ordering, both in position and in orientation. We have in mind the novel phases of liquid crystals presenting various, but recently elucidated, textures synthesized in the past few years.

Moreover, to these naturally favored kinds of ordering, we have to add the effects induced by the application of laser waves, often of very high intensity, by the experimenter. The electric field of the wave can induce effects of, for example, reorientation and electrostriction involving perturbations in the order of the microscopic entities constituting the macroscopic system under investigation.

Hence, the problem of micro- to macropassage resides, in the first place, in calculations involving *statistical averages* over assemblages of N microscopic particles in a volume V. Consider a molecular optical property, represented in the molecular system of reference (e_α, e_β, e_γ), by the tensor $t_{\alpha\beta\gamma}$ (we already know how to reduce its structural complexity). This property, expressed in laboratory coordinates (E_i, E_j, E_k), is described by Eq. II.2.3:

$$T_{ijk} = c_{i\alpha}\, c_{j\beta}\, c_{k\gamma}\, t_{\alpha\beta\gamma}$$

This equation can be generalized to order n by way of the product of n directional cosines of the transformation. Generally, we shall have to calculate *statistical orientation averages to take into account all the molecular orientations possible within the laboratory system of coordinates.* The statistical average will be denoted by $< >_\Omega$ to distinguish it from the quantum mean value $< >$ defined in the beginning of Chapter I. We thus define a *molecular average* as

$$< T_{ijk} >_\Omega = < c_{i\alpha}\, c_{j\beta}\, c_{k\gamma} >_\Omega t_{\alpha\beta\gamma}$$

or, for simplicity, as

$$T_{ijk} = < c_{i\alpha}\, c_{j\beta}\, c_{k\gamma} >_\Omega t_{\alpha\beta\gamma}$$

The property relating to the system as a whole is obtained simply on multiplying the right-hand term of the above equation by the number N of microscopic entities.

Most of the systems we shall be dealing with will *macroscopically present spherical symmetry in the absence of external stresses (the electric and magnetic fields of the laser wave, or auxiliary fields).* The system then consists of microsystems *at random orientation* (at least on the macroscopic scale). This is the case for almost all gases and liquids, obviously to the exception of certain phases of liquid crystals. We now have to distinguish two essential cases:

— No appreciable external stresses are present. The average is now referred to as the *thermal equilibrium average*.

— External stresses affecting the random orientation of the microsystems are present. Most often the following statements hold:

(i) The electric field of the laser beam induces a privileged orientation, thus destroying the isotropicity of the macroscopic properties of the system as a whole.

(ii) The experimenter deliberately causes the destruction of orientational symmetry by externally applying static or slowly varying electric or magnetic fields.

Obviously, the very nature of the optical properties—we have in mind their tensoriality—has led us to give priority to the phenomenon of reorientation, especially when caused by the electric field of a laser wave. The wave, however, can also induce many other perturbations: *electronic perturbations* (very fast and well adapted to important applications in materials of any kind, particularly semiconductors); *translational perturbations in radial ordering* which, too, are fast and consist of *molecular redistribution* on the microscopic scale keeping the mean density of the microsytems unchanged, or less fast (electrostrictive effects involving heterogeneity in density); and, finally, perturbations of thermal origin, very frequent if the wave is absorbed by the material. If necessary, we shall be discussing these perturbations in the course of this work.

Moreover, in addition to that of the statistical average, we shall have to give attention to the problem of *the internal field*. In fact, the field acting on an atom or molecule of the medium is by no means identical to the electric field conveyed by the externally applied laser beam. This, indeed, is a very old—and very badly solved—problem. We shall show how to take it into account in situations of nonlinear laser-matter interaction.

Finally, in an altogether general approach, we shall propose an expression for the macroscopic susceptibility of order n of dense fluids.

III.1 Statistical average and examples

III.1.1 Review of statistical physics

The total energy E of a conservative system (like those we deal with in this book) takes the form

$$E = T + V_F + V_I$$

where T is the kinetic energy, V_F is the potential energy due to external stresses (electric and magnetic fields, if present; gravitation; etc.), and V_I is the potential energy of interaction between the microsystems. If each of the microsystems possesses s degrees of freedom, the configurational space possesses 2s dimensions, denoted by Γ, and E is a function of the variables Γ, which we refer to as canonical variables. Consider a quantity Q dependent on the canonic variables Γ. The statistical average of Q is defined as

$$< Q(\Gamma) > = \int_\Gamma Q(\Gamma)\, f(\Gamma)\, d\Gamma \qquad \text{(III.1.1)}$$

where f (Γ) is the *Gibbs distribution function*, given by

$$f(\Gamma) = \exp\left[-E(\Gamma)/kT\right] / \int_\Gamma \exp\left[-E(\Gamma)/kT\right] d\Gamma \qquad \text{(III.1.2)}$$

Let us concentrate on the case when Q formally depends on the generalized coordinates τ only (for example, τ can stand for the variables of position **r** or the variables of orientation Ω). Then

$$< Q(\tau) > = \int_\Gamma Q(\tau)\, f(\Gamma)\, d\Gamma \qquad \text{(III.1.3)}$$

First of all, we calculate the integrals over the configurational variables whence we extract the τ. We thus obtain

$$< Q(\tau) > = \frac{\int_\tau Q(\tau) \exp[-V_F(\tau)/kT] \exp[-V_I(\tau)/kT] \, d\tau}{\int_\tau \exp[-V_F(\tau)/kT] \exp[-V_I(\tau)/kT] \, d\tau} \tag{III.1.4}$$

The above equation will serve as starting point for calculating the statistical average in the presence of external stresses. This can be the case of a gas of molecules with dipole moments **p** in a static electric field **E**. Under the action of the external field strength (the external stress) each molecule acquires potential energy V_F to an amount proportional to **pE**, destroying the initial orientational isotropicity of the system as a whole.

On assuming

$$V_F(\tau) << V_I(\tau) \text{ and } V_F(\tau) << kT \tag{III.1.5}$$

and on performing the expansion

$$\exp[-V_F(\tau)/kT] = \sum_{n=0}^{\infty} (1/n!) [-V_F(\tau)/kT]^n \tag{III.1.6}$$

Eq. III.1.4 leads to

$$<Q(\tau)> = \frac{\sum_{n=0}^{\infty} (1/n!)(-1/kT)^n < Q V_F^n(\tau) >_0}{\sum_{n=0}^{\infty} (1/n!)(-1/kT)^n < V_F^n(\tau) >_0} \tag{III.1.7}$$

where the average $< >_0$ is defined as

$$<Q(\tau)>_0 = \frac{\int_\tau Q(\tau) \exp[-V_I(\tau)/kT] \, d\tau}{\int_\tau \exp[-V_I(\tau)/kT] \, d\tau} \tag{III.1.8}$$

The value $< >_0$ is the average calculated *in the absence of external forces*, i.e., the statistical average for the unperturbed system. It characterizes the *natural properties of the system* as well as *its structure*. It is referred to as *the statistical average at thermal equilibrium*. Equation. III.1.4 is important insofar as it enables us (obviously, approximately) to express *an average in the presence of stress* by an expression where each of the terms involves *a thermal equilibrium average. The stress applied perturbs the system (later on we shall see that it simultaneously gives rise to the effect under investigation) but provides information concerning the system at equilibrium (as long as the perturbation is not excessively great).*

As done above, one can express the distribution function $f(\tau)$ in a form similar to that of Eq. III.1.7, i.e., in the form of a series expansion in growing orders of the perturbation. The following table shows the results thus obtained.

We shall now concretize some of the results derived above by invoking examples from the domain of radiation-matter interaction.

III.1.2 Examples of applications: Averages of tensorial entities

III.1.2.1 Averages at thermal equilibrium

We shall have recourse to components of the Kronecker and Levi-Civita tensors to define—when possible—scalar quantities with completely symmetric indices (we deal here with spherical symmetric macrosystems only). Here are some examples:

(a) Average of the product of two unit vector components

Consider, in orthonormal three-dimensional space, two vectors **a** and **b**, the characteristics of any microscopic system, with components a_i and b_j. We have to calculate $< a_i\, b_j >_0$, which is a tensorial entity of rank 2 symmetric in i and j. We

TABLE III.1.1 — **The first four terms of the average of a quantity Q expanded in a series in the perturbation V_F: $< Q > = < Q >_0 + < Q >_1 + \cdots$ and that of the associated distribution function f: $< f > = f_0 + f_1 + \cdots$ (the symbol 0 stands for the average at thermal equilibrium as obtained for $V_F = 0$.) Deliberately, we have omitted the τ-dependence of Q and V_F.**

Order n of the expansion	$< Q >_n$	f_n
0	$< Q >_0$	f_0
1	$(-1/kT)\left[< QV_F >_0 - < Q >_0 < V_F >_0\right]$	$-(f_0/kT)\left[V_F - < V_F >_0\right]$
2	$(-1/2k^2T^2)\left[< QV_F^2 >_0 - < Q >_0 < V_F^2 >_0 - 2\left(< QV_F >_0 - < Q >_0 < V_F >_0\right) < V_F >_0\right]$	$(f_0/2k^2T^2)\left[V_F^2 - < V_F^2 >_0 - 2\left(V_F - < V_0 >\right) < V_F >_0\right]$
3	$(-1/6k^3T^3)\left[< QV_F^3 >_0 - < Q >_0 < V_F^3 >_0 - 3\left(< QV_F^2 >_0 < V_F >_0 + < QV_F >_0 < V_F^2 >_0\right) + 6\left(< QV_F >_0 < V_F >_0 + < Q >_0 < V_F^2 >_0 - < Q_0 > < V_F >_0^2\right) < V_F >_0\right.$	$(f_0/6k^3T^3)\left[V_F^3 - < V_F^3 >_0 - 3\left(V_F^2 < V_F >_0 + V_F < V_F^2 >_0\right) + 6\left(V_F < V_F >_0 + < V_F^2 >_0 - < V_F >_0^2\right) < V_F >_0\right]$

use the components of the Kronecker tensor (a symmetric tensor of rank 2) to write

$$< a_i\, b_j >_0 = x\, \delta_{ij}$$

where x is a scalar to be determined.

On right and left multiplication by δ_{ij}, let

$$< a_i\, b_i >_0 = \mathbf{a}\, \mathbf{b} = 3x$$

We arrive at

$$< a_i\, b_j >_0 = (\mathbf{a}\, \mathbf{b} / 3)\, \delta_{ij} \qquad \text{(III.1.9)}$$

and if $\mathbf{a} = \mathbf{b}$ we obtain

$$< a_i\, a_j >_0 = (1/3)\, \delta_{ij} \qquad \text{(III.1.10)}$$

(b) Average of the product of four unit vector components

We shall have to generate the components of a completely symmetric tensor of rank 4. We shall make use of the products of two components of the Kronecker tensor. As above, we write

$$< a_i\, b_j\, c_k\, d_l > = x\, \delta_{ij}\, \delta_{kl} + y\, \delta_{ik}\, \delta_{jl} + z\, \delta_{il}\, \delta_{jk} \qquad \text{(III.1.11)}$$

On performing the necessary contractions, the reader will easily determine the three scalars x, y, and z:

$$x = (1/30)\, [4\, (\mathbf{a}\, \mathbf{b})\, (\mathbf{c}\, \mathbf{d}) - (\mathbf{a}\, \mathbf{c})\, (\mathbf{b}\, \mathbf{d}) - (\mathbf{a}\, \mathbf{d})\, (\mathbf{b}\, \mathbf{c})\,] \qquad \text{(III.1.12)}$$

y and z are obtained by circular permutation of the vectors **a**, **b**, **c**, and **d**. In particular, on equating **a** = **b** = **c** = **d**, we get

$$< a_i\, a_j\, a_k\, a_l >_0 = (1/15) \left(\delta_{ij}\, \delta_{kl} + \delta_{ik}\, \delta_{jl} + \delta_{il}\, \delta_{jk} \right)$$

$$< a_i^2\, a_j^2 >_0 = 1/15 \qquad \text{(III.1.13)}$$

$$< a_i^4 >_0 = 1/5$$

We now proceed directly to apply the results obtained above to the description of Rayleigh scattering of an optical wave on matter. We have shown (see Eq. II.2.3) that here the scattered radiation decomposes into a part called *symmetric* and a part called *antisymmetric* and that the symmetric part appears as the sum of an *isotropic* part and an *anisotropic* part. The following inset permits the determination of the latter two contributions.

(c) Average of the product of an odd number of unit vector components

All such averages vanish since symmetric tensors of odd rank do not exist.

Inset III.1 ***Rayleigh scattering (without change in frequency) by a fluid of noninteracting molecules***

The effect is due essentially to the nonresonant interaction between the light wave and the electrons of the molecules.

$\mathbf{E}_i$ and $\mathbf{E}_d$ are the incident and scattered electric fields; $\mathbf{e}_i$ and $\mathbf{e}_d$ are the respective unit vectors.

The following Rayleigh ratio, proportional to the number density ρ of the particles and inversely proportional to power 4 of the incident wavelength λ_L, has been shown to hold:

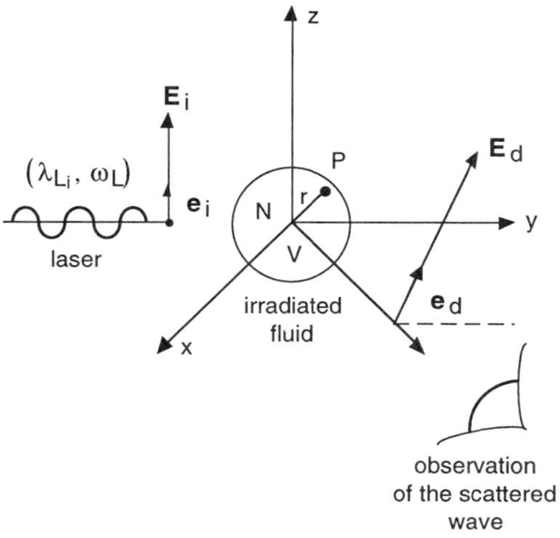

$$R = \left(64\,\pi^4 e^4 \rho / h^2 \lambda_L^4\right) \sum_m \left[\omega_{0m}^2 / \left(\omega_L^2 - \omega_{0m}^2\right)\right] \left[\left(\mathbf{r}\mathbf{e}_i\right)_{0m} \left(\mathbf{r}\mathbf{e}_d\right)_{0m}\right]^2 \quad \text{(III.1.14)}$$

where ω_L is the circular frequency of the laser wave, ω_{0m} is one of the circular eigenfrequencies of the microscopic particles (see Chapter I; the quantum ground state is denoted by the subscript 0), e is the charge of the electron, and h is Plank's constant. **r** gives the position of the particle.

The theoretical justification of Eq. III.1.14 will be given later (see Section VIII.1.1).

The expression in square brackets at the end of the right-hand term of Eq. III.1.14 can also be written as follows:

$$\left(r_\alpha r_\beta\right)_{0m} \left(r_\gamma r_\delta\right)_{0m} e_{i\alpha}\, e_{d\beta}\, e_{i\gamma}\, e_{d\delta}$$

Thus we know that in order to take into account all possible orientations of the particles (microsystems), we have to calculate $< e_{i\alpha}\, e_{d\beta}\, e_{i\gamma}\, e_{d\delta} >_0$ — the product of four unit vector components. By Eq. III.1.11, we get

$$< e_{i\alpha}\, e_{d\beta}\, e_{i\gamma}\, e_{d\delta} >_0 = (1/30)\left[\left(3\delta_{\alpha\beta}\,\delta_{\gamma\delta} + 3\delta_{\alpha\delta}\,\delta_{\beta\gamma} - 2\delta_{\alpha\gamma}\,\delta_{\beta\delta}\right)\left(\mathbf{e}_i\,\mathbf{e}_d\right)^2 + 4\delta_{\alpha\gamma}\,\delta_{\beta\delta} - \delta_{\alpha\delta}\,\delta_{\beta\gamma} - \delta_{\alpha\gamma}\,\delta_{\beta\delta}\right] \quad \text{(III.1.15)}$$

Once the directions of polarization of the incident and scattered waves are known, Eq. III.1.15 permits the calculation of the scattering process.

Isotropic Rayleigh scattering

The isotropic Rayleigh effect is obtained by calculating separately the averages $< e_{i\alpha}\, e_{d\beta} >_0$ and $< e_{i\gamma}\, e_{d\delta} >_0$:

$$< e_{i\alpha}\, e_{d\beta}\, e_{i\gamma}\, e_{d\delta} >_{0,iso} = < e_{i\alpha}\, e_{d\beta} >_0 < e_{i\gamma}\, e_{d\delta} >_0$$

Equation III.1.10 immediately leads to

$$< e_{i\alpha}\, e_{d\beta}\, e_{i\gamma}\, e_{d\delta} >_{0,iso} = (1/9)\,(\mathbf{e}_i\, \mathbf{e}_d)^2\, \delta_{\alpha\beta}\, \delta_{\gamma\delta} \qquad \text{(III.1.16)}$$

which, on insertion into Eq. III.1.14, gives

$$R_{iso} = \left(64\pi^4 e^4 \rho / 9 h^2 \lambda_L^4\right) \sum_m \left[\omega_{0m}^2 / \left(\omega_L^2 - \omega_{0m}^2\right)^2\right] (r_\alpha\, r_\alpha)_{0m}\, (r_\beta\, r_\beta)_{0m}\, (\mathbf{e}_i\, \mathbf{e}_d)^2 \qquad \text{(III.1.17)}$$

Scattering vanishes if the vectors $\mathbf{E}_i$ and $\mathbf{E}_d$ are at right angles. If they are parallel, we get $R_{iso} = 16\pi^4 \rho \overline{\alpha}^2 / \lambda_L^4$, where $\overline{\alpha}$ is the average microscopic polarizability, given by

$$\overline{\alpha}^2 = \left(4 e^4 / 9\, h^2\right) \sum_m \left[\omega_{0m}^2 / \left(\omega_L^2 - \omega_{0m}^2\right)^2\right] (r_\alpha\, r_\alpha)_{nm}^2 \qquad \text{(III.1.18)}$$

We note that $\overline{\alpha}$ is in fact proportional to the trace (i.e., to thrice the mean value) of the second-rank tensor ($r_\alpha\, r_\beta$.) Also, Rayleigh scattering—isotropic or otherwise—is fundamentally a binary quantity, involving two points in the scattering volume. The effect is thus especially well adapted to the study of interactions between scattering microsystems.

The reader is invited to take note of the following facts:

— The relation between $\overline{\alpha}$ and $\overline{\chi}$ (the mean volume susceptibility of order 1) is particularly simple (with accuracy to the choice of internal field, a problem we shall consider later on):

$$\overline{\chi} = \rho \overline{\alpha} \qquad \text{(III.1.19)}$$

— The refractive index (see Eq. II.2.16) is proportional to $\overline{\alpha}$ (and *not* to its square) and consequently should be less sensitive to intermolecular interactions.

Anisotropic Rayleigh scattering

On removal of the isotropic average (Eq. III.1.16) from Eq. III.1.15 we remain with a scattered component, to which we refer to as *anisotropic*. We now shall calculate this anisotropic component, and the result will prove that the name is well chosen. Some straightforward calculations lead to

$$< e_{i\alpha}\, e_{d\beta}\, e_{i\gamma}\, e_{d\delta} >_{0,aniso} = (1/90)\Big[\left(9\delta_{\alpha\delta}\,\delta_{\beta\gamma} - 6\delta_{\alpha\gamma}\,\delta_{\beta\delta} - \delta_{\alpha\beta}\,\delta_{\gamma\delta}\right)(\mathbf{e}_i\, \mathbf{e}_d)^2 + 3\left(4\delta_{\alpha\gamma}\,\delta_{\beta\delta} - \delta_{\alpha\delta}\,\delta_{\beta\gamma} - \delta_{\alpha\beta}\,\delta_{\gamma\delta}\right)\Big]$$

On performing the contractions we arrive at

$$\begin{aligned} R_{ani} = \left(64\pi^4 e^4 \rho / 90 h^2 \lambda_L^4\right) \sum_m \left[\omega_{0m}^2 / \left(\omega_L^2 - \omega_{0m}^2\right)^2\right] \\ \times \left[3 \left(r_\alpha r_\beta\right)_{0m}^2 - \left(r_\alpha r_\alpha\right)_{0m}^2\right] (\mathbf{e}_i \mathbf{e}_d + 3) \end{aligned} \quad \text{(III.1.20)}$$

The presence of the expression $\left[3 \left(r_\alpha r_\beta\right)_{0m}^2 - \left(r_\alpha r_\alpha\right)_{0m}^2\right]$ in Eq. III.1.20 justifies our use of the name. Depolarized anisotropic scattering corresponds to the case when the vectors $\mathbf{E}_i$ and $\mathbf{E}_d$ are at right angles. We then write

$$\gamma^2 = \left(2e^4 / h^2\right) \sum_m \left[\omega_{0m}^2 / \left(\omega_L^2 - \omega_{0m}^2\right)\right]\left[3 \left(r_\alpha r_\beta\right)_{0m}^2 - \left(r_\alpha r_\alpha\right)_{0m}^2\right]$$

where γ defines the anisotropy of first-order polarizability of the microsystem and permits the simple expression of the relation

$$R_{ani,\, dep} = 16\pi^4 \rho\gamma^2 / 15\lambda_L^4 \quad \text{(III.1.21)}$$

Also, polarized anisotropic scattering can occur (if $\mathbf{E}_i$ and $\mathbf{E}_d$ are parallel.) The depolarization ratio then amounts to

$$R_{ani,\, pol.} / R_{ani,\, dep} = 4/3$$

(d) Average of the product of two directional cosines

As previously, we write

$$< c_{i\alpha}\, c_{j\beta} >_0 = x\delta_{ij}\, \delta_{\alpha\beta}$$

and determine x on multiplying the above equation by δ_{ij}. We get

$$< c_{i\alpha}\, c_{j\beta} >_0 = (1/3)\, \delta_{ij}\, \delta_{\alpha\beta} \quad \text{(III.1.22)}$$

Application: calculation of the macroscopic volume susceptibility of a system of "anisotropic" particles. We write (deliberately avoiding the problem of the internal field)

$$\chi_{ij} = \rho < c_{i\alpha}\, c_{j\beta} >_0 \alpha_{\alpha\beta} \quad \text{(III.1.23)}$$

whence

$$\chi_{ij} = (\rho / 3)\, \alpha_{\beta\beta}\, \delta_{ij} \quad \text{(III.1.24)}$$

Starting from a nondiagonal molecular tensor $[\alpha_{\alpha\beta}]$ we have thus arrived at a diagonal and isotropic macroscopic tensor. Here, we have a case of *isotropy by compensation.*

On multiplying Eq. III.1.24 by δ_{ij}, we get

$$\chi_{ii} = \rho\, \alpha_{\beta\beta}$$

and, on dividing the above obtained trace by 3 we arrive at equality (Eq. III.1.19) between averages.

(e) Average of a product of three directional cosines

We shall make use of the product of two Levi-Civita tensors to create a true symmetric tensor of rank 6. We write :

$$< c_{i\alpha}\, c_{j\beta}\, c_{k\gamma} >_0 = x\, \varepsilon_{ijk}\, \varepsilon_{\alpha\beta\gamma}$$

We multiply both terms by ε_{ijk}. The right-hand term now gives the value $6x\varepsilon_{\alpha\beta\gamma}$ (see Inset II.3), whereas the left-hand term gives $\varepsilon_{\alpha\beta\gamma}$. Hence x = 1/6 and we get the following expression:

$$< c_{i\alpha}\, c_{j\beta}\, c_{k\gamma} >_0 = (1/6)\, \varepsilon_{ijk}\, \varepsilon_{\alpha\beta\gamma} \qquad \text{(III.1.25)}$$

Applications:

— *The optical activity tensor for a macroscopic system composed of chiral nonoriented molecules.* The macroscopic tensor η_{ijk} is, starting from the microscopic tensor $\eta_{\alpha\beta\gamma}$, defined as

$$\eta_{ijk} = < c_{i\alpha}\, c_{j\beta}\, c_{k\gamma} >_0 \eta_{\alpha\beta\gamma}$$

As already stated, $\eta_{\alpha\beta\gamma}$ is antisymmetric with respect to an interchange of its indices α and β.

By Eq. III.1.25, we have

$$\eta_{ijk} = (1/6)\, \varepsilon_{ijk} \left(\eta_{\alpha\beta\gamma}\, \varepsilon_{\alpha\beta\gamma}\right) = \eta\, \varepsilon_{ijk}$$

$$\text{with } \eta = (1/6)\, \varepsilon_{\alpha\beta\gamma}\, \eta_{\alpha\beta\gamma} \qquad \text{(III.1.26)}$$

The macroscopic tensor now presents complete antisymmetry (antisymmetry by compensation) because it is directly proportional to the Levi-Civita tensor.

— *Generation of the second harmonic of a laser wave by a random assemblage of molecules without center of symmetry.* Above [see Section II.3.1.1 (c)], we have shown that the second term of the series expansion of optical polarization in growing powers of the electric field strength of the wave was (see Eq. II.3.15)

$$P_i^{(2)}(2\omega_L) = \chi_{ijk}^{(2)}(-2\omega_L, \omega_L, \omega_L)\, E_j(\omega_L)\, E_k(\omega_L)$$

We refer to the tensor $[\beta_{\alpha\beta\gamma}]$ of rank 3 as the molecular polarizability of order 2. Its components differ from zero since the molecules lack a center of symmetry; moreover, it is naturally symmetric in its indices (β, γ.)

Our problem consists in determining whether the macroscopic tensor $\chi_{ijk}^{(2)}$ exists or not. If $\left[\chi_{ijk}^{(2)}\right]$ is nonzero, then $P_i^{(2)}(2\omega_L)$ exists and we shall show (see Chapter VIII) that the material gives rise to a harmonic wave at frequency $2\omega_L$. Now, once again having recourse to Eq. III.1.25, we may write

$$\chi_{ijk}^{(2)} = < c_{i\alpha}\, c_{j\beta}\, c_{k\gamma} >_0 \beta_{\alpha\beta\gamma} = (1/6)\, \rho\, \varepsilon_{ijk} \left(\varepsilon_{\alpha\beta\gamma}\, \beta_{\alpha\beta\gamma}\right)$$

The term $\varepsilon_{\alpha\beta\gamma}\,\beta_{\alpha\beta\gamma}$ vanishes since $\beta_{\alpha\beta\gamma}$ is symmetric in its indices β and γ whence $\left[\chi^{(2)}_{ijk}\right] = \left[\ 0\ \right]$.

Thus second-harmonic generation cannot take place, as one might expect since the macroscopic irradiated specimen exhibits a center of symmetry.

(f) Average of the product of four directional cosines

Formally, a product of this kind is a tensor of rank 8. We write its average in the form of the sum

$$< c_{i\alpha}\, c_{j\beta}\, c_{k\gamma}\, c_{l\delta} >_0 \; = x\, \delta_{ij}\, \delta_{kl} + y\, \delta_{ik}\, \delta_{jl} + z\, \delta_{il}\, \delta_{jk} \qquad \text{(III.1.27)}$$

and determine x, y, and z multiplying Eq. III.1.27 successively by $\delta_{ij}\delta_{kl}$, $\delta_{ik}\delta_{jl}$, and $\delta_{il}\delta_{jk}$. This leads to

$$x = (1/30)\left[4\, \delta_{\alpha\beta}\, \delta_{\gamma\delta} - \delta_{\alpha\gamma}\, \delta_{\beta\delta} - \delta_{\alpha\delta}\, \delta_{\beta\gamma}\right]$$

$$y = (1/30)\left[4\, \delta_{\alpha\gamma}\, \delta_{\beta\delta} - \delta_{\alpha\delta}\, \delta_{\beta\gamma} - \delta_{\alpha\beta}\, \delta_{\gamma\delta}\right] \qquad \text{(III.1.28)}$$

$$z = (1/30)\left[4\, \delta_{\alpha\delta}\, \delta_{\beta\gamma} - \delta_{\alpha\beta}\, \delta_{\gamma\delta} - \delta_{\alpha\gamma}\, \delta_{\beta\delta}\right]$$

Application: Depolarized Rayleigh scattering at the circular frequency ω_L from a random assemblage of axially symmetric molecules. We have shown (see Inset III.1) that $R_{ani,\,dep}$ is proportional to γ^2. The reader will easily find that in the case of a random assemblage of N reorientationally correlated molecules, we obtain

$$R_{ani,\,dep} \approx (3/2)\left\langle \sum_{p=1}^{N} \sum_{q=1}^{N} \left(\alpha^{(p)}_{ij} - \overline{\alpha}\, \delta_{ij}\right)\left(\alpha^{(q)}_{ij} - \overline{\alpha}\, \delta_{ij}\right) \right\rangle_0 \qquad \text{(III.1.29)}$$

However, we have shown (see Problem II.3) that

$$\alpha_{ij} = \overline{\alpha}\, \delta_{ij} + \gamma\left(c_{i3}\, c_{j3} - \delta_{ij} / 3\right)$$

whence

$$R_{aniso,\,dep} \approx \left(3\, \gamma^2 / 2V\right)\left\langle \sum_{p=1}^{N} \sum_{q=1}^{N} \left(c^{(p)}_{i3}\, c^{(p)}_{j3} - \delta_{ij} / 3\right)\left(c^{(q)}_{i3}\, c^{(q)}_{j3} - \delta_{ij} / 3\right) \right\rangle_0$$

The double sum, on separating the contributions due to the square terms (q = p) from those due to the rectangular terms ($q \neq p$), takes the following form:

Square terms: By Eq. III.1.27, we have 2N/3
Rectangular terms: We obtain $(2N/3)J_A$

where

$$J_A = (1/2) \sum_{(q \neq p)=1}^{N} \left(3 \cos^2 \theta_{pq} - 1\right) \qquad \text{(III.1.30)}$$

θ_{pq} is the angle between the axes of symmetry of molecules p and q.

To prove the above results let us first show that $c_{i3}^{(p)} c_{i3}^{(q)} = \cos\theta_{pq}$. To this aim we shall make use of the Euler angles formalism (see Inset II.1.) We have

$$\begin{aligned} c_{i3}^{(p)} c_{i3}^{(q)} &= \sin\psi^{(p)} \sin\theta^{(p)} \sin\psi^{(q)} \sin\theta^{(q)} + \cos\psi^{(p)} \sin\theta^{(p)} \cos\psi^{(q)} \sin\theta^{(q)} \\ &\quad + \cos\theta^{(p)} \cos\theta^{(q)} \\ &= \cos\theta^{(p)} \cos\theta^{(q)} + \sin\theta^{(p)} \sin\theta^{(q)} \cos\left(\psi^{(p)} - \psi^{(q)}\right) \end{aligned}$$

and Eq. II.1.1 enables us to write $c_{i3}^{(p)} c_{i3}^{(q)} = \cos\left(\theta_p - \theta_q\right) = \cos\theta_{pq}$

Accordingly, we have obtained

$$R_{ani,\,dep} \approx \rho\gamma^2 \left(1 + J_A\right) \tag{III.1.31}$$

J_A is the angular correlation parameter between axially symmetric molecules. For noncorrelated molecules, $< \cos^2\theta_{pq} >_0 = 1/3$ and $J_A = 0$: The depolarized anisotropic scattering is simply proportional to the number density ρ and to the anisotropy γ^2 of the molecules.

III.1.2.2 Averages in the presence of stresses

The present subsection will be given a much fuller development in Chapter VIII, where we shall be considering certain effects belonging to the field of nonlinear optics. Here, we shall adduce but a simple example concerning reorientation of noncorrelated polar molecules (with a *permanent* microscopic moment **p** along their z-axis) by a static electric field. This, in fact, is the static Kerr effect. The results we derive here will then serve for a comparison with a purely optical effect—the optical Kerr effect—induced by laser waves. Thus we proceed to calculate the average of the polarization **P** induced in the macroscopic system in the direction z of the reorienting electric field **E**.

We have $|\mathbf{P}| = < c_{i\alpha} > |\mathbf{p}|$, where $c_{i\alpha} = \cos\theta$ (see Inset II.1). In zeroth order, $< \cos\theta >_0 = 0$ and P vanishes.

We thus have to make use of the expansion to order 1 (see Table III.1.1). In spherical coordinates (see Section II.1.1.2) we have

$$V = -\,\mathbf{p}\,\mathbf{E} = -\,pE\cos\theta$$

$$< V >_0 = 0$$

$$f_0 = 1/4\,\pi$$

$$f_1 = \left(1\,/\,4\pi\,kT\right) pE\cos\theta$$

And with regard to Eq. III.1.1 we obtain

$$< \cos\theta > = (pE / 4\pi kT) \int_0^{2\pi} d\varphi \int_0^{\pi} \cos^2\theta \sin\theta \, d\theta = pE / 3\, kT$$

Finally, for $<P>_1$ we obtain

$$<P>_1 = p^2 E / 3\, kT \qquad \text{(III.1.32)}$$

It should, however, be stated that if **E** is the electric field of an optical wave, then the time average of V is zero and $<P>_1 = 0$. *A laser wave has no direct reorienting effect on permanent molecular dipoles.*

We have already mentioned the problem of the internal field. We shall now give it a closer look.

III.2 Internal field

In the introduction to the present chapter we said that electric polarization is essentially a nonlinear function of the externally applied electric field expressible in terms of an expansion in growing powers of the electric field strength. In vectorial form, the expansion takes the form

$$\mathbf{P} = \chi^{(1)}\, \mathbf{E} + \chi^{(2)}\, \mathbf{E}^2 + \cdots + \chi^{(n)}\, \mathbf{E}^n + \cdots \qquad \text{(III.2.1)}$$

Also, we have discussed some properties of the first two terms of the expansion. Making use of microscopic quantities (which we shall distinguish from macroscopic quantities by a subscript mic), we have

$$\mathbf{P} = \rho \left[\chi^{(1)}_{mic}\, \mathbf{E}_{int} + \chi^{(2)}_{mic}\, \mathbf{E}^2_{int} + \cdots + \chi^{(n)}_{mic}\, \mathbf{E}^n_{int} + \cdots\right] \qquad \text{(III.2.2)}$$

Here, $\mathbf{E}_{int}$ is the internal electric field, i.e., the field *in reality* acting on an individual microscopic particle. In all dense materials, $\mathbf{E}_{int}$ differs from **E**. In particular, the difference arises from the fact that each of the particles illuminated becomes the source of a supplementary electric field which adds up to the externally applied field. Very numerous models of internal electric fields have been proposed, some of them highly complicated (see the bibliography, III.5.3). We chose to assume that $\mathbf{E}_{int}$ and **E** are interrelated as follows:

$$\mathbf{E}_{int} = \mathbf{E} + \lambda\, \mathbf{P} / \varepsilon_0 \qquad \text{(III.2.3)}$$

where λ is a constant that depends on the material involved and the internal field model assumed. In the case of gases, λ is often assumed as equal to 1; we then get Eq. II.2.13 with $\mathbf{E}_{int} = \mathbf{D}/\varepsilon_0$.

Very often, in the case of cubic crystals, and liquids with "isotropic" molecules, use is made of the Lorentz-Lorenz model, so that $\lambda = 1/3$.

On replacing $\mathbf{E}_{int}$ in Eq. III.2.2 by its expression given in Eq. III.2.3, we get

$$\mathbf{P}\left[1-\rho\,\chi_{mic}^{(1)}\,\lambda/\varepsilon_0\right]=\rho\left[\chi_{mic}^{(1)}\,\mathbf{E}+\cdots+\chi_{mic}^{(n)}\left(\mathbf{E}+\lambda\mathbf{P}/\varepsilon_0\right)^{n}+\cdots\right]$$

or, after putting $1-\rho\,\chi_{mic}^{(1)}\,\lambda/\varepsilon_0=\mathrm{F}^{-1}$,

$$\mathbf{P}=\rho\left[\chi_{mic}^{(1)}\,\mathrm{F}\,\mathbf{E}+\cdots+\chi_{mic}^{(n)}\,\mathrm{F}\left(\mathbf{E}+\lambda\mathbf{P}/\varepsilon_0\right)^{n}+\cdots\right] \qquad \text{(III.2.4)}$$

Now the series expansion (Eq. III.2.1) assumes that the terms of orders 2, 3, ..., n are small compared to the term of order 1, so that we may write

$$\mathbf{P}\approx\rho\,\chi_{mic}^{(1)}\,\mathrm{F}\,\mathbf{E}$$

Equation III.2.4 thus becomes

$$\mathbf{P}\approx\rho\left[\chi_{mic}^{(1)}\,\mathrm{F}\,\mathbf{E}+\cdots+\chi_{mic}^{(n)}\,\mathrm{F}\left(1+\lambda\,\rho\,\mathrm{F}\chi_{mic}^{(1)}/\varepsilon_0\right)^{n}\mathbf{E}^{n}+\cdots\right]$$

and noting that $1+\lambda\,\rho\,\mathrm{F}\chi_{mic}^{(1)}/\varepsilon_0=\mathrm{F}$, we obtain

$$\mathbf{P}\approx\rho\left[\chi_{mic}^{(1)}\,\mathrm{F}\,\mathbf{E}+\cdots+\chi_{mic}^{(n)}\,\mathrm{F}^{n+1}\,\mathbf{E}^{n}+\cdots\right] \qquad \text{(III.2.5)}$$

This proves that, except for the term of order 1 which is multiplied by the internal field factor F, each of the terms of order n of the expansion has to be multiplied by F^{n+1}.

In particular, in the very many cases to which the Lorentz-Lorenz model is applicable, F takes the form

$$\mathrm{F}=1+\chi^{(1)}/3\varepsilon_0$$

However,

$$\chi^{(1)}\approx\varepsilon_0\,(n^2-1)\ \text{(see Eq. II.2.16)},\quad \text{whence } \mathrm{F}=(n^2+2)/3 \qquad \text{(III.2.6)}$$

Thus, the microscopic susceptibility of order n—also referred to as polarizability of order n—has to be multiplied by $[(n^2+2)/3]^{n+1}$.

III.3 Microscopic → macroscopic passage

We have by now available all the elements necessary for the passage from the microscopic to the macroscopic description (see Inset III.2).

Earlier, we introduced the concept of *"formal"* susceptibility with reference to the order of the microscopic susceptibility. What is the exact meaning to be conveyed by this adjective?

Inset III.2 ***Microscopic → macroscopic passage***

The fundamental equation describing the transition is

$$\chi^{(n)}_{ijk\ldots} = \rho F^{n+1} \langle c_{i\alpha}\, c_{j\beta}\, c_{k\gamma}\cdots\rangle\, \chi^{(n)}_{mic\ \alpha\beta\gamma\ldots} \qquad \text{(III.3.1)}$$

with $\chi^{(n)}_{ijk\ldots}$	the macroscopic volume susceptibility of order n [a tensor of rank (n + 1)]
ρ	the number density of the particles
F	the local (internal) field factor (the model is chosen according to the symmetry of the system)
$\langle c_{i\alpha}\, c_{j\beta}\, c_{k\gamma}\ldots\rangle$	the average of the product of the (n + 1) directional cosines of the transformation. In the absence of stresses the average is taken at thermal equilibrium. In the presence of stresses applied to the material under investigation one should make use of the expansions given in Table III.1.1. Usually, two types of stresses are distinguished: — Stresses *intrinsic* to laser-matter interaction if the electric field of the laser wave is sufficiently intense to induce privileged orientations in the material. — Stresses *extrinsic* to laser-matter interaction if the system is acted on by externally applied electric or magnetic fields, generally static or at low frequencies.
$\chi^{(n)}_{mic\ \alpha\beta\gamma\ldots}$	the "formal" microscopic susceptibility of order n, also referred to as molecular polarizability of order n.

We have shown that the average calculated for a product of (n + 1) directional cosines—if not performed for thermal equilibrium—could involve the emergence of higher powers of the perturbation energy V. The example considered Section III.1.2.2 showed the expression for that average to involve the amplitude of the static electric field E applied to the assemblage of dipoles. In the case of an intrinsic stress, the electric field of the laser can induce a reorientation, characterized by the average energy *per* molecule (time-average):

$$V = (-1/2)\, \alpha_{ij}\, E_i\, E_j \qquad \text{(III.3.2)}$$

where α_{ij} is the microscopic susceptibility of order 1 (equivalent to $\chi^{(1)}_{mic}$).

When averaging at order 1 of the perturbation we thus introduce the product

E_iE_j. *In this case we have to lower by 2 the rank of the microscopic susceptibility* $\chi^{(n)}_{mic}$ *occurring in Eq. III.3.1.*

The preceding example suffices to show that, to generate a microscopic susceptibility *of a given order n*, we have to choose and couple the following

— a microscopic susceptibility of *order n',* and

— a calculated order of the statistical average containing the *n"th power of the electric field conveyed by the laser*, with n" such that:

$$n = n' + n'' \qquad \text{(III.3.3)}$$

It should be stated clearly that the choice of n' and n" can be made in many ways. Thus, a "mixture" of different microscopic susceptibilities of different orders implying *totally different physical phenomena* will *jointly* combine to produce one and the same experimentally measured macroscopic susceptibility. Here resides one of the chief difficulties in the field of laser-matter interaction. We shall meet with examples of this kind and shall discuss them in full detail in Chapter VIII. Luckily, however, the different physical phenomena underlying the "mixture" are often associated with relatively different characteristic time rates. In such cases, *time-resolved experiments* can help us to separate the different contributions.

Luckily, too, lasers—to which we devote the next three chapters—provide the experimenter with the possibility of using intense light pulses with durations ranging from 10^{-15} seconds up to continuous. This remarkable diversity will be widely exploited in the course of our further work.

III.4 Problems and outlined solutions

III.4.1 Problem III.1: Elementary calculations of thermal equilibrium averages

Calculate the averages $< \cos^{2k}\theta >_0$, $< \sin^{2k}\theta >_0$, $< \cos^{2k}\varphi >_0$, $< \sin^{2k}\varphi >_0$, $< \cos^{2k}\psi >_0$, and $< \sin^{2k}\psi >_0$, where θ, φ, and ψ are Euler angles, defined in Inset II.1. Tabulate the results obtained for k = 1, 2, 3, 4, and 5.

In spherical coordinates, by Eq. III.1.4, we write

$$< \cos^n\theta >_0 = (1/4\,\pi)\int_0^{2\pi} d\varphi \int_0^{\pi} \cos^n\theta \sin\theta \, d\theta$$

Integration gives $< \cos^n\theta >_0 = [1 + (-1)^n] / 2(n+1)$. Hence

$$n = 2k \text{ (even n)}: < \cos^{2k}\theta >_0 = 1/(2k+1)$$

$$n = (2k+1) \text{ (odd n)}: < \cos^{(2k+1)}\theta >_0 = 0$$

Similarly, for the sines we obtain

$$< \sin^n \theta >_0 = (1/2) \int_0^{\pi} \sin^{n+1} \theta \, d\theta = (1/2)\, I_{n+1}.$$

Recurrential integration proceeds as follows:

$$I_{n+1} = \int_0^{\pi} \sin^{n+1} \theta \, d\theta = \int_0^{\pi} \sin^{n-1} \theta \left(1 - \cos^2 \theta\right) d\theta$$

$$= I_{n-1} - (1/n) \int_0^{\pi} \cos \theta \left[n \cos \theta \sin^{n-1} \theta\right] d\theta$$

Per parties integration of the last expression yields

$$\int_0^{\pi} (\sin \theta) \sin^n \theta \, d\theta = I_{n+1}$$

whence

$$I_{n+1} = I_{n-1} - (1/n)\, I_{n+1}$$

$$I_{n+1} = n / (n+1)\, I_{n-1}$$

Since

$$I_1 = \int_0^{\pi} \sin \theta \, d\theta = 2$$

We get $I_3 = 4/3$, $I_5 = 16/15$ …, $<\sin^2 \theta>_0 = 2/3$, and $< \sin^4 \theta >_0 = 8/15$ ….

Now calculate the averages over φ:

$$< \cos^n \varphi >_0 = (1/4\,\pi) \int_0^{2\pi} \cos^n \varphi \, d\varphi \int_0^{\pi} \sin \theta \, d\theta = (1/2\pi) \int_0^{2\pi} \cos^n \varphi \, d\varphi$$

$$< \sin^n \varphi >_0 = (1/2\,\pi) \int_0^{2\pi} \sin^n \varphi \, d\varphi$$

These two integrals are to be found in the literature (see the bibliography, III.5.2), whence

$$\int_0^{2\pi} \cos^{2n+1} x \, dx = \int_0^{2\pi} \sin^{2n+1} x \, dx = 0$$

$$\int_0^{2\pi} \cos^{2n} x \, dx = \int_0^{2\pi} \sin^{2n} x \, dx = 2\pi\, (2n-1)\,!!\, / \,(2n)\,!!$$

leading to

$$< \cos^n \varphi >_0 = < \sin^n \varphi >_0 (2n-1)\,!!\, / \,(2n)\,!!$$

$(2n-1)\,!!$ and $(2n)\,!!$ are, respectively, odd and even factorials. Similar results are obtained for ψ. The set of results thus obtained are tabulated below:

k	2k	$< \cos^{2k} \theta >_0$	$< \sin^{2k} \theta >_0$	$< \cos^{2k} \varphi >_0 = < \sin^{2k} \varphi >_0$ $= < \cos^{2k} \psi >_0 = < \sin^{2k} \psi >_0$
1	2	1/3	2/3	1/2
2	4	1/5	8/15	3/8
3	6	1/7	16/35	5/16
4	8	1/9	128/315	35/128
5	10	1/11	256/693	63/256

III.4.2 Problem III.2: Calculation of the Born contributions to the static Kerr effect in a dense fluid of axially symmetric molecules

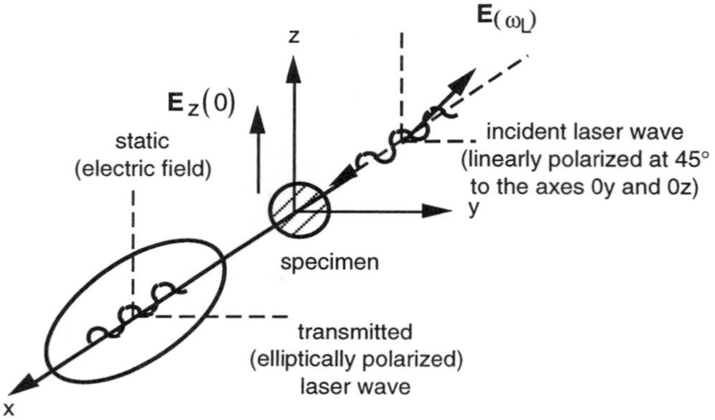

The initially isotropic specimen with mean index $n = n_{zz} = n_{yy}$ (at circular frequency ω_L of the laser wave) becomes birefringent under the action of the static electric field $\mathbf{E}_z(0.)$ The ellipticity of the transmitted laser wave, which is the quantity measured, provides a measure of the difference in index: $\Delta n = n_{zz} - n_{yy}$. The static Kerr effect concerns only the contribution to Δn proportional to $|\mathbf{E}_z(0)|^2$. Show that in the case of an isotropic fluid of axially symmetric molecules, two contributions appear, referred to as respectively as Born 1 and Born 2, involving respectively the first and second power of the permanent dipole moment $p = |\mathbf{p}|$ of the molecules.

We have recourse to Eq. II.2.15 to express the index n_{ij} and then expand the polarization in accordance with Eq. III.2.2

$$n_{ij}^2 - \delta_{ij} = (1/\varepsilon_0)\left[\chi_{ij}^{(1)} + \chi_{ijk}^{(2)} E_k(0) + \chi_{ijkl}^{(3)} E_k(0) E_l(0) + \ldots\right]$$

The Kerr effect, by definition, is described by the term of order n = 3 :

$$\begin{cases} n_{zz}^2 - 1 = (1/\varepsilon_0)\chi_{zzij}^{(3)} E_i(0) E_j(0) \\ n_{yy}^2 - 1 = (1/\varepsilon_0)\chi_{yyij}^{(3)} E_i(0) E_j(0) \end{cases}$$

Hence $\Delta n = n_{zz} - n_{yy} \approx \left(1/2\ n\ \varepsilon_0\right)\left(\chi^{(3)}_{zzzz} - \chi^{(3)}_{yyzz}\right) E_z^2(0)$

We now proceed to calculate the two Born contributions.

Born 1: Determine the order n" of the statistical average required. Since Born 1 has to be proportional to p and also because the perturbation energy due to the static electric field is $V_F = -\mathbf{p}\ \mathbf{E}_z(0)$ (see Eq. III.1.2.2) we write n" = 1 (see Eq. III.3.3). The average will be taken for order 1 of the perturbation. Hence it results that the order n' of the microscopic susceptibility chosen shall be 2. Consequently, it is the microscopic susceptibility $\chi^{(2)}_{mic}$ (which, for simplicity, we now put equal to β) that will appear in Born 1. Appendix II.8.1.2 states that the tensor β_{123}, assumed as completely symmetric, has seven nonzero components, two of which are independent: $\beta_{333} = \beta_{//}$ (03 is supposed to be the symmetry axis C_∞ of a molecule belonging to the class $C_{\infty v}$), and $\beta_{113} = \beta_{131} = \beta_{311} = \beta_{223} = \beta_{232} = \beta_{322} = \beta_\perp$. On inserting in Δn the respective tensors we get (without accuracy to the internal field factor):

$$\chi^{(3)}_{zzzz} = \rho\left[\left(\beta_{//} - 3\ \beta_\perp\right) < c^3_{z3} >_1 + 3\ \beta_\perp < c_{z3} >_1\right]$$
$$\chi^{(3)}_{yyzz} = \rho\left[\left(\beta_{//} - 3\ \beta_\perp\right) < c^2_{y3} c_{z3} >_1 + \beta_\perp < c_{z3} >_1\right]$$

The averages are to be obtained with the data of Table III.1.1 using Euler angles (see Inset II.1) and the results derived in Problem III.1 :

$$< c^3_{z3} >_1 = \left(-1/kT\right)\left[< \cos^3\theta\ V_F >_0 - < V_F >_0 / 3\right] = p\ E_z(0) / 5\ kT$$
$$< c_{z3} >_1 = \left(-1/kT\right)\left[< \cos\theta\ V_F >_0 - < V_F >_0 / 3\right] = p\ E_z(0) / 3\ kT$$
$$< c^2_{y3}\ c_{z3} >_1 = -\left(1/kT\right)\left[< \sin^2\theta \cos^2\varphi \cos\theta\ V_F >_0 - < \sin^2\theta \cos^2\varphi \cos\theta >_0 < V_F >_0\right] = p\ E_z(0) / 15\ kT$$

With the Lorentz-Lorenz internal field model we finally obtain

$$\Delta n \approx \left(\rho p\ E^2_z(0) / 15\ \varepsilon_0\ n\ kT\right)\left(\beta_{//} + 2\ \beta_\perp\right)\left[\left(n^2 + 2\right)/3\right]^4$$

Born 2: Since the contribution Born 2 is proportional to p^2, we take the statistical average at order 2 of V_F (n" = 2). This leads to n' = 1: We thus shall apply the microscopic susceptibility of order 1, i.e., $\chi^{(1)}_{mic,\ ij} = \alpha_{ij} = \overline{\alpha}\ \delta_{ij} + \gamma\left(c_{i3}\ c_{j3} - \delta_{ij}/3\right)$ (see Problem II.3), leading to

$$\chi^{(3)}_{zzzz} = \rho\left[\overline{\alpha} + \gamma\left(< c^2_{z3} >_2 - 1/3\right)\right]$$
$$\chi^{(3)}_{yyzz} = \rho\left[\overline{\alpha} + \gamma\left(< c^2_{y3} >_2 - 1/3\right)\right]$$
$$\Delta n = \left(\rho\gamma / 2\varepsilon_0\ n\right)\left(< c^2_{z3} >_2 - < c^2_{y3} >_2\right)$$

As above, we obtain

$$< c^2_{z3} >_2 = \left(p^2\ E^2_z(0) / 2\ k^2\ T^2\right)\left(< \cos^4\theta >_0 - < \cos^2\theta >_0^2\right) = 2\ p^2\ E^2_z(0) / 45\ k^2\ T^2$$
$$< c^2_{y3} >_2 = \left(p^2\ E^2_z(0) / 2\ k^2\ T^2\right)\left(< \cos^2\psi \sin^2\theta \cos^2\theta >_0 - < \cos^2\psi \sin^2\theta >_0 < \cos^2\theta >_0\right) = -\ p^2\ E^2_z(0) / 45\ k^2\ T^2$$

And finally

$$\Delta n \approx \left(\rho p^2 E_z^2(0) / 30\, \varepsilon_0\, n\, k^2\, T^2\right) \gamma \left[\left(n^2 + 2\right) / 3\right]^4$$

The two Born contributions 1 and 2 exist only in fluids consisting of polar molecules. They can be distinguished on grounds of their thermal behavior, which obeys a T^{-1} and T^{-2} law, respectively.

III.5 Bibliography

III.5.1 An introduction to statistical physics can be found in:

CHANDLER, D. *Introduction to Modern Statistical Mechanics*, Oxford University Press, Oxford, 1987.

For a complete treatment of the subject, consult:

TODA, M., KUBO, R., AND SAITO, N. *Statistical Physics I,* Springer-Verlag, New York, 1983.

KUBO,R., TODA, M., AND HASHITSUME, N. *Statistical Physics II,* Springer-Verlag, New York, 1985.

III.5.2 For calculations of the averages, see:

GRADSHTEYN, I. S., AND RYZHIK, J. M. *Table of Integrals, Series, and Products,* Academic Press, New York, 1965.

III.5.3 A general discussion of the internal field problem is given in:

KITTEL, C. *Quantum Theory of Solids,* John Wiley & Sons, New York, 1987.

PART TWO

The Laser Wave and Its Properties

CHAPTER IV

The Laser

IV.1 Generalities

IV.1.1 What is a laser?

The laser is an optical *oscillator*. In general, whatever the domain under consideration, in order to have an oscillator one needs an *amplifier* and a *feedback procedure* (a resonator) which sends all or part of the output signal back to the input with the required phase conditions. Figure IV.1.1 illustrates an example of such an oscillator. The medium described here is supposed to be an amplifying medium for the frequency ν_L belonging to the optical domain.* This amplification can be described by a gain parameter G_{ν_L}, which is a function of the intensity I_{ν_L} of the optical wave. Part of the wave coming out of the amplifier is sent back to its entrance by the four mirrors M_1 (partially reflecting) and M_2, M_3, M_4 (all three totally reflecting), in such a way that the reinjected wave is in phase with the incident wave, or "mother" wave (the problem of the "birth" of this first wave will be discussed in Section IV.3). When the wave travels through the amplifying medium the second time, it is amplified again. Of course, this procedure is repetitive.

---

* In fact, the term LASER (light amplification by stimulated emission of radiation) refers not to the optical oscillator itself, but to the physical effect of amplification which is a prerequisite to oscillation.

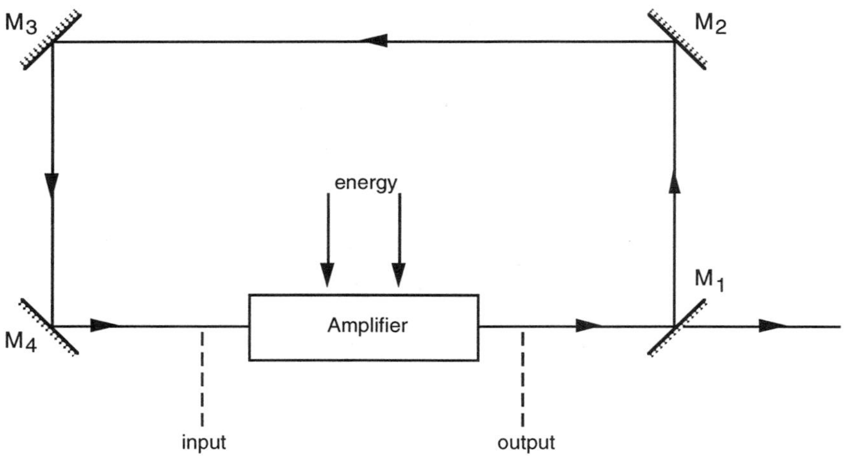

Figure IV.1.1 : Configuration of a laser. The resonator drawn here is an optical ring-cavity

Amplification occurs because outside energy is injected into the system. On the other hand, as the intensity I_{ν_L} increases, the gain G_{ν_L} decreases. Intuitively, we can understand that equilibrium is reached when the gain G_{ν_L} is equal to the losses P_{ν_L} of the resonator. The losses can be of two kinds:

— Those due to the presence of a great number of intrinsic attenuations within the resonator: diffraction of the optical parts, spurious absorptions, and the like.

— Those due to the fact that part of the radiation is let out of the cavity in order to use it, whence the voluntarily partial reflection of mirror M_1.

So a steady state is obtained for $G_{\nu_L} = P_{\nu_L}$, by saturation of the gain at the oscillation frequency of the laser.

Subsequently, we shall call "laser" the optical oscillator built on the principle of physical amplification.

Amplification	+	Resonator	→	Oscillator
(by stimulated emission)		(optical cavity)		(LASER)

The resonator, which plays the part of a photon trap, is also frequently referred to as an *optical cavity*. Here, we speak about an *optical ring cavity*. We arbitrarily chose a levogyre (left-turning) propagation direction for the wave in the cavity (see Figure IV.1.1). A dextrogyre (right-turning) wave, which is not shown here, may also propagate in the cavity. Optical amplification occurs

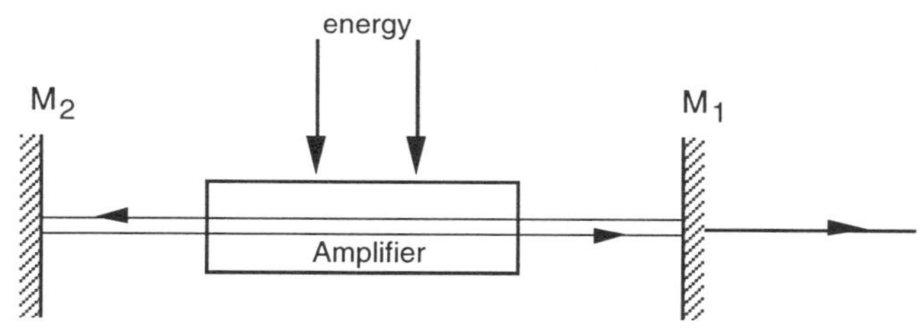

Figure IV.2.2 : Configuration of a laser. The resonator drawn here is a linear optical cavity.

whatever the direction in which the wave travels through the medium. If needed, it is possible to simplify the resonator by turning mirrors M_1 and M_4 by 45°, respectively in the levogyre and in the dextrogyre direction. This way, the wave is "folded up", and mirrors M_2 and M_3 are no longer needed. We thus end up with the traditional scheme of the *linear cavity* (see Figure IV.1.2) which is technically simpler and often used in practical experimental situations.

Let us call attention to two very important points:

— Not all mirrors can be used. In fact, they have been studied thoroughly, as will be shown in Section IV.2, and their properties are absolutely essential for the optical quality of the laser wave. A considerable amount of theoretical and experimental research is devoted to their study.

— As mentioned, the energy of the auto-oscillation is not generated spontaneously in the amplifying medium. We shall see later why a medium at thermal equilibrium cannot be an amplifying medium. Extra energy must be provided to the system. This energy can be of very different origins: electrical energy, optical energy (classical radiating sources, other lasers, solar energy), thermal, chemical, nuclear, etc. It should be said that, usually, the energy yields are very low, between 0.1% and 80% (for some molecular, chemical or solid-state lasers).

IV.1.2 The history of the laser

The laser may seem to have been invented rather late (1958). An oscillator in the microwave domain (MASER) already existed since the early 1950. Pérot and Fabry's works had shown as early as 1896 the great advantage, in spectroscopy, of plane cavities formed by placing two parallel mirrors face to face. At the start of this century, Einstein had opened the way to the experimental development of an optical amplifier, based on stimulated emission which had been discovered in 1917. However, the greatest difficulty remained: How could one go about building a resonator in the optical domain which would be comparable to the one used for microwaves? In this case, the parallelipepidal cavity is made resonant by polishing all faces with a precision very much greater then the wavelength $\lambda \approx 1$ mm. Similar polishing in the optical domain needs a precision greater than $\lambda \approx 1\ \mu$m, which was technically unthinkable. Fundamental works of Basov, Prokhorov, Townes, and Schawlow (1958) showed how the Pérot-Fabry interferometer can confine a mode of optical propagation around its axis. Only a few square millimeters of each of the two mirrors now need precision polishing, thus greatly reducing the technical problems. A year later, Maiman (Hughes) built the first ruby oscillator, and in 1960 the first gas laser was designed at MIT by Javan.

This is the start of a real revolution in optics—experimental as well as theoretical, fundamental as well as applied. Thirty years later, this impulse is still very much alive. In 1992, the annual world market of the laser represents more than a billion dollars. About 60% of this market is taken up by tooling machines (carbon dioxide laser) and by lasers used for communications (diode laser). The wide variety of existing laser sources makes an exhaustive presentation almost impossible. We shall therefore limit ourselves to a very fundamental description, thus giving the reader enough basic knowledge to enable him to study the subject more thoroughly in the specialized works cited at the end of each chapter.

IV.2 Principles of laser oscillators

IV.2.1 Amplification

IV.2.1.1 Einstein's phenomenological theory

Let us consider an isolated atom, in the absence of radiation and without interaction with any material medium. Let (1) and (2) denote the two states of this atom, characterized respectively by their energies E_1 and E_2 (see Figure IV.2.1) and by their degree of degeneracy g'_1 and g'_2 .

Let us write $E_2 - E_1 = \hbar\omega_0$. State (1) is the state of lowest energy. Let us now suppose that the atom interacts with some outside radiation, of much greater intensity than that of the thermal radiation which we shall neglect here. The volume energy density of this radiation, averaged over time and in a band of width $\delta\omega_L$ around $\omega_L = \omega_0$, is $\bar{\rho}$. $\bar{\rho}$ is measured in units of Jm $^{-3}$.$\bar{\rho}$ can be written as a function of $\bar{I}$, the mean intensity of the wave in the energy band $\delta\omega_L$ around ω_L, and of E, the amplitude of the associated electric field, in the following way:

$$\bar{\rho} / n^2 = \bar{I} / nc = \varepsilon_0 E^2 / 2 \qquad \text{(IV.2.1)}$$

ε_0 is the electric permittivity of vacuum, n is the refractive index of the medium at frequency ω_L, and c is the velocity of light in vacuum. There are two well-known processes by means of which energy can be exchanged between the atom and the light wave. They are:

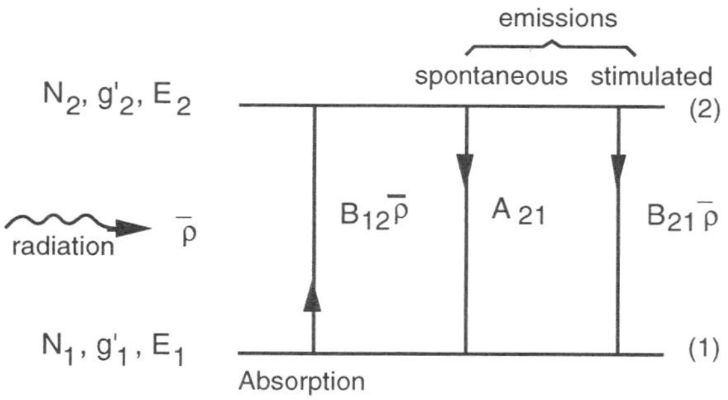

Figure IV.2.1: Radiative exchanges between a population of atoms and incoming radiatin (g'_1 and g'_2 are the degeneracies of levels 1 and 2).

— *Absorption* (on the left side of Figure IV.2.1): It consists in the absorption of a photon of energy $\hbar\omega_0$ inducing a transition $1 \rightarrow 2$. It represents a transfer of energy from the radiation to the atom. The transition probability per unit time is postulated to be equal to $B_{12}\,\bar{\rho}$. The atom goes from state (1) to state (2).

— *Spontaneous emission* (in the middle of Figure IV.2.1): The atom, which is in state (2), emits a photon of energy $\hbar\omega_0$. Energy is transferred from the atom to the radiation. After emission, the atom is in state (1) again. The photon is emitted in a random direction with random phase (with respect to an arbitrarily fixed origin of time). The probability of this transition per unit time is postulated to be equal to A_{21}.

These two processes by themselves do not account correctly for the experimental results obtained on thermal blackbody radiation. Einstein postulated the existence of a third exchange procedure, a much less intuitive one than the preceding two, called *stimulated emission* (on the right side of Figure IV.2.1). Therefore one must add a probability $B_{21}\,\bar{\rho}$ of radiation-stimulated emission to the probability A_{21} of spontaneous emission. Radiation produced by stimulated emission has two very important properties:

— The photon is emitted in the *same direction* as the exciting photon provided by the radiation.

— The associated wave vector (see Chapter I) is in *phase* with the exciting wave vector.

The three procedures are illustrated in Figure IV.2.2. In part c, it is clear that the elementary gain of stimulated emission is equal to 2. Now if instead of looking at a single atom, we look at two populations of atoms—N_1 atoms in state (1) and N_2 atoms in state (2)—the coefficients relative to states (1) and (2) must be multiplied by N_1 and N_2. N_1 and N_2 are atomic volume densities.

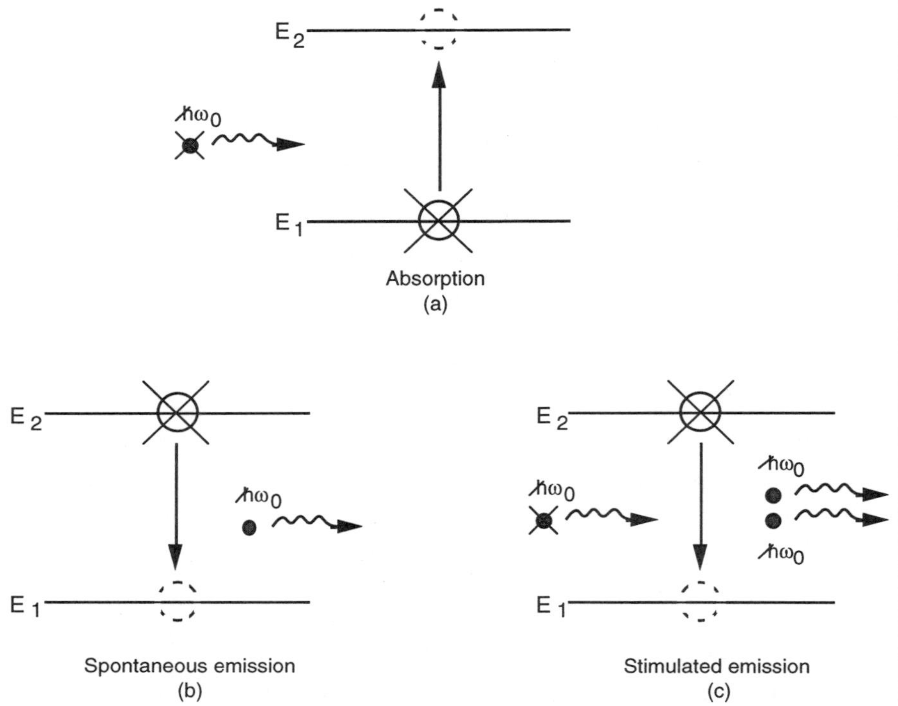

Figure IV.2.2: Illustration of the interaction processes between a light wave and an atomic medium.

Thus, we obtain the *rate equations* which describe how the populations change with time:

$$\begin{aligned} \dot{N}_1 &= N_2 B_{21} \overline{\rho} + N_2 A_{21} - N_1 B_{12} \overline{\rho} \\ \dot{N}_2 = \dot{N}_1 &= - N_2 B_{21} \overline{\rho} - N_2 A_{21} + N_1 B_{12} \overline{\rho} \end{aligned} \qquad \text{(IV.2.2)}$$

Of course, the three Einstein coefficients depend only on the atomic medium. They are absolutely independent on the outside radiation.

Starting from a classical microscopic model, we shall now proceed to show how we can find a physical meaning for coefficients B_{12} and A_{21}.

IV.2.1.2 Classical microscopic theory

At optical frequencies, light-matter interaction is usually limited to the interaction of the radiation with the electron (or electrons) which is (are) least tied to the atom or molecule. We shall describe the electron starting from the model of the one-dimensional harmonic oscillator (the parameter being x). The radiation will be described classically by the x-component of the electric field, noted E_x (t), which is such that

$$E_x(t) = \mathcal{R}[\varepsilon_x(\omega_L)\exp(i\omega_L t)] \qquad \text{(IV.2.3)}$$

ω_L is the angular frequency of the radiation, ε_x is a complex amplitude, and $\mathcal{R}$ denotes the real part of an expression. Figure IV.2.3 shows the model. m and – e are, respectively, the mass and the charge of the electron. γ is the characteristic damping coefficient of the oscillator. K is the spring constant. The term *harmonic* refers to a restoring force $-Kx_j(t)$ proportional to the elongation x_j of electron j.

In the absence of radiation:

The movement of electron j is described by the equation

$$m\ddot{x}_j(t) = -K\,x_j(t) - m\gamma\,\dot{x}_j(t) \qquad \text{(IV.2.4)}$$

which has the following solution:

$$x_j(t) = x_{j0}\exp[(-\gamma/2 + i\omega'_0)(t - t_0) + i\varphi_{j0}] \qquad \text{(IV.2.5)}$$

with $\quad \omega'^2_0 = \omega^2_0 - (\gamma/2)^2 \quad$ and $\quad \omega^2_0 = K/m$

ω_0 is the resonant angular frequency of the undamped harmonic oscillator, and we usually have $\gamma/2 << \omega_0$. In atoms, ω_0 is usually on the order of 10^{15} to 10^{16} rad s^{-1}. x_{j0} and φ_{j0} represent the amplitude and the phase at time t = 0, respectively. The movement is a damped oscillation. Indeed, the energy of electron j can be written

$$U_j(t) = \frac{1}{2}m|\dot{x}_j(t)|^2 + \frac{1}{2}K|x_j(t)|^2 = U_{j0}\exp[-\gamma(t - t_0)] \qquad \text{(IV.2.6)}$$

With time, the energy relaxes with a time constant $\tau = \gamma^{-1}$, characteristic of the energetic losses. Basically, these losses consist of:

— Radiative type losses with frequency ω_0. In this case, spontaneous emission plays the most important part. The corresponding time constant τ_1 can be estimated from classical electromagnetism ("antenna"-type radiation) and can be written

$$\tau_1 = 6\,\pi\,\varepsilon\,m\,c^3 / e^2\,\omega^2_0 \qquad \text{(IV.2.7)}$$

where ε is the permittivity of the medium.

In this microscopic model, τ_1^{-1} is the equivalent of coefficient A_{21} describing spontaneous emission in the macroscopic model we developed earlier.

— Nonradiative type losses, due to inelastic collisions of the electron, to the emission of thermal phonons in the medium, and so on. It is always difficult to identify exactly which of the various mechanisms are responsible for the nonradiative energy dissipation in a medium.

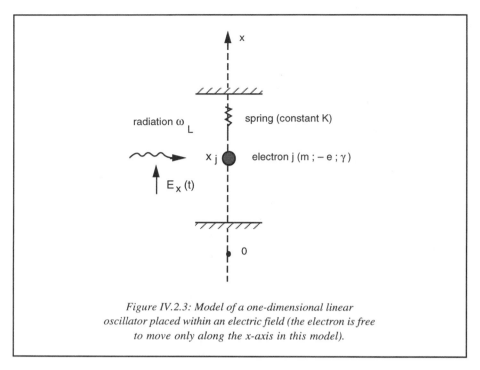

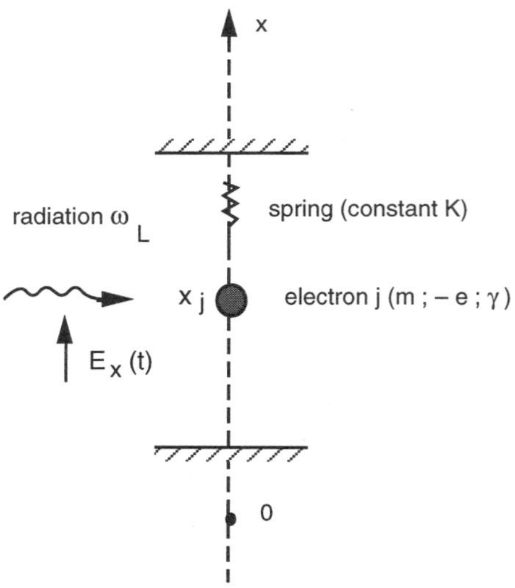

Figure IV.2.3: Model of a one-dimensional linear oscillator placed within an electric field (the electron is free to move only along the x-axis in this model).

We are now ready to introduce a fundamental physical quantity: the volume density of polarization $P_x(t)$, defined as follows:

$$P_x(t) = \sum_{j=1}^{N_1} p_{xj}(t) = -N_1 e x_j(t) \qquad \text{(IV.2.8)}$$

N_1 is the number of electrons per unit volume. In Eq. IV.2.8, we assumed the electrons to be sufficiently diluted in the medium so as to be able to neglect the interactions between different microscopic electronic dipole moments. Putting Eq. IV.2.5 into Eq. IV.2.8 and assuming that $\omega'_0 \approx \omega_0$ and that $\varphi_{j0} = \varphi_0$, we find that for all j we obtain

$$P_x(t) = N_1 p_{x_0} \exp\left[(-\gamma/2 + i\omega_0)(t - t_0) + i\varphi_0\right] \qquad \text{(IV.2.9)}$$

with $p_{x_0} = -ex_0$.

We conclude that macroscopic polarization occurs in the medium, with an oscillation frequency ω_0 and a damping constant $\gamma/2$.

However, writing the formula in this way brings about the problem of the phase of the electronic movements. We implicitly assumed that we had *total coherence* of all oscillators at time t_0. This is an extremely idealized view. By making the damping constant γ larger, we can take into account the partial incoherence of the vibrations as well as energy losses due to the nonradiative relaxation processes discussed above. Our solution becomes more general if we include these two extra damping factors (nonradiative phenomena and dephasing of the electronic movements) into the constant γ we introduced earlier. Of course, at the limit of an infinite damping constant, the polarization tends to zero.

Let us now introduce the coupling between the electron and the radiation.

In the presence of radiation:

If we take a homogeneously broadened system at steady state and, after having added the radiation-electron interaction force, we multiply both sides of Eq. IV.2.4 by $(-N_1 e)$, we obtain

---

Inset IV.1 ***Spectrum of the electronic oscillation***

The frequency width of the electronic oscillation is connected in a very fundamental way to two types of broadening:

— ***Homogeneous broadening***: This broadening occurs when all active elements of the amplifying medium have the same resonant frequency ω_0 (as was the case in our example above).

— ***Inhomogeneous broadening***: This broadening is due to the simultaneous presence of several classes (k) of oscillators, each characterized by a different frequency $\omega_0^{(k)}$ and such that the width of the distribution of the $\omega_0^{(k)}$ frequencies is greater than the homogeneous width of a single class.

1. ***Homogeneous broadening***

The oscillation is basically damped by two processes.

(a) ***Relaxation of the oscillation energy***

It is due to:

— The spontaneous transfer of the energy of some dipoles to the radiation. This is *spontaneous emission*, characterized by the constant $\gamma_{rad} = \tau_1^{-1}$.

microscopic model	$\longleftrightarrow$	phenomenological model
γ_{rad}	$\longleftrightarrow$	A_{21}

values of τ_1 for some laser transitions:
- dyes in liquids: 10^{-9} to 10^{-7} s.
- gases under low pressure: 10^{-7} to 10^{-5} s.
- doped solids: 10^{-4} to 10^{-3} s.

— The nonradiative transfer of the energy of some of the dipoles to the outside environment by way of nonelastic collisions. This is often caused by a dissipative coupling of the active dipoles with their environment.

(b) *Phase loss of some dipoles*

This is due to elastic collisions which do not modify the energy of the electron but which introduce a random term in the oscillation phase, thus increasing the partial incoherence. The corresponding constant will be called τ_2^{-1} and we write

$$\gamma = \gamma_{rad} + \gamma_{nrad} = \tau_1^{-1} + \tau_2^{-1} + \gamma_{nrad}$$

$$\gamma \approx \Delta\omega_o$$

total width of the damped oscillation

2. *Inhomogeneous broadening*

Each class of dipoles is characterized by its resonance frequency $\omega_0^{(k)}$ and by a homogeneous broadening $\Delta\omega_0$ of its oscillation. The observed global broadening consists in the envelope of all the homogeneous broadenings of the different classes.

Main causes of inhomogeneous broadening

gas laser: Doppler effect

solid laser: a variety of local environments due to imperfections of the cristal lattice

$$\mathcal{P}_x(\omega_L) / \mathcal{E}_x(\omega_L) = (N_1 e^2/m)\left(1 / \left[\left(\omega_0^2 - \omega_L^2\right) + i\omega_L \Delta\omega_0\right]\right) \quad \text{(IV.2.10)}$$

As for Eq. IV.1.3, we write:

$$P_x(t) = \mathcal{R}\left[\mathcal{P}_x(\omega_L)\exp(i\omega_L t)\right] \quad \text{(IV.2.11)}$$

and we identify the damping constant γ to the *global width* $\Delta\omega_0$ of the oscillation centered at ω_0 measured in terms of angular frequency.

After a transitory state, the radiation imposes a forced oscillation on the electrons at frequency ω_L. A complex susceptibility $\chi(\omega_L) = \chi'(\omega_L) + i\,\chi''(\omega_L)$ can be defined as

$$\chi(\omega_L) = \mathcal{P}_x(\omega_L) / \varepsilon_0 \mathcal{E}_x(\omega_L) \quad \text{(IV.2.12)}$$

By using the constitutive relations of Maxwell's equations

$$\mathcal{D} = \varepsilon_0 \mathcal{E} + \mathcal{P} = \varepsilon_0 \mathcal{E} + \varepsilon_0 \chi \mathcal{E} = \varepsilon_0 \mathcal{E}(1 + \chi) \quad \text{(IV.2.13)}$$

and by defining $\Delta\omega_L = 2(\omega_L - \omega_0) / \Delta\omega_0$ as the parameter measuring the distance between the frequency of the radiation (ω_L) and the electronic oscillation (ω_0), we finally obtain

$$\chi'(\omega_L) = -\left(N_1 e^2 / m\, \varepsilon_0\, \omega_0\, \Delta\omega_0\right)\left[\Delta\omega_L / \left(1 + \Delta\omega_L^2\right)\right]$$
$$\chi''(\omega_L) = -\left(N_1 e^2 / m\, \varepsilon_0\, \omega_0\, \Delta\omega_0\right)\left[1 / \left(1 + \Delta\omega_L^2\right)\right] \qquad \text{(IV.2.14)}$$

Using the coherent radiative relaxation parameter τ^{-1} taken from Eq. IV.2.7, this can be written as

$$\chi'(\omega_L) = -\left(3N_1 \lambda_0^3\, \tau_1^{-1} / 4\pi^2 \Delta\omega_0\right)\left[\Delta\omega_L / \left(1 + \Delta\omega_L^2\right)\right]$$
$$\chi''(\omega_L) = -\left(3N_1 \lambda_0^3\, \tau_1^{-1} / 4\pi^2 \Delta\omega_0\right)\left[1 / \left(1 + \Delta\omega_L^2\right)\right] \qquad \text{(IV.2.15)}$$

Figure IV.2.4 shows the evolution of the real and of the imaginary part of the "lorentzian" susceptibility defined above. It is possible to show that these two components are related to each other by the universal relations called Kramers–Kronig relations. The real and imaginary parts of the susceptibility describe, respectively, the dispersive and the dissipative characteristics of the linear response of the system to the electromagnetic field (see reference IV.5.5).

The amplitudes $\chi'(\omega_L)$ and $\chi''(\omega_L)$ are inversely proportional to the total width $\Delta\omega_0$ of the oscillator. The curves show a resonance occurring at $\omega_0 = \omega_L$.

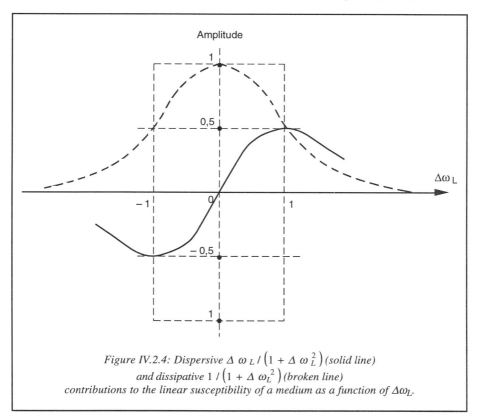

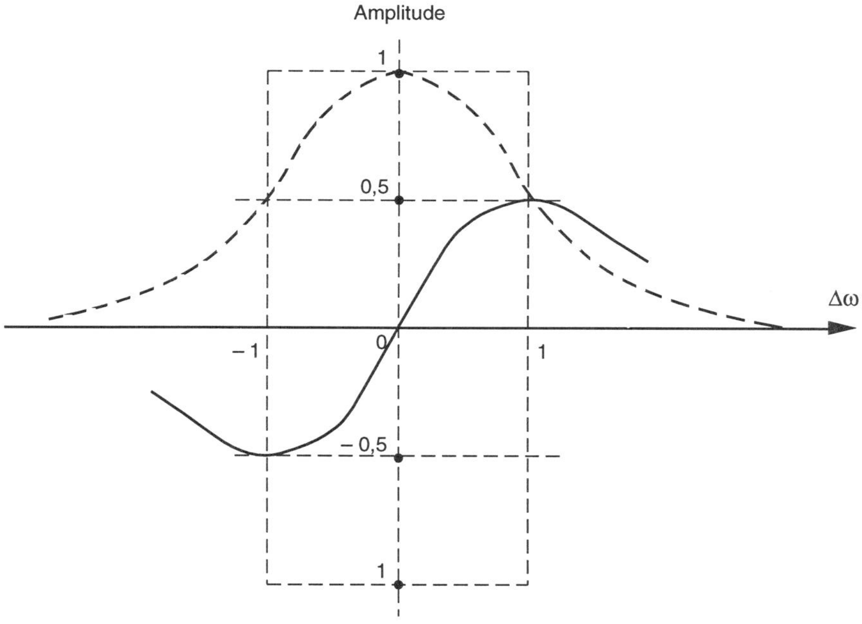

Figure IV.2.4: Dispersive $\Delta\omega_L / \left(1 + \Delta\omega_L^2\right)$ (solid line) and dissipative $1 / \left(1 + \Delta\omega_L^2\right)$ (broken line) contributions to the linear susceptibility of a medium as a function of $\Delta\omega_L$.

We now want to calculate the energy delivered to the electron by the radiation. We may write

$$du(t) = f_x(t)\, dx(t) = -e\, E_x(t)\, dx(t) \qquad \text{(IV.2.16)}$$

and the instantaneous power transferred per unit volume is

$$\dot{u}(t) = du(t)/dt = -N_1 e\, E_x(t)\, dx(t)/dt = E_x(t)\, dP_x(t)/dt \qquad \text{(IV.2.17)}$$

Let us write $E_x(t)$ and $P_x(t)$ in the following way:

$$E_x(t) = \frac{1}{2}\left[\mathcal{E}(\omega_L)\exp(i\omega_L t) + \mathcal{E}^*(-\omega_L)\exp(-i\omega_L t)\right]$$

$$P_x(t) = \frac{1}{2}\left[\mathcal{P}(\omega_L)\exp(i\omega_L t) + \mathcal{P}^*(-\omega_L)\exp(-i\omega_L t)\right] \qquad \text{(IV.2.18)}$$

averaging over time yields

$$\begin{aligned}\bar{\dot{u}}(t) &= (i\omega_L t/4)\left[-\mathcal{E}(\omega_L)\mathcal{P}^*(-\omega_L) + \mathcal{E}^*(-\omega_L)\,\mathcal{P}(\omega_L)\right]\\ &= \left(-\frac{1}{2}\right)\mathcal{E}_0\,\omega_L\,\chi''(\omega_L)\left|\mathcal{E}(\omega_L)\right|^2\end{aligned} \qquad \text{(IV.2.19)}$$

The instantaneous power delivered to the electron by the radiation is proportional to the square of the electric field of the wave (i.e. to its power) and to the imaginary dissipative part of the susceptibility. Substituting the second equation of Eq. IV.2.15 into Eq. IV.2.19 gives

$$\bar{\dot{u}}(t) = \left(3\,\varepsilon_0 N_1\,\omega_L\,\lambda_0^3\,\tau_1^{-1}/8\pi^2\,\Delta\omega_0\right)\left[1/\left(1+\Delta\omega_L^2\right)\right]\left|\mathcal{E}(\omega_L)\right|^2 \qquad \text{(IV.2.20)}$$

This result will allow us to find a physical meaning for $B_{12}\bar{\rho}$, Einstein's absorption coefficient. Indeed, in the case of $N_2 = 0$ [no spontaneous emission occurring from state (2)] and for a monochromatic wave of frequency ω_L, the second equation of Eq. IV.2.2 becomes

$$-\dot{N}_1 = N_1 B_{12}\,\bar{\rho} \qquad \text{(IV.2.21)}$$

Multiplying both sides by $\hbar\omega_0$ and setting $\bar{\dot{u}}(t)$ with $-\dot{N}_1\,\hbar\omega_0$, we end up with the expression

$$B_{12}\,\bar{\rho} = \left(3\,\varepsilon_0\,\omega_L\,\lambda_0^3\,\tau_1^{-1}/8\,\pi^2\,\hbar\omega_0\,\Delta\omega_0\right)\left[1/\left(1+\Delta\omega_L^2\right)\right]\left|\mathcal{E}(\omega_L)\right|^2 \qquad \text{(IV.2.22)}$$

and identifying $\bar{\rho}$ to $\varepsilon\left|\mathcal{E}(\omega_L)\right|^2/2$ gives

$$B_{12}/A_{21} = \left(6\,\omega_L\,\lambda_0^3/8\,\pi^2\,n^2\,\hbar\omega_0\,\Delta\omega_0\right)\left[1/\left(1+\Delta\omega_L^2\right)\right] \qquad \text{(IV.2.23)}$$

Here n is the refractive index, and the constant A_{21} can be written as

$$A_{21} = e^2\,\omega_0^2 / 6\pi\,\varepsilon\,mc^3 \qquad \text{(IV.2.24)}$$

With the help of an extremely simple microscopic model, we were able to focus on some important points concerning absorption and spontaneous emission. Specifically, absorption is described by the parameter $B_{12}\,\overline{\rho}$, which

— is proportional to the radiative energy conveyed by a volume λ_0^3,

— is inversely proportional to the oscillation band-width $\Delta\omega_0$,

— vanishes when $\Delta\,\omega_L^2 \gg 1$—in other words, when the frequency of the radiation ω_L becomes very different from the characteristic frequency of the medium ω_0.

We must remember that the classical microscopic model does not account for stimulated emission. Only by quantifying the medium—as will be done in Chapter VII—does one get a complete description. However, the phenomenological model we just studied shows a close resemblance between absorption and stimulated emission. We shall see later that these two phenomena are described by the same kind of physics and that the results obtained above are valid for stimulated emission when replacing N_1 by N_2. One of the consequences is that $B_{12} = B_{21}$ for two states (1) and (2) having the same degeneracy. But let us turn to the problem of determining the required conditions for amplification to take place in the illuminated medium.

IV.2.1.3 Steady-state amplification conditions in the case of states with the same degeneracies

Let a wave of cross section S travel through a thickness dz of a medium whose refractive index is close to 1. Neglecting spontaneous emission, the intensity $\overline{I}(z)$ is described by

$$\left[(n/c)\,\overline{I}\,(z) - (n/c)\,\overline{I}\,(z+dz)\right] Sc/n = (N_{1S} - N_{2S})\,B_{12}\,\overline{\rho}\,\hbar\omega_0\,Sdz \qquad \text{(IV.2.25)}$$

This equation, which states the conservation of energy during the light-matter interaction, may also be written as

$$-\,d\,\overline{I}\,(z) = (N_{1S} - N_{2S})\,B_{12}\,\hbar\omega_0\,\overline{I}\,(z)\,n\,dz / c$$

or again

$$d\,\overline{I}\,(z) / dz = (N_{2S} - N_{1S})\,B_{21}\,\hbar\omega_0\,\overline{I}\,(z)\,n / c \qquad \text{(IV.2.26)}$$

Amplification of the wave can only occur if $N_{2S} > N_{1S}$—in other words, if a population inversion has taken place between states (1) and (2).

Let us now examine how we can obtain such a population inversion for two simple systems.

(a) Two-level system

— *In the absence of applied radiation* (thermal equilibrium):

Populations N_{1S} and N_{2S} obey the Boltzmann statistics because the system interacts with the thermal radiation. Therefore, in the case of identical degeneracies, these populations must be such that

$$N_{2S} / N_{1S} = \exp(-\hbar\omega_0 / kT) \qquad (IV.2.27)$$

where k is the Boltzmann constant and T the temperature of the medium.

At room temperature (T = 300 K) and in the near IR region of the optical spectrum, we have $\hbar\omega_0 / kT \approx 70$ and $N_{2S} << N_{1S}$.

— *In presence of applied radiation*:

The system is described by Eq. IV.2.2, which, for the steady-state conditions $(\dot{N}_1 = \dot{N}_2 = 0)$, leads to

$$N_{2S} / N_{1S} = B_{12}\,\overline{\rho} / (B_{21}\,\overline{\rho} + A_{21}) \qquad (IV.2.28)$$

We assumed that the surrounding thermal radiation had no noticeable effect on N_{2S}, which is justified in the visible and in the near-infrared region.

Starting from Eq. IV.2.28, we can calculate the population difference $\Delta N_S = N_{1S} - N_{2S}$. Let $N_S = N_{1S} + N_{2S}$ stand for the total population. Then

$$\Delta N_S / N_S = 1 / (1 + 2\,B_{12}\,\overline{\rho} / A_{21}) \qquad (IV.2.29)$$

Figure IV.2.5 shows how $\Delta N_S/N_S$ varies with $2\,B_{12}\,\overline{\rho} / A_{21}$, a dimensionless parameter which is proportional to the radiation intensity $\overline{I}$ (see Eq. IV.2.1). Let $\overline{I}_{Sat}$ be the *saturation intensity* corresponding to $\Delta N_S / N_S = 1/2$, then

$$\Delta N_S / N_S = 1 / \left(1 + \overline{I} / \overline{I}_{Sat}\right) \qquad (IV.2.30)$$

It is easy to see that, as $\overline{I}$ tends to infinity, populations N_{1S} and N_{2S} both tend to $N_S/2$: The two populations can, at best, achieve equality, but it is impossible to invert the populations. Let us examine the consequences of this situation.

For a two-level system, we just showed the following

— At thermal equilibrium, the Boltzmann statistics make such an inversion impossible since $(E_2 - E_1) >> kT$ in the optical domain, and therefore $N_{2S} << N_{1S}$ at any temperature.

— In the presence of an absorbed electromagnetic wave, the populations may become equal, but the hoped-for inversion cannot occur.

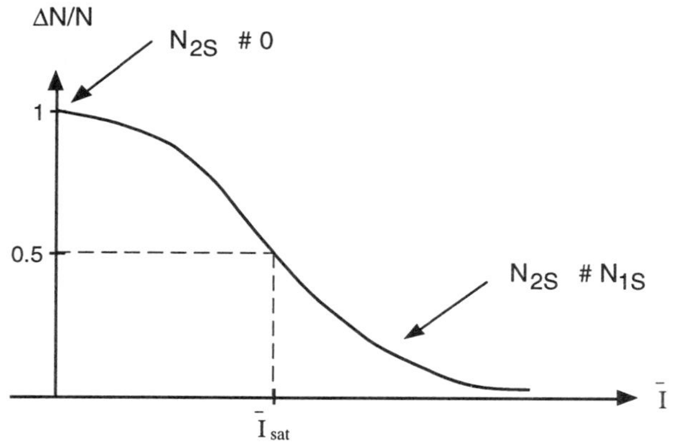

Figure IV.2.5: The relative population difference between states 1 and 2 in a two-level system, plotted as a function of the radiation intensity.

It thus clearly appears impossible to obtain a laser effect with a two-level system. The first existing laser (a ruby laser) used a three-level system. Nowadays, the amplification process of many kinds of lasers can be correctly explained by a four-level system. We shall describe only an idealized and simplified version of the latter system.

(b) Four-level system

A model of the four-level system is shown on Figure IV.2.6. The rate equations are as follows (see Section 2.2.1.1):

$$\dot{N}_4 = B_{14}\,\overline{\rho}_p\,N_1 - B_{41}\,\overline{\rho}_p\,N_4 - \gamma_{43}\,N_4 = W_p\,(N_1 - N_4) - W_{43}\,N_4$$

$$\dot{N}_3 = \gamma_{43}\,N_4 + B_{23}\,\overline{\rho}_L\,N_2 - B_{32}\,\overline{\rho}_L\,N_3 - A_3\,N_3 = W_{43}\,N_4 + W_{32}\,(N_2 - N_3) - A_3\,N_3$$

$$\dot{N}_2 = B_{32}\,\overline{\rho}_L\,N_3 - B_{23}\,\overline{\rho}_L\,N_2 - \gamma_{21}\,N_2 = W_{32}\,(N_3 - N_2) - W_{21}\,N_2 \qquad \text{(IV.2.31)}$$

We assumed that nonradiative relaxation occurs from level 4 to level 3 and from level 2 to level 1, with characteristic constants γ_{43} and γ_{21}. The term $-A_3N_3$ describes the total spontaneous emission from level 3.

At steady state, we have

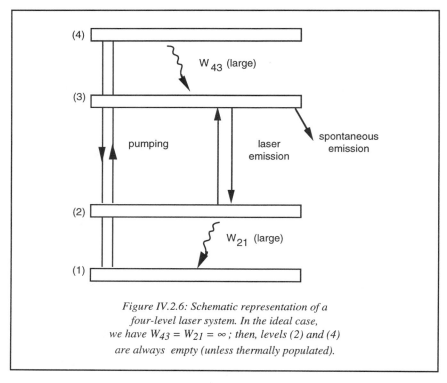

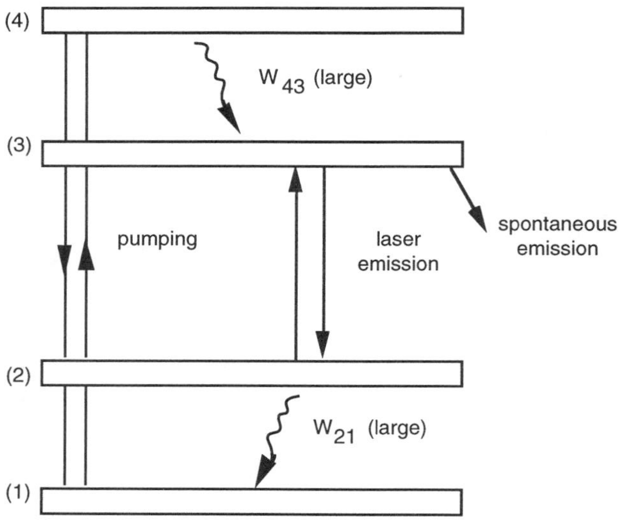

Figure IV.2.6: Schematic representation of a four-level laser system. In the ideal case, we have $W_{43} = W_{21} = \infty$; then, levels (2) and (4) are always empty (unless thermally populated).

$$N_{4S} = N_{1S}\, W_p / [W_{43}\,(1 + W_p / W_{43})]$$

$$N_{3S} = N_{4S}\, W_{43}\,(1 + W_{21} / W_{32}) / [W_{21} + A_3\,(1 + W_{21} / W_{32})] \qquad \text{(IV.2.32)}$$

$$N_{2S} = N_{4S}\, W_{43} / [W_{21} + A_3\,(1 + W_{21} / W_{32})]$$

Equations IV.2.32 can also be written as

$$N_{4S} = N_{1S}\, W_p \, \frac{1}{W_{43}\,(1 + W_p / W_{43})}$$

$$N_{3S} = N_{1S}\, W_p \, \frac{(1 + W_{21} / W_{32})}{(1 + W_p / W_{43})} \, \frac{1}{W_{21} + A_3\,(1 + W_{21} / W_{32})} \qquad \text{(IV.2.33)}$$

$$N_{2S} = N_{1S}\, W_p \, \frac{1}{(1 + W_p / W_{43})} \, \frac{1}{W_{21} + A_3\,(1 + W_{21} / W_{32})}$$

Population inversion between levels 3 and 2 is always true since we have $N_{3S} > N_{2S}$. When $W_{43} = W_{21} = \infty$ (corresponding to infinitely fast nonradiative transitions) the four-level model is said to be *ideal*, and we have

$$N_{4S} = 0$$

$$N_{3S} = N_{1S}\, W_p / (W_{32} + A_3)$$

$$N_{2S} = 0$$

Levels 4 and 2 are always empty. They play the part of *relay levels.*

The population of level 3 is proportional to the pump efficiency W_p. The total population is $N_t = N_{1S} + N_{3S}$, while the population difference between levels 3 and 2 is such that

$$N_{3S} - N_{2S} \approx N_{3S} = N_t / [1 + (W_{32} + A_3) / W_p] \qquad \text{(IV.2.34)}$$

IV.2.1.4 Description of widely used four-level amplifiers

(a) Solid amplifiers

— *Amplifying media*

Solid amplifying media usually consist of mineral salt cations included in monocrystalline matrices (the inclusion occurs during the synthesis of the crystal). The task of listing all the different cations having given rise to laser emission is an arduous one. The wide range of possible matrices already give about a hundred possible media. One must not forget that the spectroscopic properties of the doping cation depend on the chemical and physical structure of the host matrix (emission wavelength, bandwidth, and so on). However, only a few of these amplifying media have been exploited commercially.

Until a few years ago, commercial lasers with solid amplifiers emitted at fixed wavelengths. This is now changing. In the laboratories, new lasers have been built with emission wavelengths that can be adjusted in a spectral range as wide as 3500 Å. These new *tunable lasers* are starting to be commercialized and seem destined to a promising future. Table IV.2.1 gives the main characteristics of the two cations most commonly used in commercial lasers at room temperature.

— *Spectral properties of a solid*

Quantum mechanics teaches us that the energy levels of a cation isolated in space are extremely narrow. In a solid matrix, due to interatomic interactions, these levels widen to form "bands". Moreover, the vibrational modes of the lattice (phonons) may cause a homogeneous widening of the bands, often associated with an inhomogeneous broadening due to a diversity of local environment seen by the cations, a diversity related to defects in the spatial order of the matrix. This broadening of the energy levels in a solid is especially noticeable in the higher energy levels. The lowest levels remain much narrower.

TABLE IV.2.1 — **Main characteristics of solid amplifying media commonly used in optically pumped lasers.**

Active ion	Nd^{3+}	Ti^{3+}
Matrix	$Y_3Al_5O_{12}$ (YAG) $YLiF_4$ (YLF)	Al_2O_3 (sapphire)
Mean concentration of the active ion	1-2%	< 1%
Pumping region (in nm)	500-800	400-600
Operating temperature (K)	300	300
Emitted wavelength (in nm)	1060	650-1100
Emitted wavelength (in nm) after frequency doubling	530	400-550
Emitted wavelength (in nm) after two successive doublings	265	200-275

— *Amplifying medium doped with the cation Nd^{3+} (neodymium YAG and YLF lasers)*

Figure IV.2.7 represents the energy diagram of the Nd^{3+} cation.

In this case, broadening due to the phonons of the electronic levels is negligible. The most usual laser transition is found at about 1.06 µm. Pumping takes place in the visible and near-infrared domains by means of pulsed or continuous standard lamps or by means of a diode laser. Obviously, we have a system of the four-level type. Here the highest level, E_4, is level $^2S_{3/2}$. The four upper levels (from $^4S_{3/2}$ to $^2S_{9/2}$) reached by the pumping deplete in favor of level $^2S_{3/2}$. The emission is not tunable.

— *Amplifying medium doped with cation Ti^{3+} (titanium-sapphire laser)*

In this case, the matrix consists of sapphire (Al_2O_3), in which the Al^{3+} cations of the matrix are replaced by Ti^{3+} cations. An active medium is obtained by growing the crystal from a mixture of molten Al_2O_3 and Ti_2O_3. Monocrystals of sufficient optical quality have been obtained only recently, which explains why this type of laser made such a delayed appearance on the market. The energy levels of a Ti^{3+} cation in an Al_2O_3 matrix are shown on Figure IV.2.8.

In this figure we may notice that this medium pertains to the four-level system as well, and we can also see that transition from state (2) to state (1) implies the emission of a phonon. Indeed, there is a large phonon broadening of the two

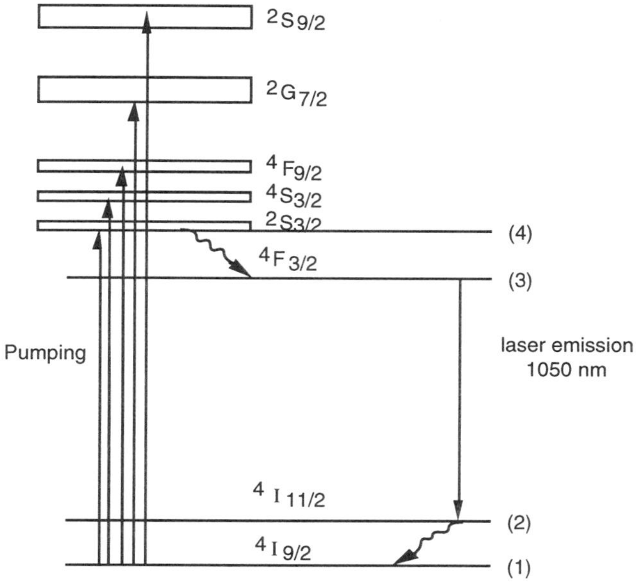

Figure IV.2.7:Energy diagram of the cation Nd^{3+} in a nonidentified solid matrix (the states are named according to the convention used in atomic spectroscopy).

electronic levels (^2E_g and $^2T_{2g}$). Therefore, the emission wavelength of a Ti^{3+}-sapphire laser can vary by changing the position of level (2) in the $^2T_{2g}$ band. To do this, one need only insert a dispersive device inside the cavity. The tuning range lies between 0.65 and 1.1 μm.

(b) Dye amplifiers

In this case, the liquid amplifying medium consists of dye molecules (rhodamines, coumarins, xanthens, and so on) dissolved in water, methanol, or ethylene glycol at a concentration of 10^{-3} to 10^{-4} mole per liter. Each electronic configuration contains a great number of states corresponding to the great number of vibrational and rotational degrees of freedom of the dye molecule. Each of these states in turn displays a large homogeneous broadening due to collisions of the molecules with their environment. The energies of all these bands overlap and form a broad continuous energy band.

Figure IV.2.9 illustrates the operating principle of this type of four level-laser.

Excitation from state (1) to state (4) is obtained by means of a standard lamp or by an auxiliary laser. The vibrational state (4) relaxes very rapidly (10^{-12} s) to state (3), whose population density is much greater than that of any vibronic excited state of band S_0, which have only a small thermal population. There is population inversion between level (3) and a continuous ensemble of levels (2). S_0 and S_1

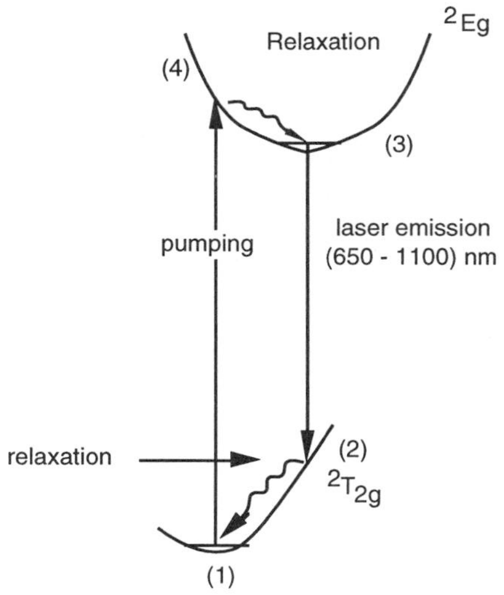

Figure IV.2.8: Structure of the energy of the Ti^{3+} cation in sapphire. (The naming of the electronic states is taken from group theory.)

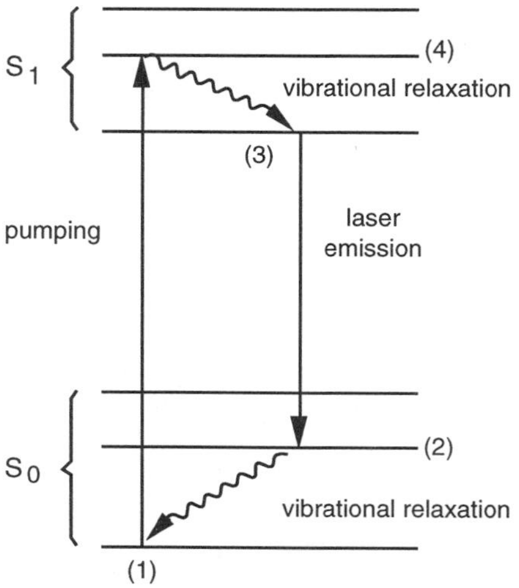

Figure IV.2.9: Principle of the dye amplifier. The vibrational structure of "bands" S_0 and S_1 is not indicated.

represent the ground (S_0) and the first excited (S_1) singulet electronic states. Of course, the laser can be tuned over a range of about 100 nm, covering almost all the fluorescent band (except for the spectral zone corresponding to the overlapping of the absorption and the fluorescence bands; in this domain, fluorescence photons are emitted to the relatively highly populated states of S_0).

IV.2.1.5 Gain of a laser amplifier

Let us now take a look at the evolution of the optical intensity as it travels through the amplifying medium. In the case of a four-level amplification, Eq. IV.2.26 can be written as

$$d\,\bar{I}(z)/\bar{I}(z)\,dz = (N_3 - N_2)\,B_{32}\,\hbar\omega_0\,n/c \qquad \text{(IV.2.35)}$$

In the case of an ideal amplifier and by neglecting spontaneous emission, the population difference between levels 3 and 2, ΔN, can be written as

$$\Delta N = N_3 - N_2 = N_3 = (N_t - N_3)\,W_p/W_{32}$$

or

$$\Delta N = N_t / (1 + W_{32}/W_p) \qquad \text{(IV.2.36)}$$

where N_t is the total population ($N_1 + N_3$).

By setting $W_{32} \approx B_{32}\,\bar{\rho} = B_{32}\,\bar{I}(z)\,n/c$ (see Eq. IV.2.1) we obtain

$$d\,\bar{I}(z)/\bar{I}(z)\,dz = G_0\left[1 + \left(\bar{I}(z)/I_{sat}\right)\right]^{-1} \qquad \text{(IV.2.37)}$$

with

$$\begin{aligned} I_{sat} &= W_p c / B_{32}\,n \\ G_0 &= B_{32}\,\hbar\,\omega_0\,N_t\,n/c \end{aligned} \qquad \text{(IV.2.38)}$$

G_0 is a positive quantity representing the *linear intensity gain* of the amplifier at weak intensities (nonsaturated gain). It measures the intensity gain per unit-length of amplifying medium, so it has the dimension of the inverse of a length. I_{sat} is a characteristic intensity, called *saturation intensity*, which allows us to differentiate between two very different limit behaviours of an amplifier illuminated by a wave of intensity $\bar{I}(z)$.

— If $\bar{I}(z) \ll I_{sat}$, the intensity of the wave increases exponentially with the distance z it traveled through the amplifying medium:

$$\bar{I}(z) = \bar{I}(0)\exp G_0 z \qquad \text{(IV.2.39)}$$

— If $\bar{I}(z) \gg I_{sat}$, the intensity increases linearly (and therefore much more slowly) with z:

$$\bar{I}(z) = \bar{I}(0) + I_{sat}\,G_0\,z \qquad \text{(IV.2.40)}$$

It is clear that for all values of $\bar{I}(z)$, the solution of Eq. IV.2.37 can be written as

$$\ln\left(\bar{I}(z)/\bar{I}(0)\right) + \left(\bar{I}(z) - \bar{I}(0)\right)/I_{sat} = G_0 z \qquad \text{(IV.2.41)}$$

Equation IV.2.37 can take the following form:

$$d\bar{I}(z)/dz = G_0 \bar{I}(z)/\left[1 + \bar{I}(z)/I_{sat}\right] = G\left(\bar{I}(z)\right)\bar{I}(z) \qquad \text{(IV.2.42)}$$

This form is similar to that of Eq. IV.2.30. This is simply due to the fact that the *saturated gain* $G\left(\bar{I}(z)\right)$, just like the population difference ΔN, is a quantity which decreases when the light intensity increases. Just like the population difference, the saturated gain of the wave is divided by two when $\bar{I}(z) = I_{sat}$.

Comment: G is the *relative linear* gain defined for the intensity. In the same way, we can define a relative linear gain for the field: $g = d\bar{E}(z)/\bar{E}(z)\,dz$. Of course, we have $G = 2g$. These linear gains should not be confused with the *relative amplification coefficients* for the intensity $a^2 = \bar{I}(z)/\bar{I}(0) = 1 + \left[\left(\bar{I}(z) - \bar{I}(0)\right)/\bar{I}(0)\right]$ and for the field

$$a = E(z)/E(0) = \left[1 + (E(z) - E(0))/E(0)\right]$$

IV.2.1.6 Steady-state laser intensity

In practice, the intensity is often large enough so that the second limit, $\bar{I}(z) >> I_{sat}$, applies. In steady-state conditions, we saw that after the laser wave traveled back and forth once through the amplifier of length L we would have

$$\bar{I}(2L) - \bar{I}_0 = I_{sat}\, G_0\, 2L \qquad \text{(IV.2.43)}$$

At steady state, the gain is equal to the losses which mainly occur during transmission $(1 - r^2)$ through the output mirror (see Figures IV.1.1 and IV.1.2) so that

$$I_{sat}\, G_0\, 2L = \bar{I}_0\left(1 - r^2\right) = \bar{I}' \qquad \text{(IV.2.44)}$$

$\bar{I}'$ represents the output power of the laser, in other words the power we can put to use. We can calculate it by replacing I_{sat} and G_0 of Eq. IV.2.44 by their expressions (see Eq. IV.2.38):

$$\bar{I}' = 2\, W_p\, N_t\, \hbar\, \omega_0\, L \qquad \text{(IV.2.45)}$$

We would like to stress the following points:

— The output intensity is independent of the reflexion coefficient of the mirror. In fact, r is usually very close to 1 (except in setups with a very high gain factor) so that when the laser is switched on, the gain is greater than the losses—a sine qua non condition for the laser to work.

— It is proportional to the volume density N of active elements.

— It is proportional to the length of the amplifying material.

— It is proportional to the parameter W_p which describes the pumping efficiency.

Of course, we depict a very idealized situation here. To be more realistic, we shall have to take into account nonradiative phenomena. Moreover, these calculations were done by taking a progressive wave which creates a steady state whose intensity does not depend on time. But in fact, the laser wave is a stationary wave and many lasers operate in a pulsing mode, where gains much larger than the cavity losses may occur. These lasers cannot be described by the same method. We shall say more about this when describing the problems which arise when time is considered in laser physics (see Chapter VI). In spite of these restrictions, the method we presented in this chapter satisfactorily describes the essential features of a great number of lasers.

IV.2.2 The optical cavity

In optics, it is a well-known fact that waves of limited dimensions diverge during propagation.The main purpose of the optical cavity (i.e., the resonator) is to confine the energy of the electromagnetic wave inside the cavity, despite diffraction which tends to disperse the wave to the outside. We can easily see that any association of a number of mirrors will not be sufficient. For instance, a linear cavity formed by two plane mirrors cannot lead to a stationary field since the diffration will cause an ever greater divergence of the wave. Later we shall see that this kind of cavity can never be used to make a continuous laser. Therefore, to find a cavity leading to the buildup of a *stationary configuration of the field*, we must use the laws of diffraction to monitor the wave from an initial field distribution on one of the mirrors, all the way through the cavity, to a reflection on the second mirror and back to the first one. Here, the final distribution must be identical to the initial one, in amplitude as well as in phase (of course, the phase may differ by a whole number of 2π). The specific shape of the field thus obtained is called an *eigenmode of the cavity*.

It is important to understand that whatever the spatial structure of the electric field initially injected into the cavity, the cavity changes this structure to leave only those modes which are eigenmodes.

It is therefore fundamental to determine the structure of these eigenmodes since these will engender the stationary state of the field outside of the cavity whose existence is the purpose of the laser. The first such calculations were done by Fox and Li (1963). Their calculations are mostly numerical and valid only for a specific cavity. We shall not reproduce them here. We shall also avoid using geometric optics to determine the stability conditions of a resonator, even though it would seem to be justified since the dimensions of the laser beam are very large compared to the wavelength of the radiation. However, in no case would this geometric analysis be able to provide the structure of the wave inside the cavity. In order to find this structure, we shall follow the procedure given by Kogelnick in 1968 and take into account the *wavelike nature of light.*

IV.2.2.1 Plane and spherical waves

In wave optics, the light beam is described as an electromagnetic wave, usually of infinite spatial dimension, in interaction with optical components of dimensions we shall suppose to be infinite so as to reduce as much as possible the diffraction losses. These beams, which may propagate either freely or within the laser cavity, are solutions of the propagation equation of waves, derived from the Maxwell equations. There are an infinite number of solutions, of which the most classical examples are the plane waves and the spherical waves. The main characteristics of these two kinds of waves are listed in Table IV.2.2.

IV.2.2.2 The laser wave

The electromagnetic wave which propagates inside a laser cavity can be neither a plane wave (with an homogeneous spatial distribution of energy) nor a spherical wave (which has a point of infinite energy density).

A laser wave must have two essential properties: (i) It must obey the laws of diffraction—since diffraction governs its propagation—and (ii) it must be identical to itself after a complete round trip through the cavity (or half a round trip in the case of a cavity consisting of two identical mirrors). This stable form will set up progressively, in a self-consistent way, taking advantage of the amplification generated by the amplifying medium.

Let us try to express this double condition by means of Huygens' diffraction theory (see bibliographic reference IV.5.5). Figure IV.2.10 defines the physical parameters used in this description. The incident wave carries a field $\mathcal{E}(x, y, z)$. z denotes the longitudinal coordinate of the diffracting object, for instance mirror M_1, inside the cavity. Each point (x, y) is considered as a source giving rise to a spherical wavelet (broken line). The distribution of the field in another plane (X, Y, Z), for instance the plane representing mirror M_2, is obtained by adding the contributions of all wavelets, and using Huygens' principle we obtain

$$\mathcal{E}(X, Y, Z) \sim (i/\lambda) \iint \mathcal{E}(x, y, z) \exp\,[-i\,\mathbf{k}\,(\mathbf{R} - \mathbf{r}_0)]\cos\theta / |\,\mathbf{R} - \mathbf{r}_0\,|\;dx\,dy \tag{IV.2.46}$$

The factor cos θ is due to the tilt angle. Except for a factor, the integrand represents the classical expression for the spherical wave (see Table IV.2.2).

If we use the paraxial (small θ) approximation, which is justified in the case of a laser, and by supposing $Z - z \approx L$, we can rewrite the above expression as follows (see also the last equation on the lower right-hand side of Table IV.2.2):

$$\mathcal{E}(X, Y, L) \sim \left[\, i \exp(-ikL) / \lambda L \right] \iint \mathcal{E}(x, y, z)$$
$$\times \exp\left\{ -ik\left[(X - x)^2 + (Y - y)^2 / 2L \right] \right\} dx\,dy$$

This equation transcribes the Fresnel approximation of Huygens' integral. Without extra approximations, it can be written in an equivalent form:

TABLE IV.2.2 — **Physical characteristics and mathematical representations of plane and spherical waves; $\mathcal{R}$ represents the real part of an expression.**

The two waves are characterized by an electric field **E**, solution of the wave equation: $\Delta \mathbf{E} - (1/v^2)\,\partial^2 \mathbf{E}/\partial t^2 = 0$

PLANE WAVES	*SPHERICAL WAVES*
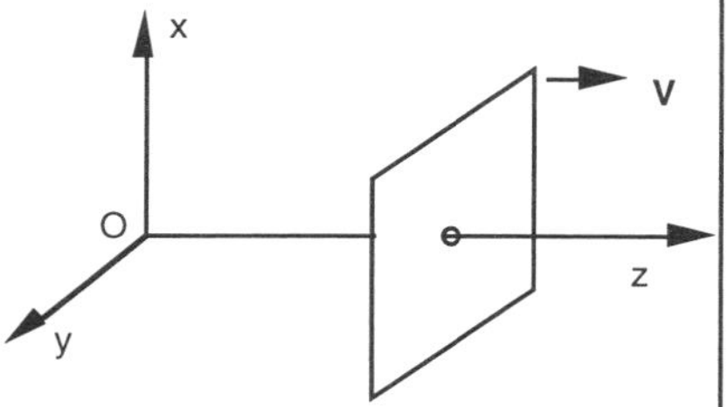	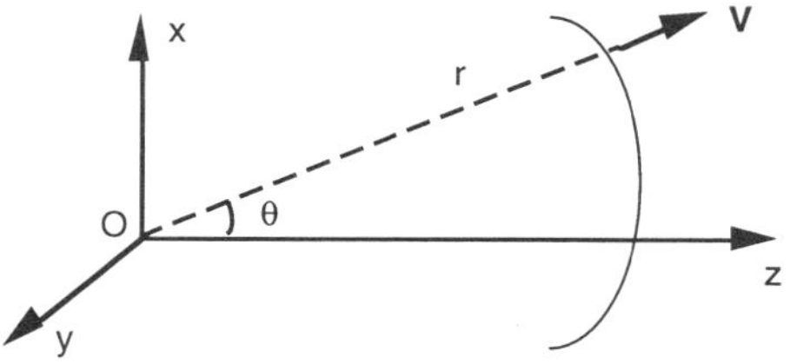
The field E lies within the plane of the wave	The field E is tangent to the wave-front
$E_x = f(z - vt)$	$E_x = \dfrac{f(r - vt)}{r}$
progressive wave moving in the direction of the positive z-axis.	divergent spherical wave moving in the direction of the positive z-axis.
$E_x = E_{0x} \cos[\omega(t - z/v) + \gamma_1]$	$E_x = (A_0/r) \cos[\omega(t - r/v) + \gamma_1]$
Monochromatic parallel plane wave	Monochromatic divergent spherical wave
$E_x = \mathcal{R}(E_{0x} \exp\{i[2\pi\nu(t - z/v) + \gamma]\})$	$E_x = \mathcal{R}((A_0/r) \exp\{i[2\pi\nu(t - r/v) + \gamma]\})$

If we define a complex amplitude $\mathcal{E}_x$ by

$\mathcal{E}_x = E_{0x} \exp[-i(2\pi\nu\, z/v - \gamma)]$	$\mathcal{E}_x = (A_0/r) \exp[-i(2\pi\nu\, r/v - \gamma)]$

we find

$$E_x = \mathcal{R}[\mathcal{E}_x \exp(-i2\pi\nu t)]$$

	For small θ, Ex can be written as $\mathcal{E}_x \approx (A_0/r) \exp\{-i[(2\pi\nu/v)(z + r_\perp^2/2z) - \gamma]\}$ with: $r_\perp^2 = x^2 + y^2$

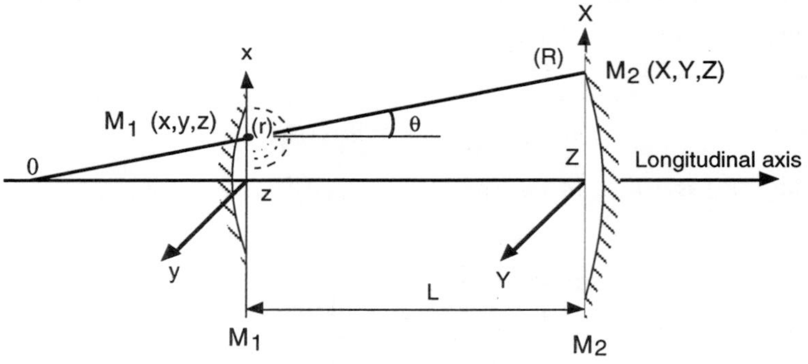

Figure IV.2.10: Huygen's principle and geometric construction to calculate Huygen's integral in Fresnel's approximation.

$$\mathcal{E}(X, Y, L) \sim (i / \lambda L) \exp\{-ik[(X^2 + Y^2)/2L + L]\} \times \iint \mathcal{E}'(x, y, z) \exp[-ik(xX + yY)/L]\, dx\, dy \qquad \text{(IV.2.47)}$$

in which we used the notation

$$\mathcal{E}'(x, y, z) = \mathcal{E}(x, y, z) \exp[-ik(x^2 + y^2)/2L]$$

Having restricted ourselves to field distributions $\mathcal{E}(x, y, z)$ with small spatial extension, we now have to take into account the reflection of mirror M_2 and then describe the trajectory back to mirror M_1 in order to calculate what the distribution $\mathcal{E}(x, y, z)$ is on M_1 after a round trip through the cavity.

For a stable state, $\mathcal{E}'(x,y)$ and $\mathcal{E}(x,y)$ must be proportional, "a" being a complex scalar:

$$\mathcal{E}'(x, y) = a\, \mathcal{E}(x, y) \qquad \text{(IV.2.48)}$$

Formally, Eq. IV.2.48 is equivalent to an equation determining the eigenfunctions of an operator. This way, we can see that solving Eq. IV.2.48 is the same thing as finding the eigenfunctions of the transformation operator

$$\iint dx\, dy \exp[-ik(xX + yY)/L]$$

which is contained in Eq. IV.2.47. Now this operator defines a two-dimensional *spatial Fourier transform* and we know its eigenfunctions are *Gaussian functions*, which may or may not be complex.

In a very simple manner we now reach an extremely important conclusion:

Diffraction controls the radial transformation of a wave in a cavity. The mathematical description of this transformation is a Fourier transform. However, the Fourier transform of a Gaussian is a Gaussian. So, if we want a wave $\mathcal{E}(x, y, z)$ which does not undergo radial distortions while building up in the cavity of the resonator, it must be a *Gaussian wave.*

Instead of trying to solve Eq. IV.2.48, we shall take a more empirical approach. We shall try to construct an approximation to a wave function, giving it *a priori* a Gaussian structure, which is a solution of the wave equation (see Table IV.2.2) derived from Maxwell's equations (which govern all electromagnetic waves considered here).

(a) Approximate solution of the wave equation: TEM structure

We shall look only at monochromatic waves of frequency ν whose time dependence is given by

$$E = \mathcal{E}(x, y, z) \exp(i2\pi\nu t)$$

Of course, $\mathcal{E}(x, y, z)$ fits the wave equation derived from Maxwell's equations:

$$\Delta \mathcal{E}(x, y, z) + k^2 \mathcal{E}(x, y, z) = 0 \qquad \text{(IV.2.49)}$$

with $k = 2\pi\nu/v$.

The reader may check that this is equivalent to the form given in Table IV.2.2. Now, let

$$\mathcal{E}(x, y, z) = \mathcal{E}_0 \psi(x, y, z) \exp(-i k z) \qquad \text{(IV.2.50)}$$

As said, the term $\exp(-i k z)$ describes a plane wave traveling in the Oz direction. Therefore, $\psi(x, y, z)$ represents the difference between a plane wave and the wave we are looking for. We suppose that it changes slowly along Oz. More exactly, we suppose that

$$\left|\partial^2 \psi / \partial z^2\right| << k \left|\partial \psi / \partial z\right| \qquad \text{(IV.2.51)}$$

Putting Eq. IV.2.50 into Eq. IV.2.49 and taking into account approximation IV.2.51, we find

$$\Delta_\perp \psi(x, y, z) - i\, 2\, k\, \partial \psi / \partial z = 0 \qquad \text{(IV.2.52)}$$

with

$$\Delta_\perp = \left(\partial^2 / \partial x^2\right) + \left(\partial^2 / \partial y^2\right) \qquad \text{(IV.2.53)}$$

Since we are looking for a wave with a Gaussian radial structure, we want a solution to Eq. IV.2.52 which writes as follows:

$$\psi(r_\perp, z) = \exp[-i P(z)] \exp\left[-i k r_\perp^2 / 2 q(z)\right] \qquad \text{(IV.2.54)}$$

defining $r_\perp^2 = x^2 + y^2$.

This specific solution to the wave equation brings about an electric field which may propagate freely inside the resonator as well as outside of it. We must now determine the complex quantities q(z) and P(z) and try to find their physical meaning. P(z) relates to a *longitudinal* variation along Oz: Its real and imaginary parts represent, respectively, the variations of phase and amplitude of the electric field. P(z) is called the *longitudinal beam parameter*. q(z) relates to the *radial* variations of the field in the plane z = constant: The real and imaginary parts of 1/q(z) represent, respectively, the variations of the wave front and that of the field amplitude distribution around the beam axis. q(z) is called the *radial beam parameter.*

Substituting IV.2.54 into Eq. IV.2.52 gives

$$-\,i\,2k/q - k^2 r_\perp^2 / q^2 - 2k\left[dP/dz - k\,r_\perp^2\,(dq/dz)/2q^2\right] = 0 \qquad \text{(IV.2.55)}$$

For Eq. IV.2.55 to be true of all $r_\perp$, we must have simultaneously

$$dq(z)/dz = 1 \qquad \text{(IV.2.56)}$$

$$dP(z)/dz = -\,i/q(z) \qquad \text{(IV.2.57)}$$

(b) Propagation law and fundamental wave structure

— *Radial beam parameter q(z)*:

Equation IV.2.56 simply implies

$$q(z) = q(0) + z \qquad \text{(IV.2.58)}$$

Let us write $q(z)^{-1}$ in the following way so as to separate its real part from its imaginary part:

$$1/q(z) = 1/R(z) - i\,\lambda/\pi\,\omega^2(z) \qquad \text{(IV.2.59)}$$

Assuming R(0) to be infinite (there is a plane wave at z = 0) we obtain

$$1/q(0) = -\,i\,\lambda/\pi\,\omega^2(0)$$

or

$$q(0) = i\,\pi\,\omega^2(0)/\lambda \qquad \text{(IV.2.60)}$$

or else, by substituting Eq. IV.2.60 into Eq. IV.2.58 we obtain

$$R(z) = z\left[1 + \left(\pi\,\omega^2(0)/\lambda z\right)^2\right] \qquad \text{(IV.2.61)}$$

$$\omega^2(z) = \omega^2(0)\left[1 + \left(\lambda z/\pi\,\omega^2(0)\right)^2\right] \qquad \text{(IV.2.62)}$$

Thus the expression $\exp\left[-\,ik\,r_\perp^2/2\,q(z)\right]$ of Eq. IV.2.54 is equal to

$$\exp\left[-\,r_\perp^2/\omega^2(z)\right]\exp\left[-\,ik\,r_\perp^2/2R(z)\right]$$

The expression exp[– iP(z)] remains to be calculated, for which we need to determine P(z).

— *Longitudinal beam parameter P(z)*:

Let us substitute Eq. IV.2.58 into Eq. IV.2.57 and integrate over z. Choosing the phase origin such that P(0) = 0, we find

$$P(z) = i\,\ln\left\{q(0)/\left[q(0) + z\right]\right\} \qquad \text{(IV.2.63)}$$

which, after an easy calculation, gives

$$\exp\left[-\,iP(z)\right] = \left[\omega(0)/\omega(z)\right]\exp\left\{i\,\arctan\left[z\,\lambda/\pi\,\omega^2(0)\right]\right\} \qquad \text{(IV.2.64)}$$

(c) Expression of the fundamental wave and physical interpretation

Combining all these elements, we find the following expression for the electric field:

$$\mathcal{E}(x, y, z) = \mathcal{E}_0 [\omega(0) / \omega(z)] \exp\left(-i \left\{k \left[z + r_\perp^2 / 2R(z)\right] - \phi(z)\right\}\right) \times \exp\left[-r_\perp^2 / \omega^2(z)\right] \qquad \text{(IV.2.65)}$$

where

$$\phi(z) = \arctan\left[\lambda z / \pi\, \omega^2(0)\right]$$

At $z = 0$, we have $\mathcal{E}(x, y, 0) = \mathcal{E}_0$. Moreover, $\omega(z)$ is at its minimum at $z = 0$ (see Section IV.2.62), which corresponds to the point where the beam has the *smallest spatial extension*, called *beam waist*. $\omega(z)$ increases hyperbolically with z (see Figure IV.2.11). Its asymptote makes an angle $\theta = \arctan [\lambda/\pi\, \omega(0)]$ with the propagation axis. θ is called the *beam divergence*. Far from the beam waist, the hyperbole may be assimilated to its asymptote, so that $\omega(z)$ varies linearly with z. The amplitude of $\mathcal{E}(x, y, z)$ then behaves like $\omega^{-1}(z)$, in other words like z^{-1}. In the same way, the phase then varies like $\left[k\left(z + r_\perp^2 / 2z\right) - \pi/2\right]$. In Table IV.2.2 the reader will notice that this behavior (of the amplitude and the phase at the same time) is characteristic of a spherical wave.

However, the last term of the second line of Eq. IV.2.65 implies a Gaussian attenuation of the amplitude around the axis with a characteristic radius $\omega(z)$ which varies with the longitudinal propagation position z.

Thus, far from the beam waist, the fundamental wave is a spherical wave with a Gaussian radial energy attenuation. This wave is also called quasi-spherical wave.

We just defined the fundamental transverse wave structure, noted TEM_{00}. In fact, many transverse wave structures with a Gaussian radial attenuation exist. They are much more complicated and can be decomposed linearly on complete orthonormal bases of waves called TEM_{mn} (m and n whole numbers; wave structure TEM_{00} corresponds to the case $m = n = 0$). Many of the lasers used today in research are characterized by the emission of the fundamental transverse wave TEM_{00}, which henceforth will be the only wave structure we shall consider in this book.

(d) Adaptation of the fundamental wave to the optical resonator; fundamental resonating modes of the cavity

In the last section, we characterized the fundamental structure of a wave which is able to propagate within an optical cavity *without deformation*. Specifically, at the longitudinal coordinate z, the radius of curvature of the wave, R(z), can be expressed with Eq. IV.2.61 and we have

$$R(z) = z + \left(\pi\, \omega^2(0) / \lambda\right)^2 / z$$

Of course, we now have to take care of two things: (i) The wave must maintain the same structure after each round trip through the cavity (which will be the case if the wave perfectly fits the geometrical shape of the resonator described in Figure

IV.2.12) and (ii) the phase condition we already mentioned must be satisfied after each transit through the cavity. Condition (i) is easy to obtain. We can either write the self-consistency conditions for q(z) during its travel back and forth through the cavity, or we can successively equate the curvatures of the wave to the corresponding curvatures of the mirrors, separated by length L. Having chosen the last method, we get a system of equations. Setting $\omega(0) = \omega_0$ we obtain

$$-\left[z_1 + \left(\pi\,\omega_0^2 / \lambda\right)^2 / z_1\right] = R_1$$

$$z_2 + \left(\pi\,\omega_0^2 / \lambda\right)^2 / z_2 = R_2 \qquad \text{(IV.2.66)}$$

$$z_2 - z_1 = L$$

This set of equations allows us to calculate the three parameters z_1, z_2, and $\pi\,\omega_0^2/\lambda$ as functions of R_1, R_2, and L. The calculation leads to the value of the beam waist ω_0 "selected" by the cavity. We find

$$\omega_0^2 = \left(\lambda L/\pi\right)\sqrt{h_1\,h_2\,(1 - h_1\,h_2)/(h_1 + h_2 - 2h_1\,h_2)^2} \qquad \text{(IV.2.67)}$$

where

$$h_1 = 1 - L/R_1$$
$$h_2 = 1 - L/R_2$$

In the same way, the values of z_1 and z_2 lead to the values of $\omega(-z_1)$ and $\omega(z_2)$—in other words, to the radius of the beam at the mirrors:

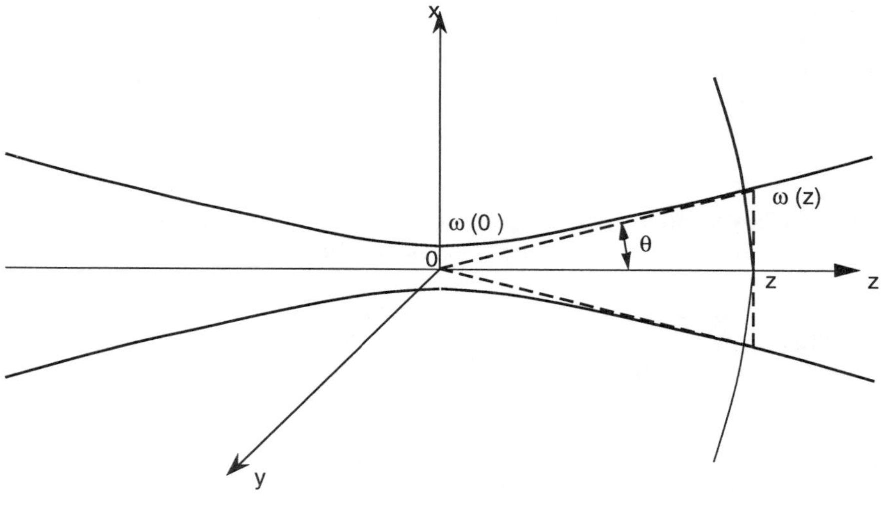

Figure IV.2.11: Transverse structure of the fundamental Gaussian propagating wave.

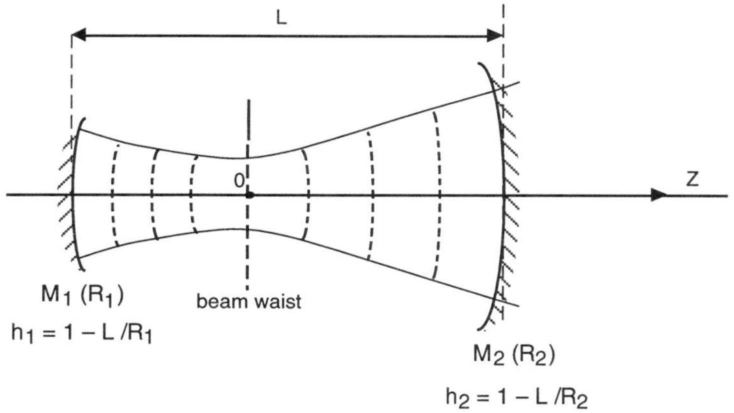

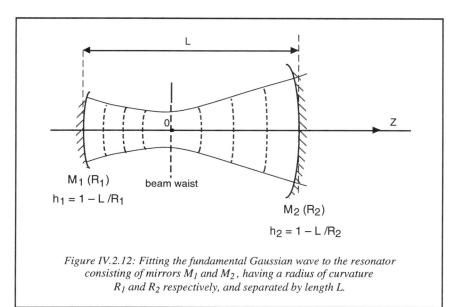

Figure IV.2.12: Fitting the fundamental Gaussian wave to the resonator consisting of mirrors M_1 and M_2, having a radius of curvature R_1 and R_2 respectively, and separated by length L.

$$\begin{aligned} \omega_1^2 &= (\lambda L/\pi)\sqrt{h_2/h_1\,(1 - h_1 h_2)} \\ \omega_2^2 &= (\lambda L/\pi)\sqrt{h_1/h_2\,(1 - h_1 h_2)} \end{aligned} \qquad \text{(IV.2.68)}$$

It is also easy to find the exact location of the beam waist. The distances z_1 and z_2 from the beam waist to the locations of each of the mirrors M_1 and M_2 are

$$\begin{aligned} z_1 &= -\,h_2\,(1 - h_1)\,L\,/\,(h_1 + h_2 - 2h_1h_2) \\ z_2 &= h_1\,(1 - h_2)\,L\,/\,(h_1 + h_2 - 2h_1h_2) \end{aligned} \qquad \text{(IV.2.69)}$$

We wish to draw attention to two things:

— Equation IV.2.67 shows that ω_0^2 can be real and finite, in other words physically acceptable, only if the product h_1h_2 is non-negative on the one hand and smaller or equal to one on the other hand. This double condition is expressed by

$$0 \le h_1\,h_2 \le 1 \qquad \text{(IV.2.70)}$$

Of course, this imposes strict conditions on the radius of curvature of the mirrors and on the distance between them. This condition is known as the *stability condition of the resonator* and is illustrated in Figure IV.2.13.

It splits the h_1, h_2 plane into two regions. The stable region is indicated in gray. In the figure, three special configurations, all three at the limit of stability, are indicated. They are the plane, the confocal, and the concentric configurations, respectively, and correspond to resonators formed by two identical mirrors.

o When $R_1 = R_2 = \infty$, in other words $h_1 = h_2 = 1$, we have the *plane*

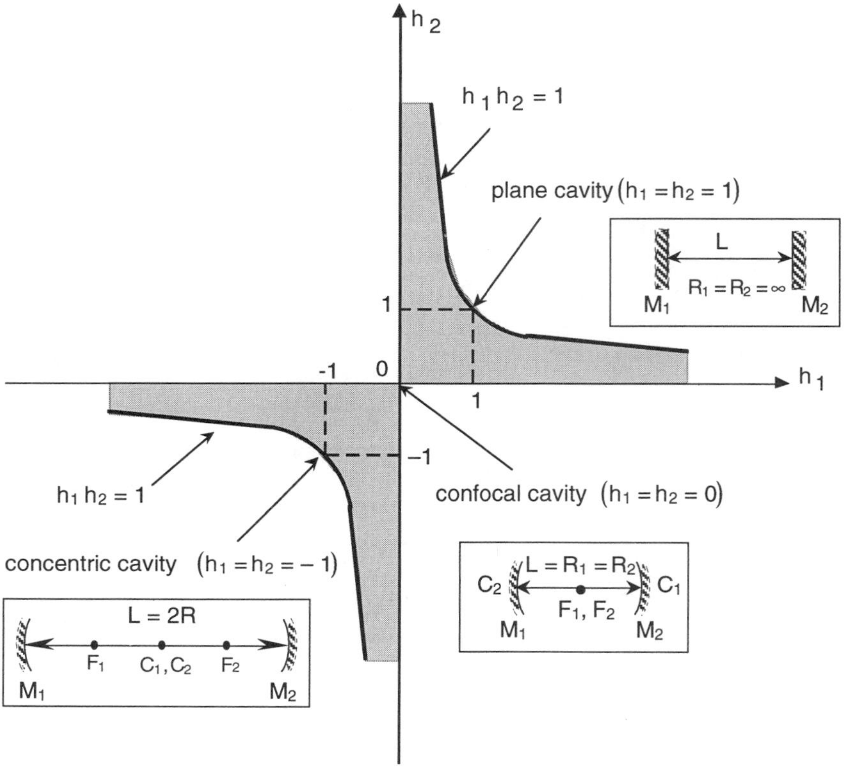

Figure IV.2.13: Stability diagram of a two mirror optical cavity. C and F, respectively, represent the centers of curvature and the focal points of mirrors M.

configuration. The cavity consists of two plane mirrors.

o When $L = R_1 = R_2$, i.e. when $h_1 = h_2 = 0$, we have the *confocal* configuration. Each mirror is placed at the focus of its partner.

o When $L = 2R_1 = 2R_2$ i.e. $h_1 = h_2 = -1$, we have the *concentric* configuration. The centers of curvature of the two mirrors merge.

This stability condition is fundamental. It guides the laser constructor. It also implies that a stable stationary resonant wave builds up progressively in the cavity after traveling back and forth many times. This condition is of course very important for the quality of the spatial properties of the generated wave.

It should be observed that the study of the resonator by means of geometric optics leads to the same stability criterion, independently of any *a priori* structure given to the wave. This will be explained in the ensuing inset.

Inset IV.2 ***Transfer matrix of an optical cavity***

From the point of view of geometric optics, and placing ourselves in the conditions where the Gaussian approximation is valid (i.e., the light rays are only slightly tilted away from the principal axis of the setup), a light ray (see Figure IV.2.14) can be described by a column vector $\begin{Bmatrix} y \\ \alpha \end{Bmatrix}$, where y and α respectively indicate the y-coordinate, and the slope of the light ray with respect to the optical axis of the system (in fact, this vector describes the curvature R = y/α of the optical wave Σ). An optical element, like any of those likely to be present in an optical cavity (plane or spherical mirrors, lenses, translations in a medium of different refractive index), will transform this column vector. This transformation can be expressed with a matrix as follows:

$$\begin{Bmatrix} y_2 \\ \alpha_2 \end{Bmatrix} = \begin{pmatrix} A_1 & B_1 \\ C_1 & D_1 \end{pmatrix} \begin{Bmatrix} y_1 \\ \alpha_1 \end{Bmatrix}$$

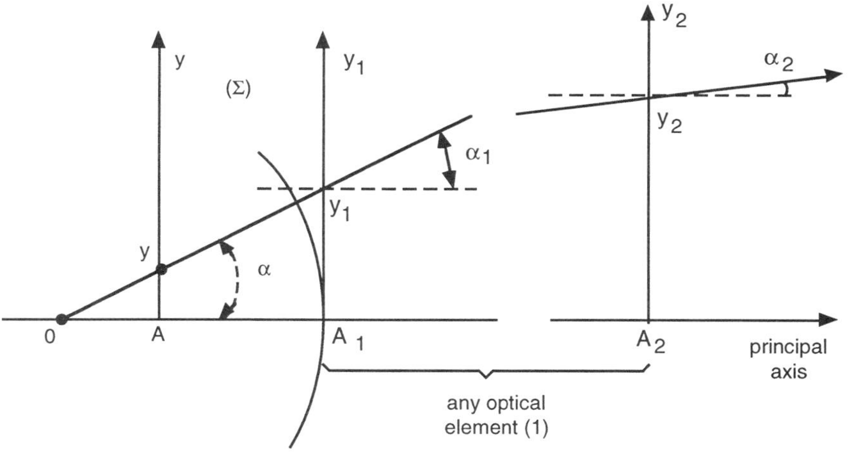

Figure IV.2.14: Geometric characterization of a ray of light before and after a nonidentified optical element.

The 2 × 2 square matrix transfers the light from plane A_1y_1 to plane A_2y_2. It is easy to see what the curvature change is

$$R_2 = (A_1 R_1 + B_1) / (C_1 R_1 + D_1)$$

More generally, a light wave traveling in a laser cavity meets a succession of n elements during a back-and-forth travel through the cavity, and the product of these n matrices (taken in the order in which the light ray meets them) forms a 2 × 2 square matrix, written $\begin{pmatrix} A & B \\ C & D \end{pmatrix}$ and which is such that

$$\begin{pmatrix} A & B \\ C & D \end{pmatrix} = \begin{pmatrix} A_1 & B_1 \\ C_1 & D_1 \end{pmatrix} \begin{pmatrix} A_2 & B_2 \\ C_2 & D_2 \end{pmatrix} \cdots \begin{pmatrix} A_n & B_n \\ C_n & D_n \end{pmatrix}$$

Matrix $\begin{pmatrix} A & B \\ C & D \end{pmatrix}$ is called the transfer matrix of the cavity.

The following table gives the transfer matrices of a few optical elements commonly seen in laser cavities.

Element	*Characteristics*	*Transfer matrix*
Translation in a medium of any refractive index.	Length: L	$\begin{pmatrix} 1 & L \\ 0 & 1 \end{pmatrix}$
Thin lens.	Focal length: f Converging lens: $f > 0$ Diverging lens: $f < 0$	$\begin{pmatrix} 1 & 0 \\ -1/f & 1 \end{pmatrix}$
Spherical mirror (for slightly slanted rays).	Radius of curvature: R Concave mirror: $R > 0$ Convex mirror: $R < 0$	$\begin{pmatrix} 1 & 0 \\ -2/R & 1 \end{pmatrix}$
Diopter (for slightly slanted rays).	Radius of curvature: R Concave diopter: $R > 0$ Convex diopter: $R < 0$ Front refractive index: n_1 Back refractive index: n_2	$\begin{pmatrix} 1 & 0 \\ (n_2 - n_1 / R) & 1 \end{pmatrix}$
Medium with a transverse gradient (along r) of the refractive index.	Length: z Refractive index such that: $n_{(r)} = n_0 - n_2 r^2/2$ We write: $\sqrt{s} = (n_2 / n_0)^{1/2}$	$\begin{pmatrix} \cos\sqrt{s}\, z & \sin\sqrt{s}\, z / \sqrt{s} \\ -\sqrt{s}\sin\sqrt{s}\, z & \cos\sqrt{s}\, z \end{pmatrix}$

The stability condition is expressed by

$$\begin{pmatrix} A & B \\ C & D \end{pmatrix} = \mathbb{1}$$

since the condition is that the light ray must be identical to itself after a back-and-forth travel through the cavity. This condition implies that we have

$$-1 \leq \mathrm{Tr}\begin{pmatrix} A & B \\ C & D \end{pmatrix} / 2 = (A + D) / 2 \leq 1$$

which is equivalent to condition IV.2.70.

And so we see that we can also come up with stability condition IV.2.70 as it was established by means of wave optics, by using geometric optics and by characterizing the cavity by matrix ABCD, as defined above. Matrix ABCD can be used more generally when studying a Gaussian beam. While in geometric optics the matrix describes the transformation of the curvature of the wave upon crossing the system, in Gaussian optics it describes the transformation of the complex parameter q(z), as defined by Eq. IV.2.59. q(z) follows the so-called ABCD law and is transformed in the following way:

$$q(z) = [A\,q(0) + B] / [C\,q(0) + D] \qquad \text{(IV.2.71)}$$

Although the demonstration of this equation falls outside the scope of this book, the reader may check that it gives the correct transformation of q(z) in a few simple cases (e.g. translation, thin lens ...).

— Now for the second point we would like to emphasize: The foregoing relations make it possible to calculate the diameter and the position of the beam waist whatever the shape of the cavity under consideration. Let us take the example of an optical cavity having one plane mirror. This is quite often the case in laser technology (at least, one often finds that one of the mirrors has a radius of curvature very much greater than that of the other).

Let us therefore set $R_1 = \infty$ (i.e., $h_1 = 1$) in Eq. IV.2.67. Then we have

$$\omega_0^2 = (\lambda L/\pi)\sqrt{h_2 / (1 - h_2)}$$

For symmetry reasons, the beam waist must be localized on the plane mirror. Clearly, if $h_1 = 1$, Eq. IV.2.69 gives the value of $z_1 = 0$.

We want to inspect the steady-state condition on the phase. The appropriate term in this case is

$$\varphi(z) = kz - \arctan\left[\lambda z/\pi\,\omega_0^2\right] \qquad \text{(IV.2.72)}$$

which describes the phase as a function of z (see Eq. IV.2.65). The phase will be stationary if, after the wave has traveled back and forth, we find that $\varphi(2L)$ has changed by a whole number, written (q + 1), times 2π (q = 0, 1, 2 ...). In this case, there is a node of the electromagnetic wave on each of the mirrors. Mathematically, we can write this condition as

$$\varphi(2L) = k_q\, 2L - 2\left[\arctan\left(\lambda z_2/\pi\,\omega_0^2\right) - \arctan\left(\lambda z_1/\pi\,\omega_0^2\right)\right] = (q+1)\,2\pi \quad \text{(IV.2.73)}$$

or, expressing z_1, z_2, and ω_0^2 using Eqs. IV.2.69 and IV.2.67, respectively, we obtain

$$\varphi(L) = k_q L - \arccos\sqrt{h_1 h_2} = (q+1)\,\pi$$

which we rewrite as

Inset IV.3 ***Summary***

1. A two-mirror optical cavity is characterized by two parameters, $h_1 = 1 - L/R_1$ and $h_2 = 1 - L/R_2$. A stationary laser wave will be able to build up inside the cavity if it is a stable one—that is, if $0 \leq h_1 h_2 \leq 1$.

2. When this condition is true, a laser wave with electric field E(x,y,z) can build up. The electric field E is given by

$$\mathcal{E}(x, y, z) = \mathcal{E}_0 [\omega(0) / \omega(z)] \exp\left[-(x^2 + y^2) / \omega^2(z)\right]$$

$$\times \exp - i \left\{ k_q \left[z + (x^2 + y^2) / 2R(z)\right] - \arctan\left[\lambda z / \pi \omega^2(0)\right] \right\}$$

where

$$\omega^2(0) = \frac{\lambda L}{\pi} \sqrt{\frac{h_1 h_2 (1 - h_1 h_2)}{(h_1 + h_2 - 2 h_1 h_2)^2}}$$

$$\omega^2(z) = \frac{\lambda L}{\pi} \sqrt{\frac{h_1 h_2 (1 - h_1 h_2)}{(h_1 + h_2 - 2 h_1 h_2)^2}} \left[1 + \frac{z^2}{L^2} \frac{(h_1 + h_2 - 2 h_1 h_2)^2}{h_1 h_2 (1 - h_1 h_2)}\right]$$

$$R(z) = z \left[1 + \frac{L^2}{z^2} \frac{h_1 h_2 (1 - h_1 h_2)}{(h_1 + h_2 - 2 h_1 h_2)^2}\right]$$

The resonance frequencies are given by the general equation

$$\nu_q = (c / 2L) \left(q + 1 + \arccos \sqrt{h_1 h_2} / \pi\right)$$

where $q = 0, 1, 2, \ldots$. The wave vector k_q is then given by

$$k_q = 2\pi \nu_q / c$$

$$\nu_q = (c/2L) \left(q + 1 + \arccos \sqrt{h_1 h_2} / \pi\right) \qquad \text{(IV.2.74)}$$

This relation shows that the only frequencies likely to give rise to stationary waves in our optical cavity are the frequencies ν_q. The frequency interval between successive modes is $\Delta\nu = c/2L$, called the *free spectral range* (FSR) of the cavity. This term is very important in laser physics. The corresponding time $\tau_t = 2L/c = \Delta\nu^{-1}$ is called the *transit time* of the cavity.

Comment: Usually, q is very large ($q \sim 10^6$), and $\left(\arccos \sqrt{h_1 h_2} / \pi\right)$ is negligible compared to $(q + 1)$. Also, in the case of a ring cavity, one must take the length of the ring L instead of 2L.

The infinite set of these propagation modes of frequencies ν_q, separated by c/2L, defines the set of longitudinal modes of the optical cavity. They are noted TEM_{00q} to remind us of their transverse structure corresponding to a fundamental Gaussian distribution of intensity m = n = 0. For a given value of q, all the possible transverse propagation modes noted TEM_{mnq}, with varying values for m and n, are called the *transverse modes* of the cavity. They correspond to the different possible transverse distributions of the electric field satisfying the self-consistency conditions for this particular cavity, i.e. they are identical to themselves after a back-and-forth run through the cavity.

IV.2.3 Laser emission

In order to describe laser emission, we now only need to combine the last two sections concerning the amplifying medium on one hand (Eq. IV.2.1) and the optical cavity on the other (Eq. IV.2.2). Figure IV.2.15 illustrates the general principle of a laser emitting continuously, shown in frequency space. In Figure IV.2.15(a), we represented the intensity of the longitudinal modes TEM_{00q}. In the figure, we can see the infinity of modes of frequencies ν_q, $\nu_{q\pm1}$, $\nu_{q\pm2}, \ldots$, separated by a frequency interval c/2L.

Can the existence of this set of modes lead to laser emission? The answer to this question can be found in our classical calculations about the amplifying medium in the foregoing pages. We showed that only if the laser frequency ν_q stayed within the characteristic band of the medium, of width $\Delta\nu_o$, did the term $B_{12}\,\overline{\rho}$ ($B_{21} = B_{12}$), responsible for the gain of spontaneous emission, take on significant values (see Eq. IV.2.22). Figure IV.2.15 (b) shows the nonsaturated gain of the amplifier and also indicates the level of the losses. As we demonstrated, amplification may take place only if

$$G_o \geq P \qquad \text{(IV.2.75)}$$

This inequality allows us to identify the frequency domain within which oscillation can occur.

Laser emission [see Figure IV.2.15 (c)] occurs only for modes whose frequencies ν_q are such that

$$\nu_{min} \leq \nu_q \leq \nu_{max} \qquad \text{(IV.2.76)}$$

which means that only a few of the modes of the cavity can give rise to laser emission, namely, those satisfying condition IV.2.76.

This is the basic way a laser operates. It is called *longitudinal multimode operation* and will be studied in Chapter VI.

IV.3 Various descriptions of the laser

In the last section, we introduced the laser, using the laws of classical physics for the wave and a phenomenological theory, due to Einstein, to describe the wave-medium interaction. In this approach, neither matter nor radiation is quantified. The *phenomenological theory* just says that atoms and molecules have

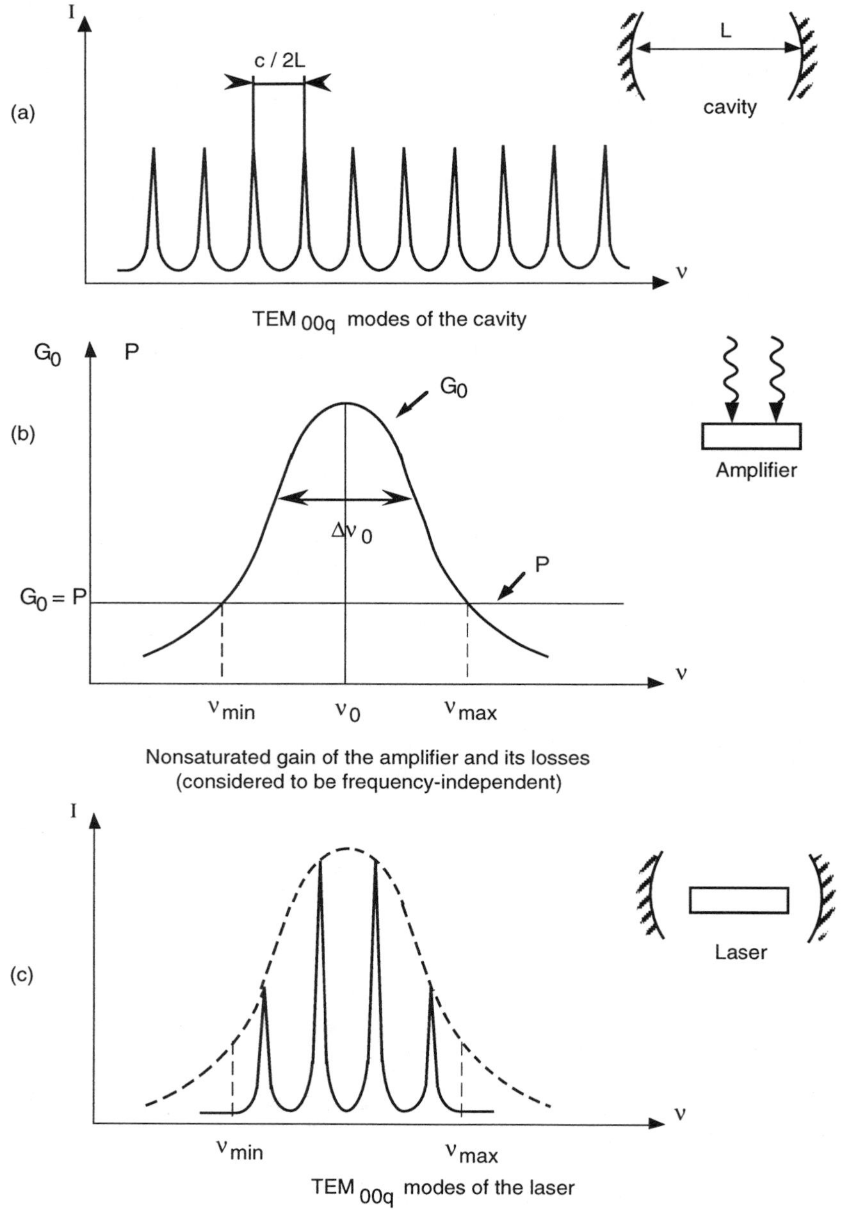

Figure IV.2.15: Diagram showing the multimode operation of a laser in frequency space.

their energy distributed in discrete levels, while, for simplicity's sake, the energy of the light wave is distributed in discrete quantities called *photons*. Moreover, the characteristic coefficients of the interaction are assumed not to depend on the orientation of the wave with respect to the medium. Yet the properties of a light beam are usually very strongly dependent on direction, so that the hypothesis of isotropy will probably have to be discarded. But in fact, when the radiation interacts with a gas, or with an amorphous liquid or solid, the random orientation of the atoms makes it possible to consider the average interaction as if it were isotropic. Einstein's coefficients must be considered as averaged coefficients, which do not depend on the geometry or on the polarization of the light beam. This is not the case when the irradiated medium is a crystallized solid in which privileged directions exist. In the latter case, the phenomenological theory must be used with care. In all cases, the nature of the theory is such that it cannot attribute a physical significance to the macroscopic coefficients it uses. Its essential quality remains its simplicity.

The *model of the classical harmonic oscillator* can predict the line shape of electronic transitions in terms of refractive indices and of absorption coefficients, and it attributes a physical meaning to Einstein's coefficient B_{12}. It gives a correct description of the interaction corresponding to a linear absorption of light by a medium. It needs some improvements (like the addition of an anharmonic term and also of higher-order terms) in order to account for the nonlinearity of laser-matter interaction. Also—and this is a fundamental limit—it cannot account for spontaneous emission.

Yet we were able to show (see Section IV.2) that by using the phenomenological and the microscopic theories simultaneously, we obtain a satisfactory description of the main characteristics of laser emission. However, a great number of very fundamental problems remain inaccessible with this method.

Lamb's semiclassical theory, in which only matter is quantified, takes the description one step further toward precision. A brief outline of this method will be given in Section VII.3.1. This semiclassical approach gives an excellent description of the problems already mentioned, and also of some completely new problems such as time-dependence of the laser intensity, saturation of the gain in the irradiated medium, competition between modes, intensity-dependence of the frequency of the modes, and so on.

With this theory, we have reached a level of description which is already very satisfying.

The ultimate step is made when both matter and light are quantified. Only a *quantum theory of the laser* can explain in a satisfying way the phenomenon of *spontaneous emission*. Now the theoretical limit to the spectral width of laser emission is an immediate consequence of spontaneous emission. Moreover, a rigorous treatment of emission is absolutely necessary if we want a correct description of the way the oscillation gets started in the laser. Last, only the *"totally" quantified theory* gives access to the statistical distribution of the photons emitted by the laser. We chose to omit this kind of theory in this book, but to initiate the reader to the quantified treatment of electromagnetic radiation we give a limited example in Exercise IV.1.

Table IV.3.1 sums up the essential characteristics of the various methods used to describe the laser, along with their assets and drawbacks.

IV.4 Problems and outlined solutions

IV.4.1 Problem IV.1: Quantum description of a coherent wave

This problem is meant to initiate the reader to the quantum description of a laser wave. The initiation is kept deliberately short and is in no way a complete treatment. The interested reader may refer to the reference in Section IV.5.4 for further initiation.

Using the quantum field theory, we wish to describe the coherent field emitted by a laser. The emission is assumed ideal; it is assumed to be a monochromatic plane wave, linearly polarized, with a constant propagation vector $\mathbf{k}$. In classic theory, its field is given by equation $\mathbf{E} = \mathbf{E}_0 \sin(\omega t - \mathbf{k}\,\mathbf{r} - \phi)$.

Mode $\mathbf{k}$ of the electromagnetic field describing the laser field is represented in the quantified field by a harmonic oscillator. For this radiation, contained in volume V, the Hamiltonian is written as $H = \hbar\,\omega\,(a^+ a + 1/2)$, where $a^+$and a are the creation and anihilation operators of an energy quantum $\hbar\omega$. It can be shown that the electric field operator associated to this representation can be written as follows:

$$\boldsymbol{E} = \mathrm{i}\,\mathrm{A}\,\mathbf{e}_0 \left\{ a \exp\left[-\mathrm{i}\left(\omega t - \mathbf{k}\,\mathbf{r}\right)\right] - a^+ \exp\left[\mathrm{i}\left(\omega t - \mathbf{k}\,\mathbf{r}\right)\right] \right\}$$

where A is a real constant and $\mathbf{e}_0$,is the unit vector in the polarization direction of field $\mathbf{E}_0$. In this expression, only a and a^+ are operators acting on the states of the field.

1. Let $|\,n >$ be a eigenstate of the Hamiltonian H. It has an associated eigenvalue E_n. Calculate E_n and derive the value of A. What do you notice about $|\,0 >$, the ground state of H? Calculate the mean value of the electric field $\mathbf{E}$ in state $|\,n >$. Compare with a actual measurement of the field, and conclude that the coherent field of the laser is not represented correctly by state $|\,n >$. Let us define a phase operator ϕ of the electric field by writing the relation between operators a and a^+ and the operator "number of photons", $a a^+$, which is $a = (aa^+)^{1/2} \exp(\mathrm{i}\phi)$, $a^+ = \exp(-\mathrm{i}\phi)\,(aa^+)^{1/2}$. Show that for a given state $|\,n >$, $< \cos\phi >$ and $< \sin\phi >$ vanish for all t. What illustration can you find for the time variations of the electric field of the wave represented by state $|\,n >$?

2. Let $|\,\alpha >$ be the eigenstate of operator a associated with eigenvalue $\alpha = |\,\alpha\,| \exp(\mathrm{i}\theta)$. Calculate $<\mathbf{E}>$, $<\mathbf{E}^2>$ for state $|\,\alpha >$, then calculate the standard deviation $\Delta E = \sqrt{<E^2> - (<E>)^2}$. Show that for large values of $|\,\alpha\,|$, the state $|\,\alpha >$ is a good representation of the coherent field of the laser. Express the measured values $\mathbf{E}$ and ϕ as a function of α. Calculate the mean value of the number of photons and the mean value of its square in state $|\,\alpha >$; find the standard deviation of this number. What do you conclude ? Write the development of the coherent states of the field on the basis of states $|\,n >$, by finding a recursion formula between coefficients n and n +1, and writing the normalization condition. Derive the probability in state $|\,\alpha >$ that the coherent field is in state $|\,n >$ with exactly n photons.

(a) We now want to show that the stationary states, which are eigenstates of Hamiltonian H and which represent a mode of the electromagnetic field, cannot describe the ideal coherent laser wave correctly. The value E_n is found by solving the eigenvalue equation:

$$H \mid n> = E_n \mid n> = \hbar\omega\left(a^+ a + 1/2\right)\mid n> = \left(n + 1/2\right)\hbar\omega \mid n>$$

Therefore $E_n = (n + 1/2)\hbar\omega$; let us remember that

$$a^+ \mid n> = \sqrt{n+1}\mid n+1> \quad \text{and} \quad a \mid n> = \sqrt{n}\mid n-1>$$

For an electromagnetic wave in a volume V, the Hamiltonian takes the form $H = V\left(\varepsilon_0 \mathbf{E}^2/2 + \mu_0 \mathbf{H}^2/2\right) = V\varepsilon_0 \mathbf{E}^2$. Using postulates P_4 anf P_7 (see Chapter I) we can calculate the mean quantum value of energy E_n:

$$E_n = <n \mid V\varepsilon_0 \boldsymbol{E}^2 \mid n> = V\varepsilon_0\left(2n + 1\right)A^2$$

Equating this to the first expression of E_n, we conclude that $A = (\hbar\omega/2\varepsilon_0 V)^{1/2}$. The ground state $\mid 0>$, which has no photon in mode $n = 0$, has an energy $E_0 = \hbar\omega/2$.

The ground state of the field is described by a fluctuating electric field with a nonvanishing mean quadratic value. In the absence of photons, each mode of the field has a nonzero energy: We have just described the electromagnetic field of vacuum, whose fluctuations explain *spontaneous emission.*

Of course we find that $<n \mid \boldsymbol{E} \mid n> = 0$, yet experimental measures give a nonzero value. Therefore we conclude that the states $\mid n>$ do not describe the laser field correctly.

We want to find a physical significance for the stationary states $\mid n>$ by calculating the quantum mean value of the phase ϕ of the field in state $\mid n>$.

We find

$$<n \mid \exp\left(\pm i\,\phi\right)\mid n> = <n \mid n \mp 1> = 0.$$

so that

$$<\cos\phi> = <\sin\phi> = 0$$

To each state, $\mid n \rangle$ corresponds a monochromatic wave with a well-defined amplitude, since $<\mathbf{E}^2>$ has a fixed value, but with a random phase.

(b) We are now going to show that the coherent states of a wave can be represented by eigenstates of a. Remembering that $\left[a, a^+\right] = \mathbb{1}$, let us write

$$<\mathbf{E}> = <\alpha \mid \boldsymbol{E} \mid \alpha> = -2\left(\hbar\omega/2\varepsilon_0 V\right)^{1/2}\mathbf{e}_0 \mid \alpha \mid \sin\left(\mathbf{k}\,\mathbf{r} - \omega t + \theta\right)$$
$$<\mathbf{E}^2> = <\alpha \mid \boldsymbol{E}^2 \mid \alpha> = A^2\left[4\mid \alpha \mid^2 \sin^2\left(\mathbf{k}\,\mathbf{r} - \omega t + \theta\right) - 1\right].$$

$\Delta E = A = (\hbar\omega/2\varepsilon_0 V)^{1/2}$. As in the classical picture, the quantum mean energy of the field, which is proportional to $\mid \alpha \mid$, oscillates. When $\mid \alpha \mid$ increases, ΔE remains constant, but the relative uncertainty decreases: The results of the measurements tend toward the mean value, and the classical and quantum representations eventually merge.

The operator "number of photons" is defined by a^+a. In state $\mid \alpha>$, the mean value of the number of photons and its square are easy to find:

TABLE IV.3.1 — **Main methods to describe the laser.**

	RADIATION: CLASSICAL	RADIATION: QUANTIFIED
MATTER — CLASSICAL — MICROSCOPIC	CLASSICAL THEORY OF THE HARMONIC OSCILLATOR o *Advantages*: Description of the interaction in terms of the refractive index and of the absorption. o *Disadvantages*: The description does not contain spontaneous emission and does not lead to the existence of laser emission.	
MATTER — CLASSICAL — PHENOMENOLOGICAL	EINSTEIN'S THEORY o *Advantages*: Simple and efficient. o *Disadvantages*: The coefficients of wave-matter interaction have to be postulated. There is no indication as to the physical meaning of these coefficients.	
MATTER — QUANTIFIED	LAMB'S SEMICLASSICAL THEORY o *Advantages*: Correct description of stimulated emission; of the time dependence of the laser intensity; of saturation; of intermode competition. o *Disadvantages*: It does not contain spontaneous emission, nor the latter's consequences on the start of oscillation; nor are its ultimate spectral limitations treated.	QUANTUM THEORY o *Advantages*: Very complete description of all phenomena of laser physics. o *Disadvantages*: The formalism is difficult to handle.

$$<\alpha\,|\,a^{+}a\,|\,\alpha> = \alpha\alpha^{*} = |\,\alpha\,|^{2}\ ;\ <\alpha\,|\,(a^{+}a)^{2}\,|\,\alpha> = |\,\alpha\,|^{2} + |\,\alpha\,|^{4}$$

The deviation Δn is then equal to $\Delta n = \left[<n^{2}> - <n>^{2}\right] = |\,\alpha\,|$, so that $\Delta n\,/<n> = |\,\alpha\,|^{-1}$. Unlike state $|\,\alpha>$, state $|\,n>$ does not correspond to a well - defined number of photons. However, as $|\,\alpha>$ increases, the uncertainty on n decreases. Indeed, $|\,\alpha>$ can be decomposed on the basis of states $|\,n>$. The result we just obtained illustrates the relative decrease of the width of the distribution as the number of photons, i.e. the intensity of the wave, increases. Therefore, when dealing with lasers in a quantum way, state $|\,\alpha>$ is commonly referred to as a state $|\,n>$.

To obtain the explicit coefficients of the decomposition, let us write $|\,\alpha> = \sum_{n} c_n(\alpha)\,|\,n>$ and apply operator a:

$$a\,|\,\alpha> = \sum_{n} c_n(\alpha)\,n^{1/2}\,|\,n-1> = \alpha \sum_{n} c_n(\alpha)\,|\,n>$$

which leads to the recursion formula on the coefficients:

$$c_{n+1} = \alpha\,(n+1)^{-1/2}\,c_n \text{ and therefore } c_n = \alpha^{n}\,(n!)^{-1/2}\,c_0$$

By normalizing $|\,\alpha>$, we can calculate c_0 and we find $c_0 = \exp\left(-|\,\alpha\,|^{2}/2\right)$. (Since we can arbitrarily choose the phase of the wave function, we were able to choose a real positive c_0). At last, $|\,\alpha>$ decomposes on states $|\,n>$ in the following way:

$$|\,\alpha> = \exp\left(-|\,\alpha\,|^{2}/2\right) \sum_{n} \alpha^{n}\,(n!)^{-1/2}\,|\,n>$$

Using postulate P_4 (see Part I), we find probability P_n to be

$$P_n = |<n\,|\,\alpha>|^{2} = \exp\left(-|\,\alpha\,|^{2}\right)|\,\alpha\,|^{2n}/\,n! = <n>^{n}\exp\left(-<n>\right)/\,n!$$

This kind of photon statistics, characterizing a coherent state of the field like the one produced by a laser, is fundamentally different from the statistics monitoring an incoherent state like the one which is found in a classical source based on spontaneous emission. P_n for instance shows a maximum when the number of photons is equal to $<n>$, while in the classical case, the probability is highest when $n = 0$ (see bibliographical reference IV.5.4).

IV.4.2 Problem IV.2: Study of the gain of a laser

1. Let us consider the amplifying medium of a laser. The pumping sequence is drawn on the diagram. λ_2 and λ_1 are the population rates (see Chapter I, Section IV.4.3). They represent the number of molecules per unit volume and per unit time which are "pumped", respectively, into the lower state (1) and into the upper state (2). These two states correspond to a laser transition of frequency ν. γ_2 and γ_1 are the relaxation constants, i.e. the probabilities per unit time that an atom leaves, respectively, state (2) or state (1) for a state other than (1) or (2). A_{21} and $B_{21} = B_{12}$ are the Einstein coefficients for transition (1) $\rightarrow$ (2).

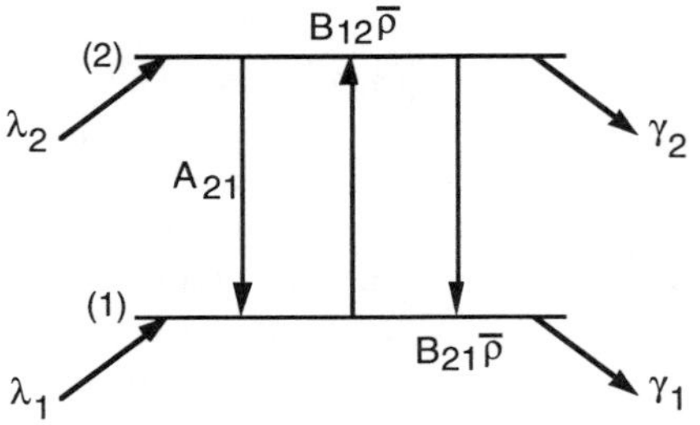

(a) We assume that the energy density $\bar{\rho}$ for transition (2) → (1) is negligible. Determine the population difference $(N_{2S} - N_{1S})_0$ and find at what condition the medium will be an amplifying medium.

(b) Let the medium be immersed in a plane wave of frequency ν and of energy density $\bar{\rho}$. Always assuming a steady state, calculate $(N_{2S} - N_{1S})$ and use it to derive the relative linear gain G of the medium at frequency ν_0, defined by $G = d\bar{I}/\bar{I}\,dz$, where $\bar{I}$ is the intensity of the wave. Show that G can take the form $G = G_0/\left(1 + \bar{I}/I_{sat}\right)$; calculate the constants G_0 and I_{sat}.

2. A length d of this medium is placed in a cavity formed by one plane mirror M_1 with a reflection coefficient of 1, and one spherical mirror M_2 with a radius of curvature R_2 and a transmission coefficient $\mathcal{T}_2$. The mirrors are separated by a distance L.

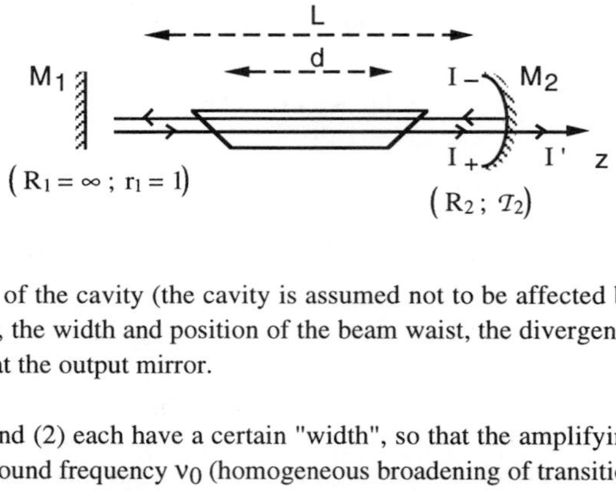

(a) Determine the stability domain of the cavity (the cavity is assumed not to be affected by the presence of the amplifying medium), the width and position of the beam waist, the divergence of the beam, and its radius of curvature at the output mirror.

(b) As a matter of fact, levels (1) and (2) each have a certain "width", so that the amplifying medium can amplify in a small region around frequency ν_0 (homogeneous broadening of transition (2) → (1)). This shows up as a Lorentzian shape for the nonsaturated gain G_0 as a function of frequency: $G_0(\nu) = G_0(\nu_0)\left(\Delta\nu/2\right)^2/\left[\left(\nu - \nu_0\right)^2 + \left(\Delta\nu/2\right)^2\right]$.

Calculate the number of laser modes which can build up a self-sustained oscillation in the cavity at very low wave intensities $\bar{I}$ ($\bar{I} \ll I_{sat}$) as a function of $G_0(\nu_0)$, $\Delta\nu$, d, $\mathcal{T}_2$, L, and c. If there is no mode selector, can all these modes oscillate simultaneously?

3. Let us assume that only the central mode, of frequency ν_0, can build up a self sustained oscillation. To calculate the intensity in this mode, we have to take into account the fact that the intensity $\bar{I}$ which saturates the gain of the amplifying medium consists of the sum of two intensities : intensity $\bar{I}_+(z)$ of the wave propagating in the positive, + **0z**, direction, and intensity $\bar{I}_-(z)$ of the wave propagating in the negative, – **0z**, direction : $\bar{I}(z) = \bar{I}_+(z) + \bar{I}_-(z)$.

(a) Let us consider the case when $\mathcal{T}_2 << 1$. Calculate $\bar{I}'$, the output intensity of the laser. In this case, we can assume that $\bar{I}_+(d) - \bar{I}_+(0) << \bar{I}_+(0)$ and that $\bar{I}_-(z) \cong \bar{I}_+(z)$.

(b) Now, $\mathcal{T}_2$ can no longer be considered as small compared to 1, and the gain of the wave after crossing the medium once is no longer very small compared to unity. Starting from the two gains $G_+ = d\bar{I}_+(z) / \bar{I}_+(z)dz$ and $G_- = d\bar{I}_-(z) / \bar{I}_-(z)dz$, show that the product $\bar{I}_+(z)\,\bar{I}_-(z) = \text{constant} = A$ throughout the amplifying medium. Integrate the gain G_+ over one passage through the amplifying medium and find the relation which exists between $\bar{I}_+(d)$ and $\bar{I}_+(0)$ as a function of A, G_0, and d. Write the boundary conditions at mirrors M_1 and M_2 explicitly; these give rise to the relations between $\bar{I}_+(0)$ and $\bar{I}_-(0)$ on one hand and between $\bar{I}_+(L)$ and $\bar{I}_-(L)$ on the other. Conclude by calculating intensity $\bar{I}_+(L)$ and the output intensity $\bar{I}'$ of the laser.

4. Indicate qualitatively how one could calculate the intensity of the laser for a frequency $\nu \neq \nu_0$ in a gaseous amplifying medium in which the different velocity classes v_z of the atoms are used to amplify waves $\bar{I}_+$ and $\bar{I}_-$. The Doppler displacement of the resonant frequency ν for class v_z is given by $v_z/c = \pm(\nu_0 - \nu)/\nu_0$. It leads to inhomogeneous broadening of the transition. What happens if the selected mode is mode $\nu \cong \nu_0$?

1. (a) Absorption and stimulated emission are negligible. The rate equations are

$$\dot{N}_2 = \lambda_2 - N_2(A_{21} + \gamma_2) \qquad \dot{N}_1 = \lambda_1 + N_2 A_{21} - N_1\gamma_1$$

At steady state, $\dot{N}_1 = \dot{N}_2 = 0$ and therefore

$$(N_{2S} - N_{1S})_0 = \lambda_2(\gamma_1 - A_{21}) / [(A_{21} + \gamma_2)\gamma_1] - \lambda_1/\gamma_1$$

Amplification will occur within the medium, provided that $(N_{2S} - N_{1S})_0 > 0$ which means

$$\lambda_2/\lambda_1 > (A_{21} + \gamma_2)/(\gamma_1 - A_{21})$$

(b) Now the rate equations take the form

$$\dot{N}_2 = \lambda_2 - N_2(A_{21} + \gamma_2) + B_{21}\bar{\rho}\,(N_1 - N_2)$$
$$\dot{N}_1 = \lambda_1 + N_2 A_{21} - N_1\gamma_1 - B_{21}\bar{\rho}\,(N_1 - N_2)$$

Writing the steady-state conditions and using Eq. IV.2.1, we find $N_{2S} - N_{1S} = (N_{2S} - N_{1S})_0 / \left[1 + (\bar{I}/I_{sat})\right]$ with

$$I_{sat}^{-1} = n B_{21}\left[1 + (\gamma_1 - A_{21})/(\gamma_2 - A_{21})\right] / c\gamma_1$$

Now we saw that $d\bar{I}/dz = (N_{2S} - N_{1S}) B_{21}\hbar\omega \bar{I}n/c$ (see Eq. IV.2.26) from which we conclude $d\bar{I}/\bar{I}dz = (N_{2S} - N_{1S})_0 B_{21}\hbar\omega n / c\left(1 + \bar{I}/I_{sat}\right)$ and, comparing to Eq. IV.2.42:

$$G = G_0\left(1 + \bar{I}/I_{sat}\right) \quad \text{with} \quad G_0 = B_{21}\hbar\omega n\left(N_{2S} - N_{1S}\right)_0 / c$$

2. (a) Here we can use condition IV.2.70 with $h_1 = 1$ and $h_2 = 1 - L/R_2$. This leads to $0 < L < R_2$; in other words, the radius of curvature of the output mirror must be greater than the length of the cavity. From Eq. IV.2.69 we conclude that $z_1 = 0$, meaning that the beam waist is on the plane mirror. Equation IV.2.67 leads to $\omega_0^2 = \lambda L \sqrt{h_2/(1-h_2)} / \pi$ or again $\omega_0^2 = \lambda \sqrt{L(R_2 - L)} / \pi$. The divergence of the beam is arc tan $(\lambda / \pi\omega_0)$, while the radius of the beam on the output mirror is $\omega(L) = \omega_0 \sqrt{1 + (\lambda L / \pi \omega_0^2)^2}$ (see Eq. IV.2.62).

(b) Equation IV.2.39 shows that at weak intensities we have $\bar{I}(2d) = \bar{I}_0 \exp(2G_0 d)$. Since the gain is equal to the losses, which are mostly due to the transmission of mirror M_2, we have $\bar{I}_0[\exp(2G_0 d) - 1] = I_0 \mathcal{T}_2^2 \exp(2G_0 d)$ or $G_0 = \mathrm{Ln}[1/(1 - \mathcal{T}_2^2)] / 2d$.

Replacing G_0 by this expression, we obtain a total width of

$$2(\nu - \nu_0) = \Delta\nu \sqrt{-2\,G_0(\nu_0)\,d / \mathrm{Ln}(1 - \mathcal{T}_2^2) - 1}$$

giving a number of modes equal to

$$2\,L\,\Delta\nu \sqrt{-2\,G_0(\nu_0)\,d / \mathrm{Ln}(1 - \mathcal{T}_2^2) - 1} / c$$

Since here we have a homogeneous broadening, each mode interacts with the total population of the amplifying medium, so that all modes compete together. This means that they cannot oscillate simultaneously.

3. (a) Since $\mathcal{T}_2^2 << 1$, $\bar{I} \approx \bar{I}_0$, and $\bar{I}' \approx \mathcal{T}_2^2 \bar{I}_0$ we obtain

$$d\bar{I}/dz = \bar{I}\,G_0 / [1 + (\bar{I}_+ + \bar{I}_-)/I_{sat}] \approx G_0 \bar{I}_0 / [1 + 2\bar{I}_0 / I_{sat}]$$

$$\mathcal{T}_2^2 = \Delta\bar{I} / \bar{I}_0 = 2\,G_0\,d / [1 + 2I_0 / I_{sat}]$$

from which we conclude $\bar{I}' = I_{sat}(2\,G_0\,d - \mathcal{T}_2^2)/2$

(b) Inverting the propagation direction of waves (+) and (–) imposes $d\bar{I}_+/\bar{I}_+ dz + d\bar{I}_-/\bar{I}_- dz = 0$, which gives $\bar{I}_- d\bar{I}_+/dz + \bar{I}_+ d\bar{I}_-/dz = 0$ and therefore $\bar{I}_+(z)\,\bar{I}_-(z) = A = \text{constant}$. So now we write

$$d\bar{I}_+/\bar{I}_+ dz = G_0 / [1 + \bar{I}_+/I_{sat} + A/(\bar{I}_+ I_{sat})]$$

Integrating this equation gives us

$$\ln[\bar{I}_+(d)/\bar{I}_+(0)] + [\bar{I}_+(d) - \bar{I}_+(0)]/I_{sat} - A\{[1/\bar{I}_+(d)] - [1/\bar{I}_+(0)]\}/I_{sat} = G_0 d$$

(set $\nu = \nu_0$ in the expression for G_0). We can express the boundary conditions by

$$\bar{I}_+(0) = \bar{I}_-(0)\,;\ \bar{I}_+(d) = \bar{I}_+(L) = \bar{I}_-(L)/(1 - \mathcal{T}_2^2)$$

and, finally, $\bar{I}'_+ = \mathcal{T}_2^2\,\bar{I}_+(L) = I_{sat}[G_0 d + \ln(1 - \mathcal{T}_2^2)]$

4. When the broadening is inhomogeneous, waves $\bar{I}_+$ and $\bar{I}_-$ are amplified independently so that G_+ depends only on $\bar{I}_+$ and G_- only on $\bar{I}_-$. For $\nu \neq \nu_0$, we must integrate each wave separately. For $\nu = \nu_0$, however, the same molecules (namely those of velocity class $v_z = 0$) contribute to gains G_+ and G_-, so that the output power of the central mode, of frequency ν_0, will be weaker. This phenomenon is called the "Lamb dip".

IV.4.3 Problem IV.3: Study of a three-level laser: Ar^+ laser

1. We take an Ar^+ laser ($\lambda = 514.5$ nm) in which an amplifying medium of length d is placed inside an optical cavity. This cavity consists of a plane mirror M_1 with a reflection coefficient $r_1 = 1$, a spherical output mirror M_2 with a radius of curvature $R_2 = 6$ m, and a reflection coefficient (for the intensity) $r_2^2 = 0.95$. The optical path between the mirrors has a length L (see diagram of Problem IV.2).

Find the stability domain for the cavity. Let L = 1.5 m. What are the coordinates of the point representing the cavity on the stability diagram? Give the dimension (ω_0) and the position of the beam waist. Give θ, the divergence of the laser beam at the output. The output power of the laser varies between 0 and 2 W. Give the maximum value of the power of the radiative field inside the cavity.

2. Population inversion within the amplifying medium is obtained by sending an electric discharge into the argon atoms (see figure below).

As a very rough approximation, we can say that the argon atoms are selectively sent from their ground state (3) into the excited state (2) of the Ar^+ (by collision with the electrons of the discharge). W_{32} represents the probability per unit time for one atom to undergo such an excitation. Once in level (2), the Ar^+ ions can spontaneously relax to state (1) of the Ar^+ ion. The corresponding Einstein coefficient is A_{21}. In the presence of radiation of energy density $\bar{\rho}$ and with frequency ν of transition (2) $\rightarrow$ (1), the absorption probability $B_{12}\,\bar{\rho}$ is equal to the probability of stimulated emission $B_{21}\,\bar{\rho}$ ($B_{12} = B_{21}$). W_{13} is the probability per unit time for an ion in state (1) to fall back into the ground state (3) of the atom. W_{31} on the other hand is the probability per unit time for an atom in state (3) to be excited (by collision with electrons) to state (1). We shall assume that the number of atoms participating in the pumping sequence is small compared to the total number N_t of atoms ($N_3 \cong N_t$).

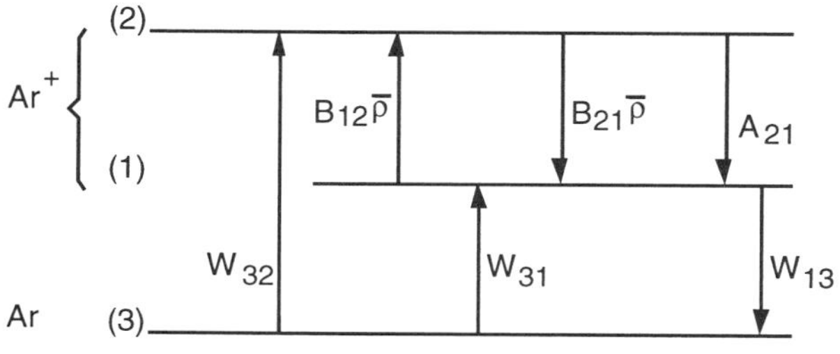

(a) Calculate the populations N_{2S} and N_{1S} of levels (2) and (1) in the absence of laser effect ($\bar{\rho} = 0$) and at steady state. At what conditions can population inversion occur?

(b) We now assume this condition is satisfied. Calculate the population difference ($N_{2S} - N_{1S}$) in the presence of an energy density $\bar{\rho}$. What do you notice about the population N_{1S} of level (1)? Explain the result.

(c) Calculate the relative linear gain $G = d\bar{I} / \bar{I}\,dz$ of the medium. Show that it can be written in the form $G = G_0 / 1 + (\bar{I} / I_{sat})$. Specify constans G_0 and I_{sat} (n being the refractive index of the medium).

(d) We assume that the self-sustaining wave oscillating at frequency ν is a plane wave and that the gain of a wave propagating in the +, respectively the –, direction is exclusively saturated by intensity $\bar{I}_+$, respectively $\bar{I}_-$, of the wave propagating in the same direction. Calculate intensity $\bar{I}_+$ and the output intensity $\bar{I}'$ of the laser at steady state, setting $r_2^2 = 1 - \mathcal{T}_2^2$ for the reflection coefficient of mirror M_2. Simplify these expressions for $\mathcal{T}_2^2 << 1$.

1. Applying Eq. IV.2.70 gives $0 < L < R_2$. The coordinates of the characteristic point are $h_1 = 1$ and $h_2 = 1 - (1.5 / 6) = 0.75$ (see Figure IV.2.13). The beam waist is located on mirror M_1, and its dimension is

$$\omega_0 = \left[\lambda \sqrt{L(R_2 - L)} \,/\, \pi\right]^{1/2} \approx 650\ \mu\text{m}$$

The beam divergence is $\theta \approx \lambda / (\pi\, \omega_0) \approx 0.25$ mrd. The maximum power found inside the cavity is 40 W.

2. (a) The rate equations take the form

$$\dot{N}_2 \approx W_{32} N_t - A_{21} N_2$$
$$\dot{N}_1 \approx A_{21} N_2 + W_{31} N_t - W_{13} N_1$$

At steady state : $N_{2S} \approx N_t W_{32} / A_{21}$; $N_{1S} \approx N_t (W_{31} + W_{32}) / W_{13}$

The condition for a population inversion is $N_{2S} > N_{1S}$ in this case:

$$W_{32} / A_{21} > (W_{31} + W_{32}) / W_{13}$$

(b) In the presence of laser emission, the rate equations change to

$$\dot{N}_2 = W_{32} N_3 - A_{21} N_2 - B_{12}\bar{\rho}\,(N_2 - N_1)$$
$$\dot{N}_1 \approx W_{21} N_2 - W_{13} N_1 + W_{31} N_t + B_{12}\bar{\rho}\,(N_2 - N_1)$$

And at steady state we find

$$N_{1S} \approx N_t (W_{32} + W_{31}) / W_{13};\quad N_{2S} \approx N_t W_{32} / A_{21} - B_{12}\,\bar{\rho}\,(N_{2S} - N_{1S}) / A_{21}$$
$$N_{2S} - N_{1S} = N_t \left[(W_{32} / A_{21}) - (W_{32} + W_{31}) / W_{13}\right] / \left(1 + B_{12}\,\bar{\rho} / A_{21}\right)$$

We find that N_{1S} does not depend on $\bar{\rho}$. The result was predictable since the number of atoms lifted from state (3) to state (1), by way of relay state (2), per unit time, is equal to the number of atoms relaxing from (1) to (3). The energy density $\bar{\rho}$ plays no part in this process.

(c) The gain is found by an equation similar to Eq. IV.2.35:

$$G = (N_2 - N_1)\, B_{12}\, h\nu n/c = G_0 / (1 + \bar{I} / I_S)$$

where

$$I_S = cA_{21} / n\, B_{12}$$

and

$$G_0 = N_t B_{12} nh\nu [(W_{32} / A_{21}) - (W_{32} + A_{31}) / W_{13}] / c$$

(d) By integrating the equation which defines the gain, we obtain the following relation:

$$\ln [\bar{I}_+ (0) / \bar{I}^- (0)] + [\bar{I}_+ (0) - \bar{I}_- (0)] / I_S = 2 G_0 d$$

But $\bar{I}^- (0) = \bar{I}^+ (0) (1 - \mathcal{T}_2^2)$. Therefore $\bar{I}^+ (0) = I_S [2G_0 d + \ln (1 - \mathcal{T}_2^2)] / \mathcal{T}_2^2$, which gives $\bar{I}' = I_S [2G_0 d + \ln (1 - \mathcal{T}_2^2)]$. If $\mathcal{T}_2^2 << 1$, a limited series expansion gives $\bar{I}' \approx I_S (2G_0 d - \mathcal{T}_2^2)$.

IV.4.4 Problem IV.4: Dye laser; study of the part played by the triplet state

A schematic representation of the amplifying medium of a laser is shown in the diagram. Pumping lifts the molecules from the lower state of the ground energy band S_0 (level 1) to the highest level (level 2) of the energy band of the first excited electronic singlet state S_1. The molecules relax very rapidly (in a time range on the order of the picosecond) from levels 2 and 4 to levels 3 and 1, respectively. These relaxations are fast enough so that we can consider that the populations of levels 3 and 4 remain at thermal equilibrium (the temperature T is kept constant).

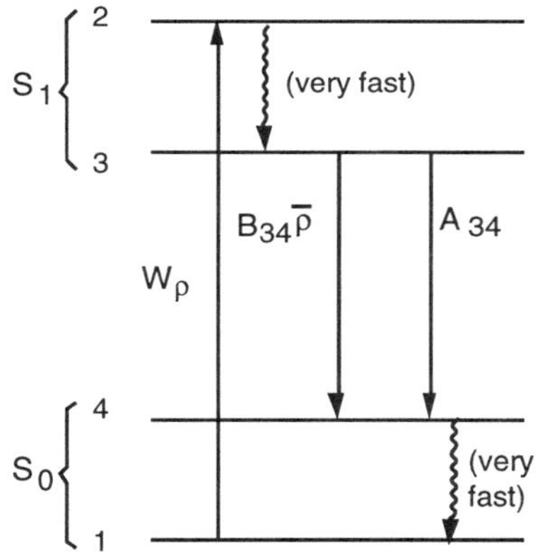

1. Assuming a purely spontaneous relaxation of level 3 toward level 4, show that population inversion between level 3 and a certain set of levels (which we ask you to determine) can occur at steady state (also assuming that the pumping intensity is weak enough so as to have $N_{3S} << N_{1S}$). Calculate the following values as a function of W_p, A_{34}, and T: the frequency difference between the lowest frequency of the pump beam $\nu_{p,\,min}$ ($h\nu_{p,min} = E_3 - E_1$) and the highest frequency ν_{max} ($h\nu_{max} = E_3 - E'_4$) for which population inversion is possible.

2. Let us now introduce some radiation into the medium with a mean energy density $\bar{\rho}$ and a frequency $\nu = \nu_3 - \nu_4$, which activates stimulated emission between levels 3 and 4. The steady-state population of level 4 is assumed to be negligible at thermal equilibrium ($N_{4S} << N_{3S}$). Calculate the new value of N_{3S} as a function of $\bar{\rho}$.

Show there is a gain at frequency ν. Find the longitudinal variation (i.e., along the z axis) of the intensity $\bar{I}$ of a beam propagating in this medium of refractive index n.

3. The ideal amplifying medium we just described is now perturbed by an intersystem conversion process involving the two first excited triplet states T_1 and T_2, as shown in the accompanying diagram. This means we have to take account of the probability per unit time k_{ST} that level (3) de-excites to T_1, i.e. level (5), which we assume to have an infinite lifetime. N_5 therefore increases with time. Moreover, in the presence of radiative energy density $\bar{\rho}$, a

$T_1 \rightarrow T_2$ transition is possible, immediately followed by a very fast nonradiative $T_2 \rightarrow T_1$ relaxation.

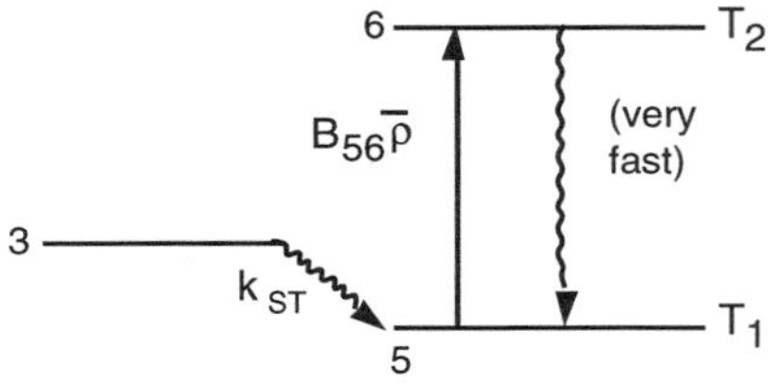

(a) Write the rate equations relative to levels (3) and (5). Show that for a given W_p, the value of N_3 at steady state tends to a maximum for $\bar{\rho} = 0$. Write the differential equation governing $\bar{I}$ as a function of N_3 and N_5. At what condition is the medium an amplifying one?

(b) This amplifying medium is used in a cavity in which the losses are assumed to be distributed uniformly: Let us call $\alpha\bar{I}$ the losses per unit length along the 0z axis. Since the population leak from S_1 to T_1 is very slow, the oscillation can be considered to be at steady state at all time, so that at each point within the medium the gain must be exactly compensated by the losses. Determine the evolution of N_5 in this steady state of the oscillation. Show that the laser oscillation can last no longer than an interval of time Δt; calculate this interval Δt as a function of W_p, α, and the spectroscopic constants of the medium.

4. It is clear that the laser we just described can operate only in pulses. Indicate qualitatively how one could proceed to avoid the disastrous effect of the triplet state, so as to obtain a dye laser able to operate in a continuous mode.

1. The relative rate equation for level (3) is $\dot{N}_3 = N_1 W_p - A_{34} N_3$. At steady state, $N_{3S} = N_{1S} W_p / A_{34}$. Population inversion will occur between level (3) and all levels (4) for which $N_3 > N_4$. The level corresponding to energy E'_4 is the level for which $N_{4S} = N_{3S} = N_{1S} \exp[-(E'_4 - E_1) / kT]$. However, we know that $E'_4 - E_1 = (E_3 - E_1) - (E_3 - E'_4) = h(\nu_{p,min} - \nu_{max})$ so that

$$\nu_{p,\,min} - \nu_{max} = -\,kT \ln(W_p / A_{34}) / h$$

2. Adding stimulated emission to the rate equation of level (3) gives $N_{3S} = N_{1S} W_p / A_{34} [1 + (B_{34}\bar{\rho} / A_{34})]$.

There will be a gain at frequency ν if $N_{3S} > N_{4S}$. Now since $N_{4S} \approx 0$, amplification always occurs and is described by $d\bar{I} / dz \approx N_{3S} B_{34} \bar{I} \hbar\omega\ n/c$.

3. (a) As soon as pumping starts, N_3 increases very quickly. Stimulated emission reaches a maximum, then decreases while N_5 increases. It eventually falls to zero; the populations do not reach an equilibrium (steady state).

The population changes are governed by the equations

$$\dot{N}_3 = N_1 W_p - N_3 A_{34} - N_3 B_{34}\bar{\rho} - N_3 k_{ST}$$

and

$$\dot{N}_5 = k_{ST} N_3$$

As $\bar{\rho}$ decreases, N_3 increases until it reaches its maximum value, which is its steady-state value too: $N_{3S, max} = N_{1S} W_p / A_{34} [1 + (k_{ST} / A_{34})]$. $\bar{I}$ fits the equation $d\bar{I} / \bar{I} dz = (N_3 B_{34} - N_5 B_{56}) \hbar\omega n/c$. The medium will be an amplifying one provided that $N_3 B_{34} > N_5 B_{56}$.

(b) We have $\alpha = d\bar{I} / \bar{I} dz$, in other words,

$$N_3 = \dot{N}_5 / k_{ST} = N_5 B_{56} / B_{34} + \alpha c / \hbar\omega n B_{34}.$$

This gives

$$N_5 (t) = (\alpha c / \hbar\omega n B_{56}) [\exp (k_{ST} B_{56} t / B_{34}) - 1]$$

$$N_3 (t) = \dot{N}_5 (t) / k_{ST} = \alpha c \exp (k_{ST} B_{56} t / B_{34}) / \hbar\omega n B_{34}$$

By writing $N_3 (\Delta t) = N_{3S, max}$ we obtain the time interval:

$$\Delta t = B_{34} \ln [N_{1S} W_p \hbar \omega n B_{34} / \alpha c (A_{34} + k_{ST})] / k_{ST} B_{56}.$$

4. In order to reduce the disastrous effect of the triplet state, dye laser technology "eliminates" those molecules which are in a triplet state by having the dye solution flow at a high enough speed. The speed of the flow must be on the order of the ratio between the diameter of the pump dye (a few tens of microns) and the time Δt needed to fill the triplet state (on the order of the microsecond). This gives a speed of about 10 meters per second.

IV.4.5 Problem IV.5: The Ti/sapphire laser

1. The energy diagram of the Ti^{3+} cations in a sapphire matrix is shown in the accompanying diagram, with standard notations. A_2 represents the total spontaneous emission from level (2). The ideal model of the four-level laser is assumed to hold in this case with $N_{4S} = N_{1S} = 0$. N_t represents the total volume density of the Ti^{3+} ions. The number of cations participating in the pumping sequence is assumed to be negligible as compared to N_t.

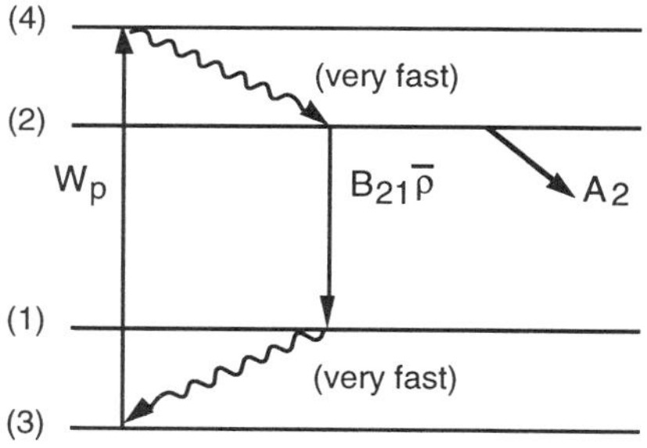

Calculate N_{2S} when $\bar{\rho} = 0$ and when $\bar{\rho} \neq 0$. Neglecting spontaneous emission, find the amplification in intensity $a^2 = \bar{I} (d) / \bar{I} (0) = 1 + 2\varepsilon$ ($\varepsilon << 1$) for a slab of thickness d.

2. This slab of amplifying medium is placed inside an optical cavity consisting of (a) two mirrors M_1 and M_2 and (b) optical elements F which are placed in the cavity so as to induce a monomode operation of the laser and so as to stabilize its wavelength (see diagram below).

The intensity transmission coefficient of elements F is $\mathcal{T}^2$, which is close to 1 ($\mathcal{T}^2 = 1 - \alpha$ with $\alpha << 1$). Mirrors M_1 and M_2 have the same radius of curvature R. M_1 is a dichroic mirror with (a) a reflection coefficient $r_1^2 = 1$ for the laser wave intensity at wavelength (λ) and (b) a transmission coefficient of 1 for the pump wave. r_2^2, the reflection coefficient of M_2, is also close to 1 ($r_2^2 = 1 - \beta$ with $\beta << 1$) at wavelength λ.

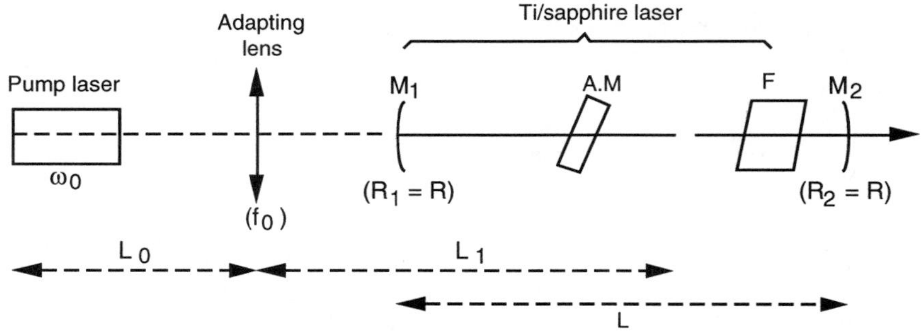

(a) Disregard the Gaussian structure of the beams. Calculate $W_{p,min}$, the smallest pumping energy-density giving rise to self-sustained oscillation of the laser. Now assume that W_p takes on a value greater than $W_{p,min}$. Calculate the intensity $\bar{I}(d)$ inside the A.M. (the Amplifying Medium) at steady state. Derive from this the output intensity $\bar{I}'$ of the laser.

(b) Let L be the distance between the two mirrors. Find the position and the dimension ω_1 of the beam waist in the cavity for wavelength λ. To adjust beam waist ω_0 of the Ar^+ pump laser to beam waist ω of the cavity, a lens of focal length f is used. At what distance L_0, measured from beam waist ω_0, must the lens be placed to have adapted beam waists? What then is the distance L_1 from the lens to the A.M. which is placed at the beam waist of the cavity? What practical conditions must L_0 and L_1 satisfy? What is the distance $\delta\nu$ between two successive longitudinal modes in mode TEM_{00}? The refractive index of the A.M. and the optical elements is assumed to be equal to 1.

Numerical calculation: R = 50 cm ; L = 50 cm. Calculate ω for $\lambda = 0.8\ \mu m$.

What do you think of the cavity's stability? To get the laser going, an energy of intensity $\bar{I}_p > 100\ kW/cm^2$ must exist within the A.M. Calculate the minimum total power of the pump laser. Can an Ar^+ laser provide such energy? Comment on this result.

(c) In fact, the cavity actually used is the cavity depicted here:

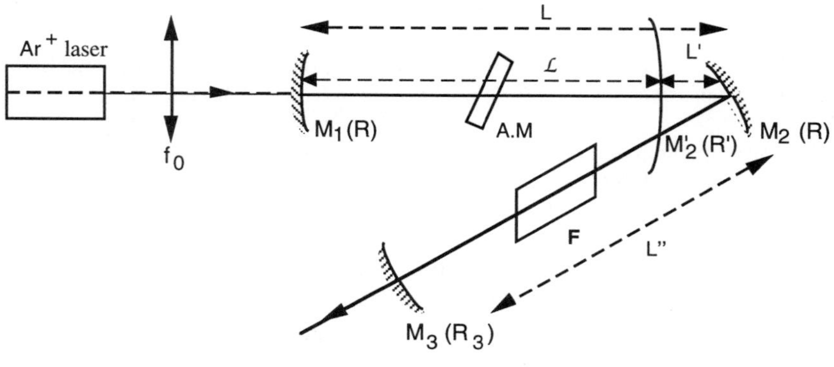

Mirror M_3 is a plane mirror. M_1 and M_2 are mirrors with radius of curvature R. M_1 and M_2 have a reflection coefficient of 1 at the laser wavelength λ. M_1 is a dichroic mirror which transmits the pump wavelength. M_3 is the output mirror. L is the distance between M_1 and M_2, while L" is the distance between M_2 and M_3. The laws of geometric optics show that for the study of the resonant mode of the cavity in the A.M., the set of mirrors (M_3, M_2) is equivalent to a mirror M'_2 having a radius of curvature $R' = -f^2 / (L'' - f)$, and placed between mirror M_2 and the A.M. at a distance $L' = f L'' / (L'' - f)$ from M_2. $f = R/2$ is the focal length of mirror M_2. Let L" and f be fixed; determine the limit values (L_{max} and L_{min}) between which L may vary while remaining within the stability domain of the cavity (we always assume $L'' > f > 0$). Choose L in the middle of the stability domain. Calculate the value ω of the beam waist at wavelength λ. What is the distance $\delta\nu$ between two successive longitudinal modes in TEM_{00} mode?

Numerical calculation: R = 20 cm, L" = 1.1 m, $\lambda = 0.8$ µm.

Calculate ($L_{max} - L_{min}$), ω, and $\delta\nu$. What is the advantage of this type of cavity?

1. When $\bar{\rho} = 0$, the rate equation for level (2) takes the form $\dot{N}_2 = N_3 W_p - N_2 A_2$ so that $N_{2S} = N_{3S} W_p / A_2 \approx N_t W_p / A_2$. If $\bar{\rho} \neq 0$, this equation changes to $\dot{N}_2 = N_3 W_p - N_2 A_2 - N_2 B_{21} \bar{\rho}$ so that $N_{2S} \approx N_t W_p / (A_2 + B_{21} \bar{\rho})$ and the amplification is written as $a^2 = \bar{I}(d) / \bar{I}(0) = 1 + N_t W_p \hbar\omega \, nd / c \bar{\rho} = 1 + 2\varepsilon$.

2. (a) At the threshold of oscillation, we have $r_2^2 a^2 \mathcal{T}^2 \approx 1$; in other words, $2\varepsilon \approx \alpha + \beta$. In this case, spontaneous emission can no longer be neglected. However, $\bar{\rho} \approx 0$ so that $W_{p,min} \approx (\alpha + \beta) c A_2 / N_t B_{21} \hbar\omega \, nd$. Far above the threshold of self-sustained oscillation, we have $\bar{\rho} \approx N_t W_p \hbar\omega \, nd / c (\alpha + \beta)$. But since $\bar{\rho} = n \bar{I} / c$, we conclude that $\bar{I}' = \beta \bar{I}(d) \approx N_t W_p \hbar\omega \, d\beta / (\alpha + \beta)$.

(b) The beam waist is located in the middle of the symmetrical cavity, at a distance L/2 of each mirror. Its dimension is $\omega_1 = \left[\lambda \sqrt{L(2R - L)} / 2\pi \right]^{1/2} \approx 250$ µm. Using relation V.2.7 we find $L_0 = f_0 + \sqrt{\omega_0^2 f_0^2 / \omega_1^2 - (\pi \omega_0^2 / \lambda)^2}$ and $L_1 = f_0 + (L_0 - f_0) \omega_1^2 / \omega_0^2$. L_0 must be greater than the length of the pump laser; L_1 must be greater than L/2. $\Delta\nu = c / 2L$. Since the cavity is a confocal one, it is at the limit of stability (see Figure IV.2.13). The minimal total starting power must be

$$\bar{P}_{p,min} = \int_0^{\infty} \bar{I}_{p,min} \exp\left[-2r^2 / \omega_1^2\right] 2\pi r \, dr$$

$$\bar{P}_{p,min} = \bar{I}_{p,min} \pi \omega_1^2 / 2 = 100 \text{ W}$$

Such power cannot be provided by a commercial laser. The solution is to reduce ω_1, as is suggested by question c.

c) The stability condition states $0 < (1 - \mathcal{L} / R)(1 - \mathcal{L} / R') < 1$ with $\mathcal{L} = L - L'$. This gives us

$$L_{min} = R - R^2 / 2(2L'' - R) + RL'' / (2L'' - R)$$
$$L_{max} = R + RL'' / (2L'' - R).$$

Numerical calculation: $L_{max} - L_{min} = R^2 / (4L'' - 2R) = 1$ cm. To be in the middle of the stability domain, we need to take $L^* = (L_{min} + L_{max}) / 2$, so that $\mathcal{L}^* = (8 L'' R - 5R^2) / 4 (2 L'' - R)$. Using Eq. IV.2.67, we find

$$\omega^4 = \lambda^2 R^4 (8 L'' - 5R)^3 (8 L'' - 3R) / 1024 \pi^2 (2L'' - R)^2 (R^2 + 8L''^2 - 9L'' R)^2$$

The numerical calculation gives a value $\omega \approx 32$ μm. The diameter of this beam waist is very small, which is what the Ti/sapphire laser technology imposes (as a matter of fact, dye laser technology also does). In this way we can avoid the use of very short focal-length mirrors, which, though they can lead to equally small beam waists, are very delicate to use in practice. The frequency distance between modes is $\delta\nu = c/2 (L^* + L'')$. Numerically, this gives $\delta\nu \approx 107$ MHz.

IV.4.6 Problem IV.6: Study of a ring cavity

Let us consider the laser represented on the diagram below. It consists of a ring cavity containing an amplifying medium which is pumped by an electromagnetic wave with electric field Ep.

1. The three mirrors M_1, M_2, and M_3 are assumed to be plane mirrors. In a first approximation, the two self-sustaining waves oscillating in the cavity are assumed to be plane waves (one travels in the + direction, the other in the – direction). M_2 and M_3 are totally reflecting. M_1 is slightly transparent, and its reflection and transmission coefficients r_1 and $\mathcal{T}_1$ are assumed to be real numbers. $u_+(0)$ and $u_-(d)$ are the complex amplitudes of the plane waves as they enter in the amplifying medium located between $z = 0$ and $z = d$. These waves are amplified upon crossing the medium with an amplification coefficient of, respectively, a_+ and a_-:

$$v_+(d) = a_+ u_+(0) \text{ and } v_-(0) = a_- u_-(d).$$

The total length of the cavity is L. $\mathcal{L} = (L - d)$ is the length of the path traveled by the waves outside of the amplifying medium.

(a) By writing the variation of $u_+(0)$ or of $u_-(d)$ during one transit through the cavity and then expressing the self-consistency of the field in a steady-state, find the condition on a_+, r_1 and $\mathcal{L}$ so that the (+) wave can oscillate in a stable self-sustaining mode; or find these conditions on a_-, r_1 and $\mathcal{L}$ so that the (–) wave may oscillate in a stable self-sustaining mode.

(b) Now we assume that a_+ and a_- are not coupled to each other in the amplifying medium. They saturate in the following way:

$$a_+ = a_0 / \sqrt{1 + (\bar{I}_+ / I_{sat})} \qquad a_- = a_0 / \sqrt{1 + (\bar{I}_- / I_{sat})}$$

where a_0 is the amplification of the nonsaturated medium. a_0 is a real number and it is a function of ν, the frequency of the self-sustained oscillation. I_{sat} is the saturation intensity of the medium; $\bar{I}_+$ and $\bar{I}_-$ are, respectively, the intensities of waves (+) and (–) within the medium.

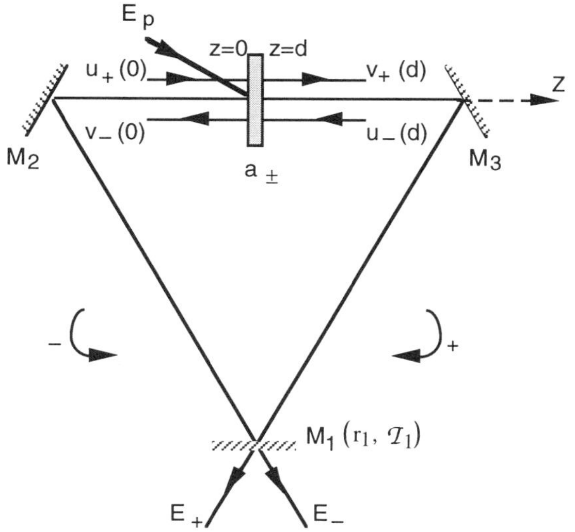

At what conditions is the assumption that the two waves are not coupled a realistic one? To find these conditions, assume that in this medium of low gain ($a_0 - 1 << 1$, $1 - r_1 << 1$), the intensities $\bar{I}_+$ or $\bar{I}_-$, respectively equal to $|u_+(0)|^2$ or to $|u_-(d)|^2$, are practically constant throughout the whole length of the medium.

By writing explicitly the conditions for self-sustained oscillations in a steady state found in question 1 (a), find as a function of $\mathcal{L}$ the frequencies ν_q which are likely to lead to a self-sustained oscillation. Determine at what conditions self-oscillation can occur. Write the intensities I_+ and I_- of the waves at these frequencies as a function of I_{sat}, r_1, and a_0. Compare the absolute values of the output fields of the laser $|E_+|$ and $|E_-|$. Is there a phase correlation between these fields?

2. Say why it is not realistic to assume that the waves in the cavity are plane waves. Show that this cavity with three plane mirrors cannot give rise to stable continuous self-sustained oscillations. Now a spherical mirror of negligible astigmatism and with a radius of curvature R_1 is used instead of plane mirror M_1. Note that mirrors M_2 and M_3 and the amplifying medium do not change the transverse structure of the wave, and therefore do not play any part in determining the structure of the cavity. Determine the structure of the equivalent cavity, its stability domain, and the characteristics of the fundamental Gaussian beam: position and dimension of the beam waist, resonance frequencies.

1. (a) We have successively

$$v_+(d) = a_+ u_+(0)\ ;\ v_+(L) = r_1 a_+ u_+(0) \exp(ikL)$$

The self-consistency of the electric field imposes that $v_+(L) = u_+(0)$, which means $r_1 a_+ \exp(ikL) = 1$. In the same way $r_1 a_- \exp(ikL) = 1$.

(b) The hypothesis that the two waves are not coupled is realistic when amplification of wave (+) and that of wave (–) are due to a different set of molecules (for instance, a medium consisting of a gas subject to Doppler effect). The equations of 1 (a) lead to

$$r_1 a_0 \exp\left(ik\mathcal{L}\right) / \sqrt{1 + \left(\bar{I}_\pm / I_{sat}\right)} = 1$$

Frequencies ν_q are such that $\nu_q \approx (q + 1)\, c/\mathcal{L}$ where q is a whole number. There is self-sustained oscillation provided that $r_1 a_0 (\nu_q) \geq 1$. The intensities are such that $r_1 a_0 / \sqrt{1 + \left(\bar{I}_\pm / I_{sat}\right)} = 1$, and thus $\bar{I}_+ = \bar{I}_- = I_{sat}\left(r_1^2 a_0^2 - 1\right)$.

The electric fields at the output of the laser are $E_+ = a_+ \mathcal{T}_1 u_+ (0)$; $E_- = a_- \mathcal{T}_1 u_- (d)$; $|E_+| = |E_-|$. The phases of the two fields, which are not coupled through the amplifying medium, are not correlated.

2. In Section IV.2.2 we explained why the hypothesis of plane waves in a laser cavity is not realistic. In the same way, a cavity consisting of plane mirrors is at the limit of the stability domain and cannot lead to the development of a stable oscillation. Following Chapter 15 (p. 581) of A. Siegman's book (see bibliographical reference IV.5.2), we shall deal with the cavity calculations using the matrix technique. The two plane mirrors and the amplifying medium do not modify the wave propagation so that the cavity is described by matrix $\begin{pmatrix} 1 & \mathcal{L} \\ 0 & 1 \end{pmatrix}$ describing a translation $\mathcal{L}$ and by matrix $\begin{pmatrix} 1 & 0 \\ -2/R_1 & 1 \end{pmatrix}$ describing reflection on spherical mirror M_1 of radius R_1. The product of these matrices gives matrix $\begin{pmatrix} 1 - 2\mathcal{L}/R_1 & -\mathcal{L} \\ -2/R_1 & 1 \end{pmatrix}$. The trace of this product matrix must lie between -1 and $+1$, which gives us the stability condition $L < 2\, R_1$. Since $h_1 = h_2 = 1 - L/R_1$ and using Eq. IV.2.67, we find $\omega_0^2 = \lambda_q \sqrt{\mathcal{L}\left(2R_1 - \mathcal{L}\right)} / 2\pi$. The cavity is symmetrical, and therefore the beam waist is exactly in the middle between mirrors M_2 and M_3. The resonance frequencies are $\nu_q \approx c\,(q + 1) / L$.

IV.4.7 Problem IV.7: Storing energy in an optical cavity

1. A resonant ring cavity consists of plane mirrors and a converging lens L of focal length $f > 0$. The total length of the ring is $L = 2\,(2L_1 + L_2)$.

Determine the two-mirror cavity equivalent to this one. Find its stability domain and the characteristics of the fundamental resonant Gaussian mode (position and diameter ω_0 of the beam waist, resonance frequencies). Numerical calculation: f = 20 cm; λ = 514.5 nm; L_1 = 5 cm; L_2 = 10 cm.

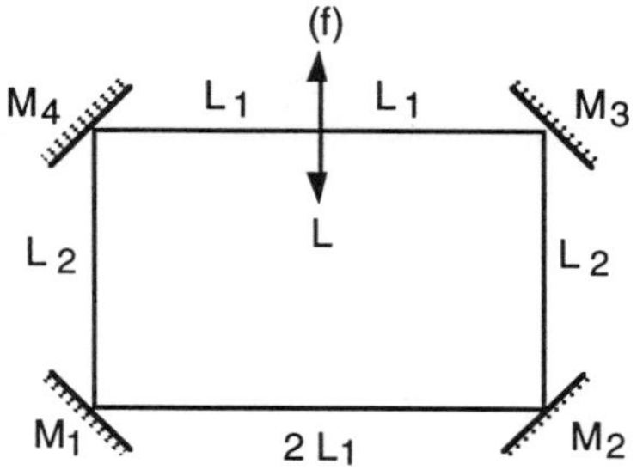

2. We wish to store the energy from a continuous monomode Ar^+ laser inside this cavity.

(a) Mirror M_1 has no losses and it introduces no dephasing. It is used to send field E_i into the cavity. r_1 and $\mathcal{T}_1$ are its real field-reflection and transmission coefficients $\left(r_1^2 + \mathcal{T}_1^2 = 1\right)$. Mirrors M_2, M_3, and M_4 also have no losses and they have transmission and reflection coefficients equal to unity. The field-transmission coefficient of the lens is $\mathcal{T}$.

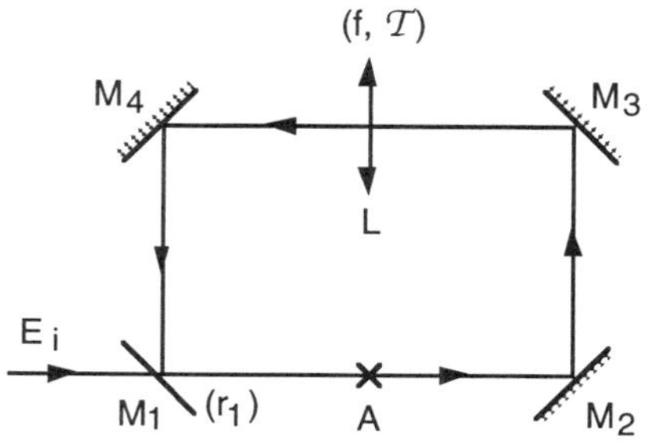

Assuming that the incident wave of amplitude E_i and the wave stored in the cavity are monochromatic plane waves, and that the exciting frequency ν_i is in resonance with an eigenfrequency ν_q of the cavity, determine the value E taken by the resulting field at point A at steady state. Write the field at point A and at time t as the sum of two terms: the first corresponding to the incident wave E_i directly transmitted through M_1; the second corresponding to the wave present at point A at time $(t - \tau_t)$ (τ_t = the transit time through the cavity), and which has just completed a transit through the ring. Calculate, as a function of $\mathcal{T}$, the reflection coefficient $r_{1,\ max}$ of mirror M_1 leading to a maximum of the energy stored inside the cavity. Calculate the storage coefficient Q defined as the ratio between the intensity in the cavity at point A and the intensity of the incident wave. Numerical calculation: $r_1 = 95\ \%$.

(b) The exciting laser providing field E_i is an Ar^+ laser operating in monomode regime. It has a resonator consisting of a plane mirror and a spherical output mirror with radius R = 4 cm. The optical path L between the two mirrors measures 2 m. Give the position and dimension of the beam waist of the laser. Now we want to excite only the fundamental transverse TEM_{00} mode of the storing cavity. To do this, the Gaussian field coming out of the laser must be changed by means of a lens into a Gaussian field whose beam waist has the same position and the same dimension as the resonant mode of the storage cavity. Determine the position of this lens with respect to the laser and with respect to the cavity so as to fulfill the above conditions (assume that the Gaussian mode of the laser beam is not deformed upon crossing the output mirror). Numerical calculation: f' = 0.90 m.

1. Let's calculate the transfer matrix:

$$\begin{pmatrix} 1 & 0 \\ -1/f & 1 \end{pmatrix}\begin{pmatrix} 1 & L \\ 0 & 1 \end{pmatrix} = \begin{pmatrix} 1 & L \\ -1/f & 1 - L/f \end{pmatrix}$$

The value of half the trace must lie between -1 and 1. This gives $-1 < (1/2)\,(2 - L/f) < 1$, i.e. $L < 4\,f$. Under these conditions, the cavity is equivalent to a symmetrical cavity consisting of two mirrors with radius of curvature $R = 2\,f$, separated by a distance L. The beam waist is located at a distance L/2 of the lens, and its diameter is $\omega_0^2 = \lambda \sqrt{L\,(4\,f - L)}\ /\ 2\pi$. The resonance frequencies ν_q are such that $\nu_q \approx c\,(q + 1)\,/\,L$. The numerical calculation gives $\omega_0 \approx 181\ \mu m$.

2. (a) $E = \mathcal{T}_1 E_i + \mathcal{T} r_1 E \exp[-i 2\pi \nu_i L / c]$. The phase term must be equal to one (we are at resonance) so that $E = \mathcal{T}_1 E_i + \mathcal{T} r_1 E$; in other words, $E = E_i \mathcal{T}_1 / (1 - \mathcal{T} r_1)$ and $E = E_i \sqrt{1 - r_1^2} / (1 - \mathcal{T} r_1)$. At the energy maximum, we must have $dE / dr_1 = 0$, so we find $r_{1,\, max} = \mathcal{T}$. The storage coefficient Q can be expressed as follows:

$$Q = E^2 / E_i^2 = (1 - r_1^2)/(1 - \mathcal{T} r_1)^2 = 1/(1 - r_1^2)$$

The numerical calculation gives $Q \approx 10$.

(b) The cavity under consideration is half a symmetric cavity, therefore the beam waist must be on the plane mirror. Its diameter is $\omega_1^2 = \lambda \sqrt{2L(2R - 2L)} / 2\pi \approx 572\ \mu m$. Now the equation

$$\omega_1^2 f'^2 / \left[(\delta_1 - f')^2 + (\pi \omega_1^2 / \lambda)^2\right] = \omega_0^2$$

must be true, where δ_1 is the distance from beam waist ω_1 to the adapting lens. This leads to $\delta_1 \approx 2.9$ m. On other hand, the lens is at a distance δ_2 from beam waist ω_0 (point A) with $(\delta_2 - f') = (\delta_1 - f')\, \omega_0^2 / \omega_1^2$, so we find $\delta_2 \approx 1.1$ m.

IV.4.8 Problem IV.8: Thermal lens in a YAG laser

The amplifying medium of the YAG laser consists of an aluminum and yttrium garnet crystal doped with Nd^{3+} ions. Population inversion between the energy levels of the Nd^{3+} ions is obtained by pumping at frequency ν_p by means of continuous or pulsed lamps placed around the cylindrical crystal rod.

1. *Propagation of a Gaussian beam in the amplifier*

Part of the pumping energy is dissipated in the rod as heat. The rod is cooled by water. The heat gives rise to a temperature gradient $T(r) = T(0) - \alpha r^2$, where r is the distance to the axis of the rod. T (0) is a constant, proportional to the energy density of the pump W_p, assumed to be uniform throughout the rod. T(0) depends on the geometry of the system. α is a positive number proportional to W_p: $\alpha = aW_p$, where a is inversely proportional to the conductivity of the medium. This leads to an inhomogeneous refractive index: $n(r) = n_0 - (dn/dT)\, \alpha r^2$, where n_0 is the value of the refractive index at T(0) and (dn/dT) is a positive parameter characteristic of the medium.

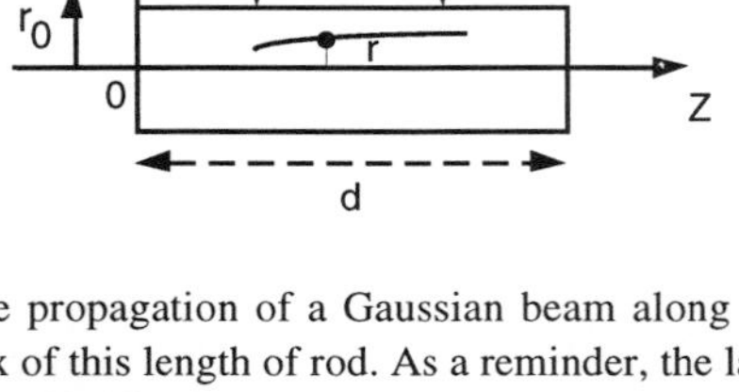

(a) We want to study the propagation of a Gaussian beam along a length z of the rod by determining the transfer matrix of this length of rod. As a reminder, the laws of geometrical optics in an inhomogeneous medium give the following propagation equation for a light ray traveling along the 0z axis, or slightly tilted away from it: $d^2r / dz^2 = dn(r) / n(r)\, dr$. r is the distance from any point M of the light ray to the axis, n(r) is the refractive index at M, and dn(r)/dr is its derivative.

Write the differential equation corresponding to a light ray propagating in the amplifying medium. To do this, assume that α (dn / dT) $r^2 << r_0$, where r_0 is the radius of the rod. Integrate this equation over a length z of the rod. Determine the transfer matrix corresponding to a length z and the transfer matrix corresponding to the whole length d of the pumped rod. Show that for small values of W_p, the rod is equivalent to a thick lens—in other words, to a thin lens of focal length f_p followed by a translation of length d_p through air. Show that f_p can take the form $f_p = b/W_p$ and calculate b. Calculate d_p and explain at what condition on W_p this "thick lens" equivalence is a good approximation.

(b) Let a Gaussian beam propagate along the positive Oz axis with a beam waist located at the entrance of the rod at z = 0. We do not want to restrict ourselves to small values of W_p, therefore we shall not use the above approximation. Use the transfer matrix $(M) = \begin{pmatrix} A & B \\ C & D \end{pmatrix}$ to calculate q(z), the complex parameter of the beam, at any location z of the rod. Separate the imaginary and the real part of $q(z)^{-1}$ and calculate the diameter $\omega(z)$ of the beam at z. Show that $\omega(z)$ is a periodic function and give its period Δz. At what conditions on d do we have a maximum or a minimum ω at the back face of the rod? Give the values of ω in these cases.

(c) The Gaussian beam has its beam waist ω_0 located at z' = 0, at a distance z_0 of the front (the entrance face) of the rod. Calculate the diameter $\omega(z')$ as a function of z' in the rod for this new geometry. Did the spatial period Δz change?

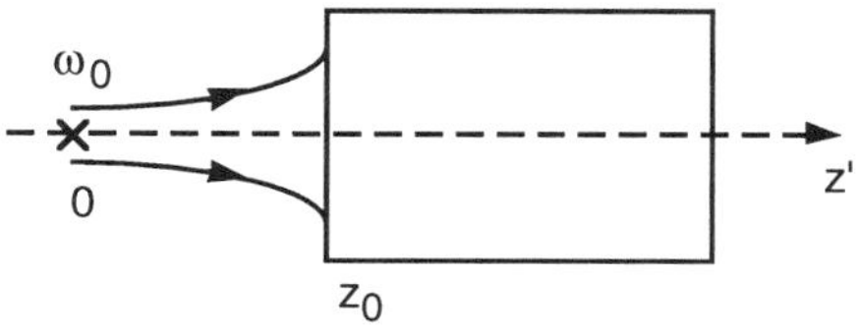

2. *Study of the optical cavity with amplifying rod*

(a) The laser cavity consists of two plane mirrors separated by a distance L. The amplifying rod is placed between them. W_p is small enough so that the rod can be assimilated to a thin lens of focal length $f_p = b / W_p$, placed exactly between the two plane mirrors (at distance L/2 away from each). Give a justification for this approximation using the results of 1(a). Conclude that this cavity is equivalent to a standard cavity and give its stability domain in terms of W_p. What are the position and the diameter of the beam waist?

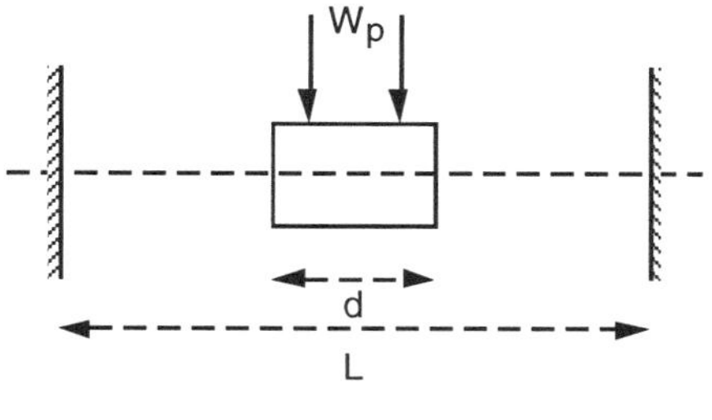

(b) Now, the rod is no longer equivalent to a thin lens, but W_p is such that d is a multiple of the period Δz as defined in 1 (b). The two mirrors are spherical and have the same radius of curvature R. They are placed symmetrically with respect to the rod. Determine the stability domain of the cavity, the positions and the diameters of the beam waists. In a second step, the rod is moved along its axis so that it come closer to one of the mirrors. How does this affect the stability domain of the cavity? The positions, and dimensions of the beam waists? The rod is put back in the center of the cavity and now we let W_p vary around the value defined just now. Describe qualitatively how this affects the parameters of the cavity.

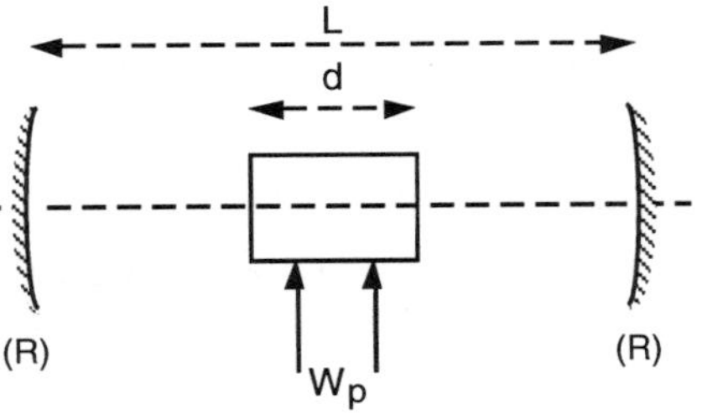

1. (a) The equation is $d^2r/dz^2 + s\,r \approx 0$ with $s = 2\,\alpha dn / n_0\, dT$. The initial conditions are $r(0) = r_0$ and $(dr/dz)_0 = r'_0$. We conclude:

$$r = r_0 \cos(\sqrt{s}\,z) + (r'_0/\sqrt{s}) \sin(\sqrt{s}\,z);\quad r' = -\sqrt{s}\,r_0 \sin(\sqrt{s}\,z) + r'_0 \cos(\sqrt{s}\,z)$$

The transfer matrices take the form:

$$\begin{pmatrix} \cos(\sqrt{s}\,z) & \sin(\sqrt{s}\,z)/\sqrt{s} \\ -\sqrt{s}\sin(\sqrt{s}\,z) & \cos(\sqrt{s}\,z) \end{pmatrix} \text{ and } \begin{pmatrix} \cos(\sqrt{s}\,d) & \sin(\sqrt{s}\,d)/\sqrt{s} \\ -\sqrt{s}\sin(\sqrt{s}\,d) & \cos(\sqrt{s}\,d) \end{pmatrix}$$

For small pumping values, we have $\sqrt{s}\,d \ll 1$. When $d_p \ll f_p$, the last matrix simplifies to $\begin{pmatrix} 1 & 0 \\ -1/f_p & 1 \end{pmatrix}\begin{pmatrix} 1 & d_p \\ 0 & 1 \end{pmatrix}$ = [thin lens with $f = f_p$] + [translation $d_p = d$]. We have $d_p = d$. On the other hand, $f_p = 1/sd$ or $f_p = n_0 / 2\,ad\,W_p\,(dn/dT)$, so that at last we obtain $b = n_0 / 2\,ad\,(dn/dT)$.

For the approximation to be valid, we need to have $d \ll n_0 / 2\,ad\,W_p\,(dn/dT)$ which means $W_p \ll n_0 / 2\,a\,(dn/dT)\,d^2$.

(b) Using Eqs. IV.2.60 and IV.2.71, we obtain

$$q(z) = \left[i\pi\omega_0^2 \cos(\sqrt{s}\,z)/\lambda + \sin(\sqrt{s}\,z)/\sqrt{s}\right] / \left[-i\pi\omega_0^2 \sqrt{s}\sin(\sqrt{s}\,z)/\lambda + \cos(\sqrt{s}\,z)\right]$$

and using Eq. IV.2.59 we obtain $\omega(z) = \omega_0\left[\cos^2(\sqrt{s}\,z) + \beta\sin^2(\sqrt{s}\,z)\right]^{1/2}$ with $\beta = \lambda^2/\left(\pi\omega_0^2\right)^2 s$. $\omega(z)$ is a periodic function of period $\Delta z = \pi/\sqrt{s}$. For $d = \Delta z$ (where p is a whole number) a minimum of ω occurs on the back face of the rod with $\omega(d)_{min} = \omega_0$. A maximum occurs for $d = (2p+1)\,\Delta z/2$ (p is a whole number), with

$\omega(d)_{max} = \omega_0 \sqrt{\beta}$.

(c) Let us do the above calculations on the new axis 0Z'; we thus obtain the similar result:

$$\omega(z') = \omega_0 \left\{ \cos^2\left[\sqrt{s}\,(z' - z_0)\right] + \left[\sqrt{\beta}\,\sin\left[\sqrt{s}\,(z' - z_0)\right] + \gamma\cos\left[\sqrt{s}\,(z' - z_0)\right]\right]^2 \right\}^{1/2}$$

with $\gamma = z_0\,\lambda / \pi\,\omega_0^2$. The period remains unchanged.

2. (a) The cavity is equivalent to a symmetrical cavity consisting of two spherical mirrors with a radius of curvature 2f and separated by a distance L. Its transfer matrix takes the form

$$\begin{pmatrix} 1 & 0 \\ -1/f_p & 1 \end{pmatrix} \begin{pmatrix} 1 & L \\ 0 & 1 \end{pmatrix} = \begin{pmatrix} 1 & L \\ -1/f & 1 - L/f \end{pmatrix}$$

The stability condition gives $-1 < 1 - L/2f < 1$ or $4\,f_p > L$, or again $W_p < 2\,n_0 / a\,L_d\,(dn/dT)$. The beam waist is at the center of the YAG rod, and its diameter is $\omega_0^2 = \lambda \sqrt{L\,(4\,f_p - L)} / 2\pi$.

(b) The beam waist is at the entrance (the front) of the rod. The output diameter at the back face plays no part at all. The equivalent cavity consists of two spherical mirrors of radius R separated by a distance (L – d). The transfer matrix is as follows:

$$\begin{pmatrix} 1 & L - d \\ -2/R & 1 - 2\,(L - d)/R \end{pmatrix}$$

The stability condition says $-1 < 1 - (L - d)/R < +1$, giving $L - d < 2R$. The beam waist is still at the center of the rod. Its diameter is

$$\omega_0^2 = \lambda \sqrt{(L - d)(2R - L + d)} / 2\pi.$$

Since d is a multiple of Δz, moving the rod inside the cavity does not change the calculations. However, changing W_p changes the stability domain, so the condition lies somewhere between $L < 2R$ and $L < 2R + d$. In the same way, the value of the diameter of the beam waist lies between $\lambda \sqrt{L\,(2R - L)} / 2\pi$ and $\lambda \sqrt{(L - d)(2R - L + d)} / 2\pi$.

IV.5 Bibliography

IV.5.1 General presentation of the laser for nonspecialists:

LENGYEL, B. A. *Lasers,* fourth edition, John Wiley & Sons, New York, 1964.

BEESLEY, M. J. *Lasers and their Applications,* Taylor and Francis, London, 1971.

IV.5.2 More specialized general presentations of the laser:

BIRNBAUM, G. *Optical Lasers,* Academic Press, New York, 1966.

VERDEYEN, J. T. *Laser Electronics,* Prentice-Hall, Englewood Cliffs, NJ, 1981.

SIEGMAN, A. *Lasers,* University Science Books, Mill Valley, 1986.

IV.5.3 A description of Lamb's theory is found in:

SARGENT, M., III, SCULLY, M. O., AND LAMB, W. E. *Lasers Physics,* Addison-Wesley, London, 1974.

IV.5.4 Quantum theory of radiation and quantum theory of spontaneous emission:

LOUDON, R. *Quantum Theory of Light,* second edition, Oxford Science Publications, Oxford, 1983.

COHEN-TANNOUDJI, C., DUPONT-ROC, J., AND GRYNBERG, G. *Atom-Photon Interaction,* John Wiley & Sons, New York, 1992.

IV.5.5 The diffraction of light is studied in:

BORN, M., AND WOLF, E. *Principles of Optics,* second edition, Pergamon Press, Oxford, 1964, pp. 370-609.

CHAPTER V

Spatial Structure of a Laser Wave and Its Consequences

V.1 Spatial structure and coherence

We showed in Chapter IV that the spatial structure of the fundamental laser wave can be described in a condensed way as follows:

$$\mathcal{E}(x, y, z) = [\mathcal{E}_0 / \omega(z)] \times \exp\left(-\left\{i\left[k_q z - \phi(z) + k_q (x^2 + y^2)/2R(z)\right] + (x^2 + y^2)/\omega^2(z)\right\}\right) \quad \text{(V.1.1)}$$

where $\omega(z)$, $\phi(z)$, and $R(z)$ are the functions of z given by (see Eqs. IV.2.61, IV.2.62 and IV.2.72)

$$\begin{aligned} \omega^2(z) &= \omega^2(0)\left[1 + \left(\lambda z / \pi \omega_0^2\right)^2\right] \\ \phi(z) &= \arctan\left[\lambda z / \pi \omega_0^2\right] \\ R(z) &= z\left[1 + \left(\pi \omega_0^2 / \lambda z\right)^2\right] \end{aligned} \quad \text{(V.1.2)}$$

These functions refer to an origin z = 0 of the longitudinal coordinate of the beam waist and to a value (ω_0) which depends on the geometry of the optical cavity. For a symmetrical cavity for instance (consisting of two identical mirrors), the beam waist is of course in the center of the cavity and has a value

$$\omega_0^2 = \lambda \sqrt{L(2R - L)} / 2\pi$$

This equality is easily obtained from Eq. IV.2.67.

In a similar fashion, the plane-spherical cavity, which we already discussed in a problem, can be considered as the equivalent of half a symmetrical cavity. Its beam waist is located on the plane mirror. Far from the beam waist, i.e. for $z >> \pi \omega_0^2 / \lambda$ which is where the beam is mostly put to use, we have

$$\begin{aligned} \omega(z) &\approx \lambda z / \pi \omega_0 \\ \phi(z) &\approx \pi / 2 \\ R(z) &\approx z \end{aligned} \qquad \text{(V.1.3)}$$

Here, the laser beam is equivalent to a spherical wave springing forth from the beam waist. Around its axis of course, the beam possesses the Gaussian amplitude attenuation mentioned earlier.

Traditionally, the property of *spatial coherence*, implied by the very structure of the beam, is expressed by a formula based on a complex correlation function. In our case, it takes the form

$$\gamma_{12}(0, \Delta x) = \overline{E(t,x)\, E^*(t, x + \Delta x)} \Big/ \sqrt{\overline{|E(t, x)|^2 |E(t, x + \Delta x)|^2}} \qquad \text{(V.1.4)}$$

x is a radial coordinate and Δx is the length separating the two points of the beam under consideration. 0 means that the two fields are considered at the same instant t. The model we use leads to $|\gamma_{12}(0, \Delta x)| = 1$ for all values of x and Δx since all points on the wave front are in phase with each other. However, we shall see later why the spatial coherence of the real laser is imperfect.

Comment: Notice that Eq. V.1.4 leads in a natural way to define a *characteristic coherence length* Δx_c.

The question now, and a very important one for those who use the laser, is whether this particular spatial structure can be modified—for instance, in order to concentrate the energy of the wave inside a small volume, or to try to reduce the angular divergence so as to be able to transmit energy over long distances.

V.2 Some consequences of spatial coherence: Spatial concentration of a laser beam

V.2.1 Transformation of a laser beam by a lens

This will be the first idea of the experimental physicist. Optics tells us that a thin lens with a thickness at the center of 2ε and a focal length f placed at $z = 0$ on the z axis (see Figure V.2.1) affects a wave only in that it changes its phase around the axis. Indeed, in the case of a converging thin lens, the axial part of the wave ($x = y = 0$) has to travel through a thicker section of the lens than those parts of the wave which are further away from the axis. In case of a divergent lens, the axial part of the wave travels through a thinner section of the lens.

Thus a phase difference appears which is inversely proportional to the focal length of the lens. Mathematically, the transformation can be written

$$\mathcal{E}(x, y, z_0 + \varepsilon) = \mathcal{E}(x, y, z_0 - \varepsilon) \exp\left[k\left(x^2 + y^2\right) / 2f\right] \qquad (V.2.1)$$

Following the usual conventions, f is positive for a converging lens and negative for a diverging lens.

Applying this relation to a Gaussian wave as described by Eq. IV.2.54 gives

$$\begin{aligned} &\mathcal{E}(x, y, z_0 - \varepsilon) \sim \exp\{-i\left[P(z) + k\left(x^2 + y^2\right) / 2q(z)\right]\} \\ &\mathcal{E}(x, y, z_0 + \varepsilon) \sim \exp\left(-i\left\{P(z) + (k/2)\left(x^2 + y^2\right)\left[q^{-1}(z) - f^{-1}\right]\right\}\right) \end{aligned} \qquad (V.2.2)$$

We can see that the lens modifies the radial parameter q(z) of the beam at point z_0 in the following way:

$$1 / q(z_0 + \varepsilon) = 1 / q(z_0 - \varepsilon) - 1 / f \qquad (V.2.3)$$

We are now ready to describe more precisely the transformation of a Gaussian beam traveling through a lens. Let q_1 and ω_1 be the values of the radial parameter and of the characteristic diameter of the beam at the beam waist before traveling through the lens (see Figure V.2.2). Of course, they are related by (see Eq. IV.2.60)

$$q_1 = i \pi \omega_1^2 / \lambda \qquad (V.2.4)$$

We successively find

$$\begin{aligned} q(0 - \varepsilon) &= q_1 + d_1 \\ 1/q(0 + \varepsilon) &= 1/q(0 - \varepsilon) - 1/f \\ q_2 &= q(0 + \varepsilon) + d_2 \end{aligned} \qquad (V.2.5)$$

From which we can derive q_2 as a function of q_1:

$$q_2 = \left[q_1(1 - d_2 / f) + d_1 + d_2 - d_1 d_2 / f\right] / (4 - q_1 / f - d_1 / f) \qquad (V.2.6)$$

We wish to find the position d_2 of the new beam waist. Substituting Eq. V.2.4 into Eq. V.2.6 and expressing the fact that q_2 is a purely imaginary number lead to the values of d_2 and q_2:

$$\begin{aligned} d_2 - f &= (d_1 - f) f^2 / \left[(d_1 - f)^2 + \left(\pi\omega_1^2 / \lambda\right)^2\right] \\ \omega_2^2 &= \omega_1^2 f^2 / \left[(d_1 - f)^2 + \left(\pi\omega_1^2 / \lambda\right)^2\right] \end{aligned} \qquad (V.2.7)$$

We conclude that a laser beam traveling through a lens does not alter its Gaussian structure, but does change the position and the diameter of the beam waist as well as the geometrical characteristics depending on it, like for instance the divergence of the beam. These transformations are illustrated in Figure V.2.3.

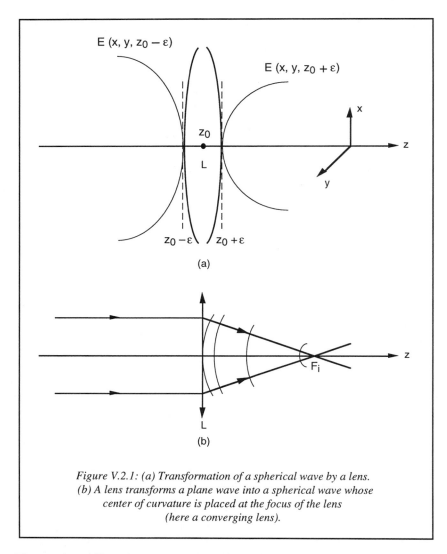

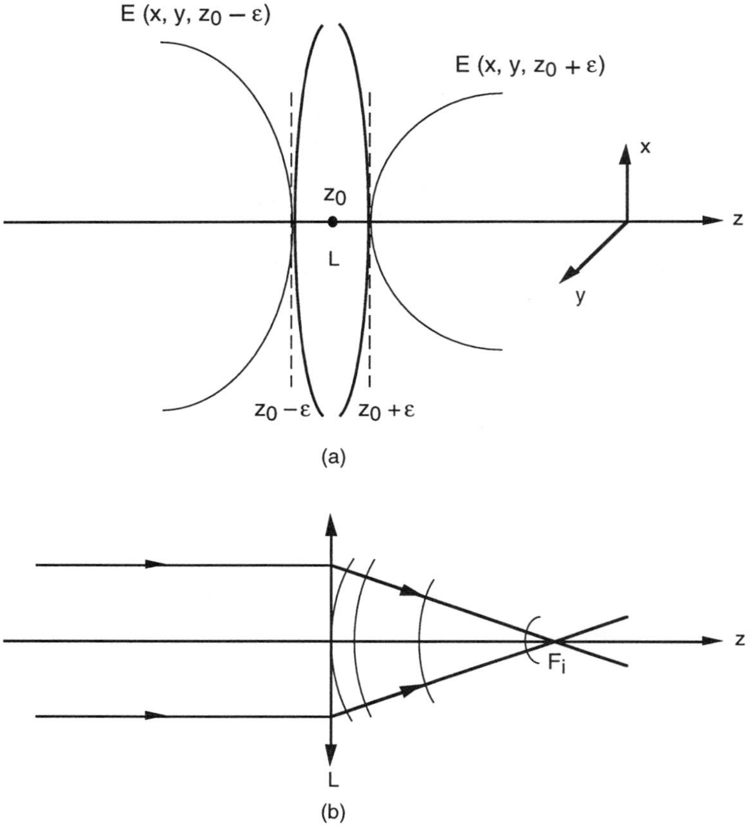

Figure V.2.1: (a) Transformation of a spherical wave by a lens. (b) A lens transforms a plane wave into a spherical wave whose center of curvature is placed at the focus of the lens (here a converging lens).

The reader will notice some very important behavior differences between a laser wave and a classical wave (indicated by a broken line). These differences are especially noticeable when the beam waist is at a distance on the order of, or less than, $\pi \omega_1^2 / \lambda$ from the focus of the lens.

— If the "object" beam waist is at the object focal point of the lens, the "image" beam waist is formed at the image focal point (and not at infinity, as would have been the case in classical optics).

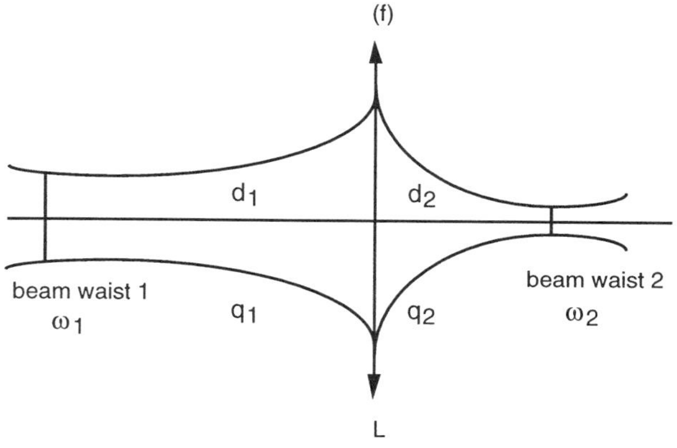

Figure V.2.2: Transformation of a Gaussian beam by a lens.

— It is no longer possible to obtain infinite magnification as is the case in classical optics when the object is at the object focus. The largest image beam waist is obtained for $d_1 = f$ and its size is $f\lambda/\pi\omega_1$.

V.2.2 Surface concentration of the energy

It is easy to see that, using the appropriate optical system, one can obtain extremely small beam waists in a simple way. For instance, starting from an intracavity object beam waist of $\omega_1 = 100\ \mu m$ placed at a distance $d_1 = 1$ m from a lens (a microscope lens for instance) having a focal length of 0.5 cm, we can theoretically obtain an image beam waist with a diameter $\omega_2 = 0.5\ \mu m$, i.e. a size comparable to the wavelength.

So, theoretically, all the energy carried by a Gaussian wave can be concentrated within a surface whose dimension is comparable to the wavelength of the beam. However, the approximations which were made to characterize Gaussian propagation (see Eq. IV.2.51) no longer holds for a beam having a size comparable to or smaller than the wavelength. Most of the foregoing expressions are then irrelevant. Yet, experimental evidence shows that beams a few microns wide can be obtained, provided that some precautions, discussed later, are taken. Let us note that very small images can be obtained in classical optics too, but only with an enormous loss of energy. Now a laser wave with a power of 1 W, focalized on 1 μm^2, gives a surface energy density of 100 MW cm^{-2}. The electric field carried by this wave has a value of about 2×10^7 V m^{-1} in vacuum. Even higher energy densities can be obtained. Practical applications of such surface concentrations of energy are of great importance:

— in fundamental research, where they generated nonlinear optics,

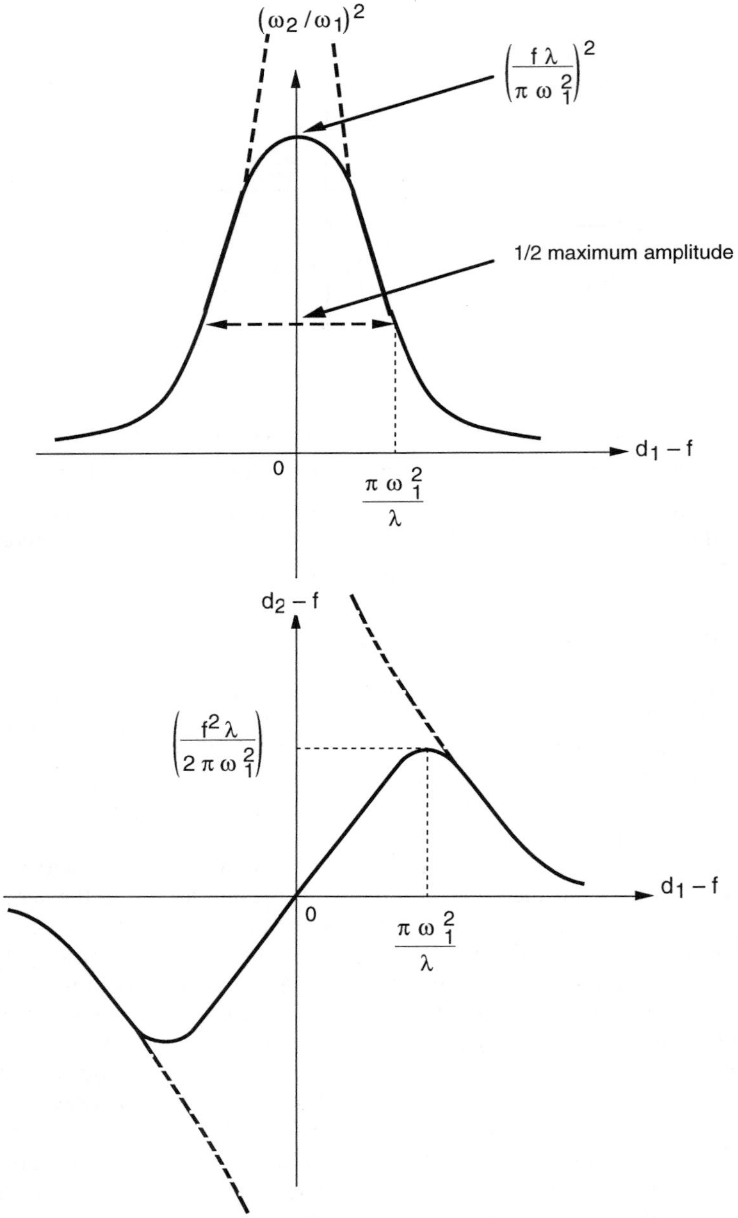

Figure V.2.3: Graphic illustration of the transformation formulae concerning the transformation of a Gaussian beam by a lens (solid line represents the Gaussian wave; broken line represents the classical wave).

— in applied research, where they are often used for mechanical actions (cutting, vaporizing, soldering) and for thermal actions on materials (surface treating); in medicine, where the use of the laser is wide spread; and in nuclear physics (inertia confinement and controlled fusion).

First comment: We described the surface energy concentration in the frame of a TEM_{00q} Gaussian laser wave. In this model, the phase ϕ_q does not depend on its location within the source. Any two points of the wave are perfectly correlated (see Eq. IV.2.65). The wave is said to have perfect spatial coherence. As a matter of fact, a perfect TEM_{00q} mode cannot exist, be it only because of index fluctuations within the cavity and because of the finite dimensions of the mirrors and of the amplifying medium: The actual propagation mode must satisfy boundary conditions which the TEM_{00q} mode could never satisfy. Since these boundary conditions are themselves a function of various fluctuating conditions within the laser, the actual transverse propagation mode itself fluctuates with time. And so a fundamental limit appears to spatial coherence. One must try to reduce it, though there is no hope of ever doing away with it completely.

In order to improve the spatial coherence of the actual wave, an experimental *spatial low-pass filtering* setup, like the one shown in Figure V.2.4, is often used.

A Gaussian wave afflicted with noise (on the left on the figure) springs forth from an "object" beam waist located far away from a lens (or set of lenses) which in turn makes an "image" beam waist in the immediate neighborhood of its image focal point (see Figure V.2.3). A pinhole of small diameter plays the part of spatial low-pass filter. For simplicity's sake, let's reduce the problem to a monodirectional filtering [along axes 0x and 0X; see Figure V.2.5 (a)] and let $\mathcal{E}(x)$ be the distribution (with noise) in the object focal plane of the lens. The phase difference induced by the lens at point X of the image focal plane is expressed by $X \sin\theta \approx X\theta = Xx / f$.

It follows that distribution $\mathcal{E}(X)$ takes the form

$$\mathcal{E}(X) = \int \mathcal{E}(x) \exp(i\,k\,x\,X / f)\,dx \qquad \text{(V.2.8)}$$

$\mathcal{E}(X)$ turns out to be the Fourier transform of $\mathcal{E}(x)$.

Figure V.2.5 (b) shows the shape of this Fourier transform which sends the noise of the polluted distribution $\mathcal{E}(X)$ far back into the wings of distribution $\mathcal{E}(x)$. The low-pass filter, consisting of a pinhole, cleans out the incident wave and thus gives rise to a much purer radial intensity profile, as shown on the right-hand side of Figure V.2.4.

Second comment: Theoretically, it is possible to imagine that the fluctuations we just mentioned, and which are in fact a fundamental limit to spatial coherence, are eliminated. Also, we can assume that the problems due to the infinite dimensions of the mirrors are somehow solved. We would then reach the ultimate

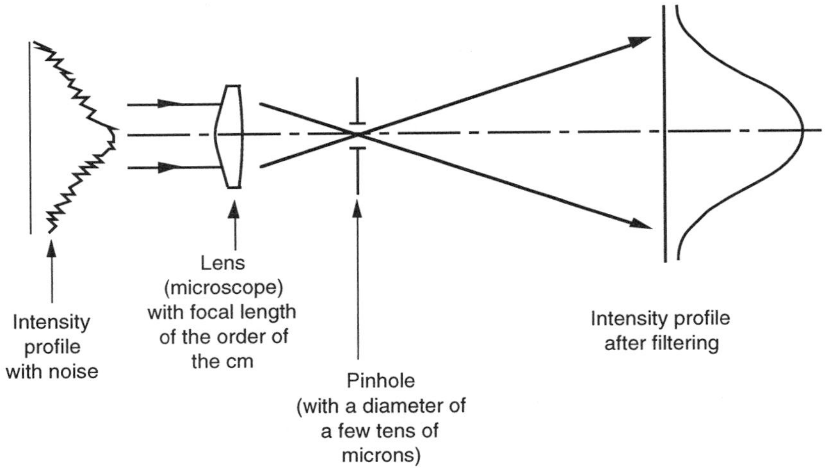

Figure V.2.4: Spatial filtering: experimental setup.

fundamental theoretical limit to spatial coherence, namely the one due to spontaneous emission, which until now was ignored in this text.

V.2.3 Angular concentration of the energy

In Chapter IV, we established that far away from the beam waist, the laser wave is characterized by its divergence $\theta = \arctan[\lambda / \pi\, \omega\,(0)]$. Laser beam waists usually have values of a few hundreds of microns. For $\omega\,(0) \approx 500\ \mu m$ at a wavelength of $\lambda = 0.5\ \mu m$, the divergence θ is close to 4×10^{-4} radians, or about 1.2 angular minutes. This is a very small divergence and it is one of the unusual and extremely fertile aspects of laser optics. However, a great number of practical applications require an even smaller divergence. Of course, such a reduction of θ can only occur by enlarging the original beam waist since, in the small angle approximation, θ varies as the inverse of ω. However, we were able to show that the maximum size of the image beam waist is obtained for $d_1 = f$ and that its maximum value is

$$\omega_2 / \omega_1 = f\,\lambda / \pi\, \omega_1^2 \qquad \text{(V.2.9)}$$

With $f = 1$ m, $\lambda = 0.5\ \mu m$, and $\omega_1 = 100\ \mu m$, we have $\omega_2 / \omega_1 \approx 16$, which gives $\omega_2 \approx 16$ mm. The divergence has been reduced about 16 times and reaches a value of about 20 angular seconds. This result is obtained at the price of a very bulky setup because the lens converges so weakly. It seems difficult to obtain smaller divergences with only one lens. It is therefore common to use quasi-afocal systems consisting of two lenses: one, lens L_1, with a short focal length f_1, either converging [see Figure V.2.6 (a)] or diverging [see Figure V.2.6 (b)]; and the

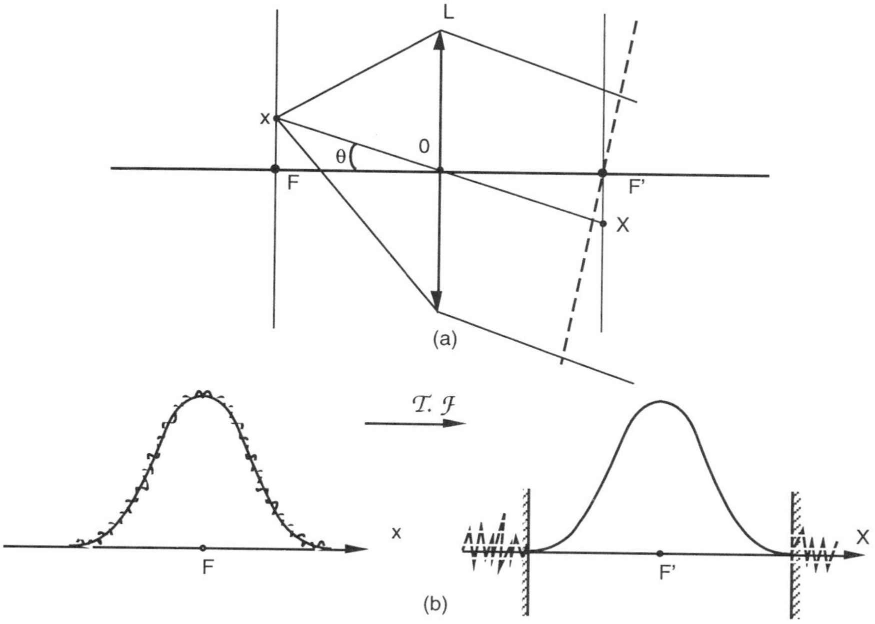

Figure V.2.5: Spatial filtering: theoretical view.

other, lens L_2, converging, but with a much longer focal length f_2. The original beam waist ω_1 must be placed far from L_1 ($d_1 >> f_1$) and at such a distance from lens L_2 that $d_3 = f_2$. Applying Eq. V.2.7 gives

$$\omega_3 / \omega_1 \approx f_2 \lambda \sqrt{d_1^2 + (\pi \omega_1^2 / \lambda)^2} / \pi \omega_1^2 | f_1 | \qquad \text{(V.2.10)}$$

If we put in the following values: $f_1 = 1$ cm, $f_2 = 20$ cm, $d_1 = 50$ cm, $\omega_1 = 100$ μm, we obtain $\omega_3 / \omega_1 \approx 160$. This setup is much more compact, and the enlargement has been multiplied by 10.

So for this case, $\omega_3 \approx 1.6$ cm. As you can see in Figure V.2.6, the intermediary beam waist ω_2 is extremely small (just a few microns). When, as is often the case in nonlinear optics, the wave carries an electric field of great intensity, a voltage breakdown can occur at the real (as opposed to virtual) image focal point of the setup using a converging L_1 lens. This explains why laser builders usually prefer to use a divergent L_2 lens in their setup, since in this case no surface concentration of energy occurs and the inconvenience of a voltage breakdown is avoided. By using optical elements with very large diameters, beam waists measuring several tens of centimeters can be obtained, leading to divergences as small as a few microradians. Here, the limit is a purely technical one: It is reached at the greatest possible diameter of the lenses.

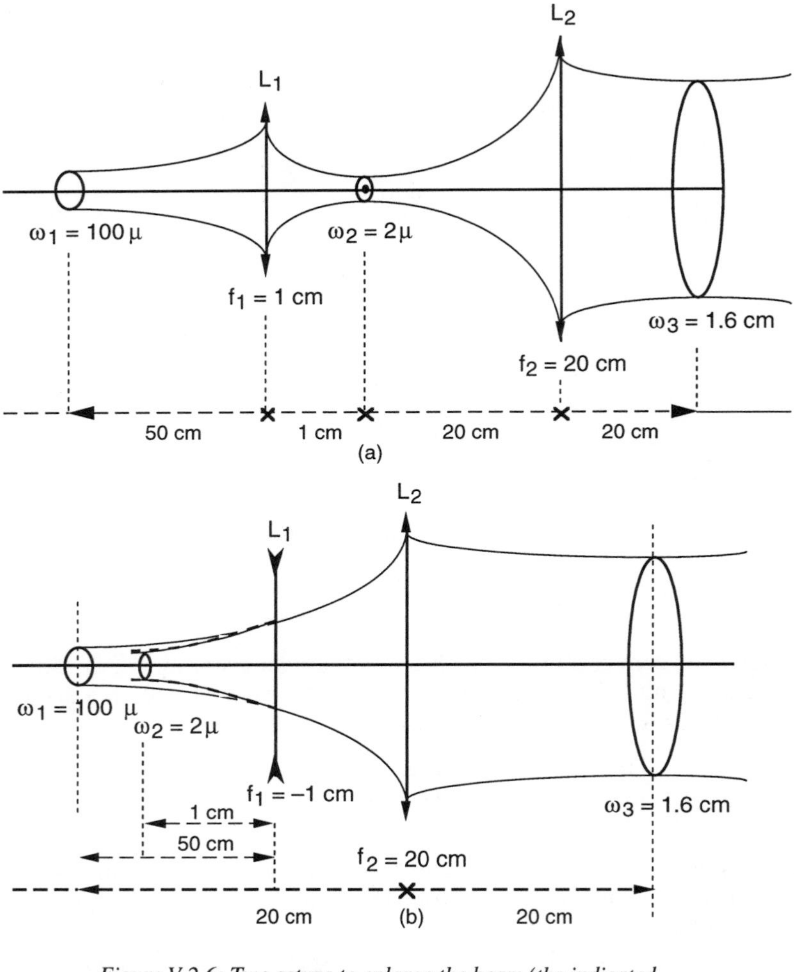

Figure V.2.6: Two setups to enlarge the beam (the indicated dimensions are only approximate, and the drawing is not to scale).

There are many practical applications to the angular energy concentration:

— In fundamental research the experimenter can use a laser beam over long distances, sometimes needed when a large number of optical elements have to be inserted between the source and the object of interest, without losing too much energy.

TABLE V.2.1 — **Comparison between a classical and a laser light source.**

<table>
<tr><th>CLASSICAL SOURCE</th><th>LASER SOURCE</th></tr>
<tr><td colspan="2">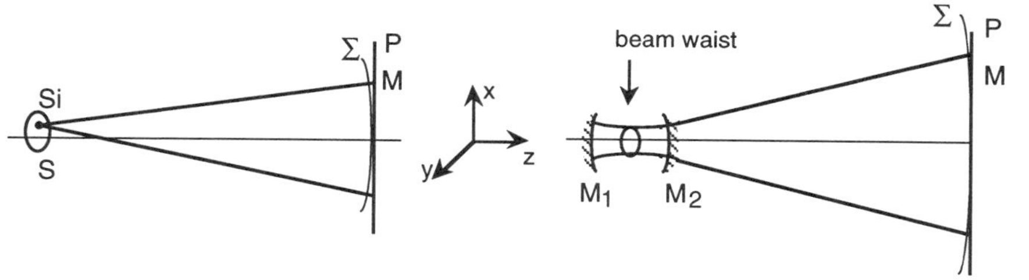
</td></tr>
<tr><td colspan="2">Emission:</td></tr>
<tr><td>Spontaneous</td><td>Stimulated</td></tr>
<tr><td colspan="2">Propagation: $\Delta\mathbf{E} - (1/v^2)\,\partial^2\mathbf{E}/\partial t^2 = 0$</td></tr>
<tr><td>Initial conditions: $E_S = \sum_i E_{Si}$</td><td>Initial conditions: E_{M2} satisfies the self-consistency condition of the optical cavity</td></tr>
<tr><td colspan="2">Shape of the wave on P:</td></tr>
<tr><td>$E_M = \sum_i E_{Mi}$</td><td>E_M</td></tr>
<tr><td>$E_{Mi} = \mathcal{E}_i(x, y, z) \exp\{-i[2\pi\nu_i t - \varphi_i(t)]\}$
$\mathcal{E}_i(x, y, z) = \frac{\mathcal{E}_0}{z}$
$\times \exp\left[-i2\pi\nu_i\left(\frac{z}{v} + \frac{1}{v}\frac{x^2+y^2}{2z}\right)\right]$</td><td>$\mathcal{E}_M = \mathcal{E}(x, y, z) \exp(-i2\pi\nu_L t)$
si $z \gg \pi\,\omega_0^2/\lambda$
$\mathcal{E}(x, y, z) \approx \frac{\mathcal{E}_0}{z}\exp(i\pi/2)$
$\times \exp\left[-i\,2\pi\nu_L\left(\frac{z}{v} + \frac{1}{v}\frac{x^2+y^2}{2z}\right)\right]$
$\times \exp\left(-\frac{x^2+y^2}{\omega^2_{(z)}}\right)$</td></tr>
<tr><td>Classical wave: a superposition of spherical waves which (i) spring forth from different points of the source and (ii) have random phase</td><td>Laser wave = quasi-spherical wave
= spherical wave × Gaussian term</td></tr>
<tr><td colspan="2">Surface concentration (1μ²) limited by diffraction:</td></tr>
<tr><td>A diameter on the order of λ, but very little energy on the concentration surface.</td><td>A diameter on the order of λ, but all the energy is on the concentration surface.</td></tr>
<tr><td colspan="2">Angular concentration (1 μrd):</td></tr>
<tr><td>It is limited by defects of the lenses. Very little energy left in the quasi-plane wave obtained.</td><td>Limited by the diameter of existing optical elements. All the energy is in the quasi-plane wave obtained.</td></tr>
</table>

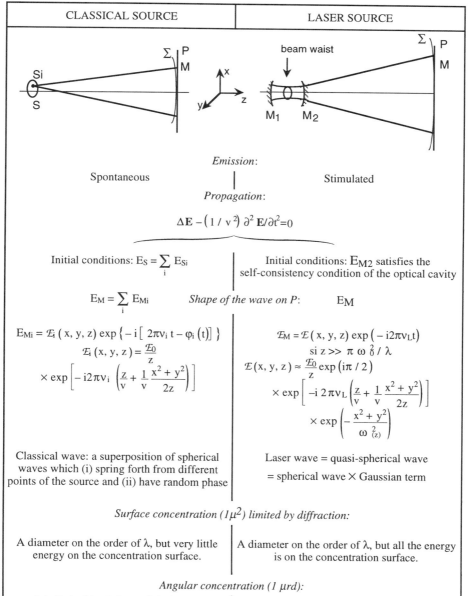

— In applied research the small divergence of the laser beam and the possibility of reducing it even more, opened the way to telemetric measurements of long distances (like the distance from the earth to the moon) and intermediate distances (from the earth to a satellite for instance).

Last, we would like to mention that the high spatial coherence of laser sources (which is the direct cause of the surface and the angular energy concentrations of the beam) makes it possible to guide laser beams through optical fibers. This technique has given rise to spectacular developments in integrated optics.

V.2.4 Comparison of the spatial properties of classical and laser light sources

This comparison is given in the form of a table (see Table V.2.1) summarizing this whole chapter.

V.3 Problem and its outlined solution

V.3.1 Problem V.1: Ring laser and injected laser technologies

1. Let us take a three-mirror optical cavity, as shown on the diagram, to confine light waves with a mean wavelength λ. Mirrors M_1 and M_2 are plane mirrors. M_1 has a reflection coefficient of 1 at λ, and M_2 has a reflection coefficient r_2 for the field at wavelength λ.

The transmission $\mathcal{T}_2$ is such that $r_2^2 + \mathcal{T}_2^2 = 1$. r_2 is very close to one, so that we can write $r_2 = 1 - \beta$ with $\beta << 1$. M_3 is a spherical mirror with radius of curvature R_3 and a reflection coefficient of unity at λ. The ring, which is shaped like an equilateral triangle, has a total optical length of L. We shall neglect any problems due to the astigmatism of mirror M_3, which are in fact compensated by clever positioning and tilting of the amplifying medium which is not shown here.

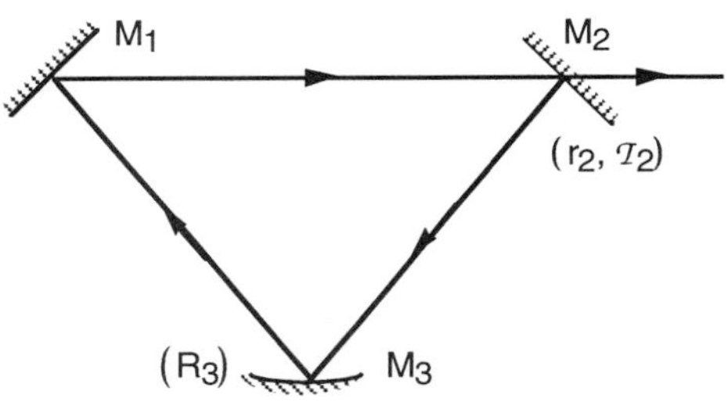

(a) Find the transfer matrix of this cavity.

(b) Using this matrix, find the stability domain of the cavity and also the characteristics of the fundamental Gaussian mode: position and size ω_0 of its beam waist, and its resonance frequencies. Numerical calculation: $R_3 = 10$ cm, L = 19 cm, $\lambda = 0.6\ \mu$m.

Show that the resonator is stable with this configuration. Calculate ω_0 and the distance between adjacent longitudinal modes. Is there a simple way to reduce ω_0 for this configuration of the cavity?

(c) A very thin plate of dye "solution" is placed at the beam waist of the cavity. It is pumped by an Ar^+ laser; the pump beam reaches the plate through mirror M_1 which is transparent for the pump wavelength λ_p. This plate plays the part of the amplifying medium for the cavity. A lens L, with focal length f, is responsible for the shaping and positioning of the beam waist of the pump beam, which must match the cavity beam waist exactly. Let ω_p be the beam waist of the pump laser; find distances d_0 and d_p. What are the conditions on d_0? Numerical calculation: $\omega_p = 300$ µm, f = 50 cm, $\lambda_p = 0.5145$ µm.

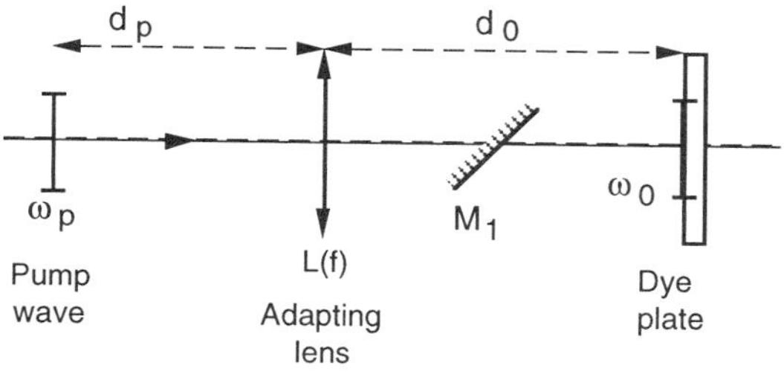

Comment on the advantages of having adapted beam waists. Could the same lens L be used in case of a very small cavity beam waist ω_0?

(d) In order to calculate the amplification of the dye plate at wavelength λ (and ignoring the Gaussian structure of the beams), we assume that the pump beam has the same intensity $\bar{I}_p$ at all points of the amplifying medium and that the intensity of the laser beam varies very little while traveling through the dye plate of thickness d. Let $\bar{I}(0)$ be the intensity of the laser beam at the entrance (front face) of the plate, n the refractive index of the medium, and N_t the number of dye molecules per unit volume. The pumping sequence is indicated schematically in the accompanying diagram. Relaxations 4 → 2 and 1 → 3 are assumed to be instantaneous. We also assume that the number of molecules participating in the sequence is small compared to N_t. Calculate the intensity amplification of the plate, defined by the ratio $a^2\left[\bar{I}_p, \bar{I}(0)\right] = \bar{I}(d)/\bar{I}(0)$. a^2 is close to one and can be written $a^2 = 1 + 2\varepsilon$ with $\varepsilon \ll 1$. Calculate the function $\varepsilon\left[\bar{I}_p, \bar{I}(0)\right]$ and derive from this the amplification $a\left[\bar{I}_p, \bar{I}(0)\right]$ of the field traveling through the amplifying medium. Discuss all the approximations made during this simplified calculation of the amplification.

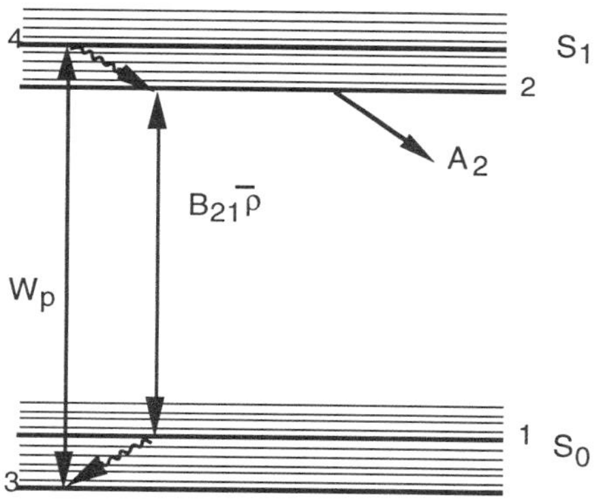

(e) To obtain a monomode operation of the laser, as well as a tunable wavelength λ, one must place a few extra optical elements inside the cavity. Specify these extra elements and discuss their function. Let $\mathcal{T}$ be the transmission coefficient for the field for the whole set of these elements at wavelength λ. Derive from the self-consistency condition of the field within the cavity the minimum value of the pump intensity $\bar{I}_{p\,min}$ (pumping threshold) allowing self-sustained oscillation of the laser in presence of these extra elements. Calculate the output intensity $\bar{I}'$ of the laser when $\bar{I}_p \gg \bar{I}_{p\,min}$. Comment on this result. What is the advantage of using a ring cavity if one wishes to obtain a powerful monomode laser?

2. The laser of 1.(e) is limited in power because the elements inside the cavity do not withstand excessively high energy densities, so this first tunable monomode laser is used *to inject*

a second laser, identical to the first one except that it has no elements in its cavity apart from the amplifying medium. This second dye laser is pumped by a second Ar^+ laser. We want to show that the injected laser operates at the wavelength λ imposed by the pilot laser, with the same property of monochromaticity, provided that the injected intensity $\bar{I}'$ is high enough.

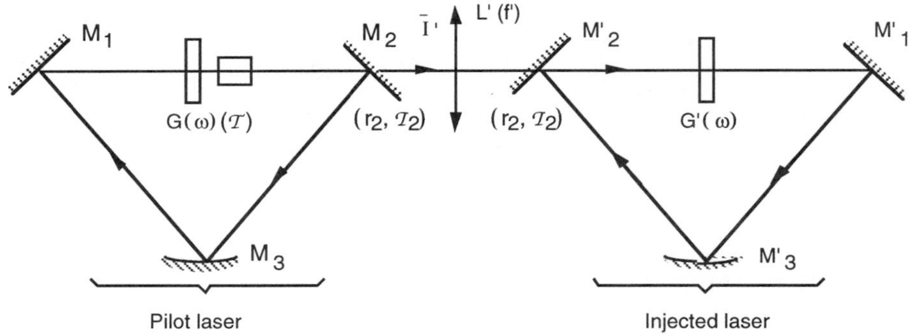

(a) To link the two lasers together, a lens L' of focal length f' is used. What is the purpose of this lens? How must one choose its focal length f'?

(b) In the injected laser, a self-sustained oscillation builds up at wavelength λ_{max}. λ_{max} is the wavelength corresponding to a maximum of the stimulated emission probability between level 2 and one of the levels, λ_{max}, of electronic band S_0 at a pump intensity $\bar{I}'_p > \bar{I}'_{p\,min}$ when $\bar{I}' = 0$. Let $B_{21\,max}$ be the corresponding Einstein coefficient. At weak injected intensity $\bar{I}'$ and at steady state, two waves coexist therefore inside the injected cavity: the self-sustained oscillation at λ_{max}, and of intensity $\bar{I}_{max}$, and the wave resulting from the injection at λ and of intensity $\bar{I}$. Using the new pump diagram drawn here, calculate the amplifications a'^2 and a'^2_{max} as functions of the pump intensity $\bar{I}'_p$ for a plate of thickness d at wavelengths λ and λ_{max}. Use the same approximations as in 1.(d) and assume that the relaxation processes of level 2 are negligible compared to stimulated emission ($A \to 0$). Now calculate the corresponding field amplifications

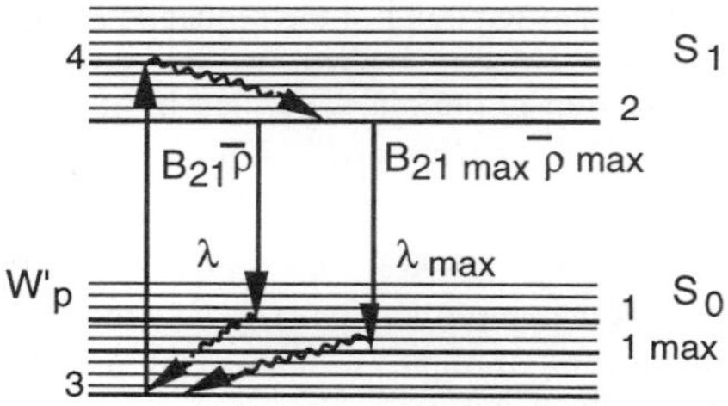

a' and a'_{max}. Set $a' = 1 + \varepsilon$ and $a'_{max} = 1 + \varepsilon_{max}$, where ε and ε_{max} are very small compared to unity and are functions of $\bar{I}'_p$, $\bar{I}$, and $\bar{I}_{max}$. $\bar{I}'_p$, $\bar{I}$, and $\bar{I}_{max}$ in turn are proportional to each other; find the proportionality coefficients (which must be independent of these intensities).

(c) Write the steady-state equations for E_{max}, the field of the self-sustained oscillation at λ_{max}, and E, the injected field at λ. Using the first equation, show that one can find the values taken by amplification a' and a'_{max} when the two waves oscillate simultaneously in the cavity; these values do not depend on the values $\bar{I}'_p$, $\bar{I}$ and $\bar{I}_{max}$. Assume that r_2 and $\mathcal{T}_2$ are independent on λ and that the optical length of the cavity L' is resonant for the injected wavelength λ.

(d) Now let $\bar{I}'_p$ be large enough ($\bar{I}'_p > \bar{I}'_{p\,min}$) so that the laser self-oscillates even without injection (i.e., $\bar{I}' = 0$). Let $\bar{I}'$ increase, starting from 0. Show that the intensity of the self-oscillation decreases and falls to zero at $\bar{I}' = \bar{I}'_i$. Calculate $\bar{I}'_i$. $\bar{I}'_i$ is called the *injection threshold*. Explain this name.

(e) Calculate $\bar{I}$ and $\bar{I}_{max}$ for $\bar{I}' < \bar{I}'_i$. For $\bar{I}' > \bar{I}'_i$, show that $\bar{I}$ is solution of a second-degree equation, and calculate $\bar{I}$. Explain the intensity variations thus found. To obtain these results, the cavity was assumed to be resonant for the injected wavelength. How can such a condition be brought about in practice?

1.(a) Proceeding as in Problem IV.6 for inspiration, we write the transfer matrix of the cavity:

$$\begin{pmatrix} 1 & 0 \\ -2/R_3 & 1 \end{pmatrix}\begin{pmatrix} 1 & L \\ 0 & 1 \end{pmatrix} = \begin{pmatrix} 1 & L \\ -2/R_3 & 1-2L/R_3 \end{pmatrix}$$

(b) The stability condition is written as $-1 < 1 - L/R_3 < 1$ or $0 < L < 2R_3$. The equivalent cavity is a symmetrical cavity consisting of two mirrors of curvature R_3 separated by a distance L. The beam waist is in the middle of segment M_1M_2. Its dimension is $\omega_0^2 = \lambda\sqrt{L(2R_3 - L)}\,/\,2\pi$. The resonance frequencies are given by $\nu_q \approx c\,(q+1)\,/\,L$. The numerical calculation gives $\omega_0 \approx 64.5$ μm and $\Delta\nu \approx c/L \approx 1.58$ GHz. To reduce ω_0, one must reduce the difference $(2R_3 - L)$. The cavity is a stable one since L (19 cm) is smaller than 2R (20 cm).

(c) Relation V.2.7 leads to $d_0 = f + \sqrt{f^2\omega_0^2/\omega_p^2 - (\pi\omega_0^2/\lambda)^2}$ and $d_p = f + (d_0 - f)\,\omega_p^2/\omega_0^2$. Since the triangle $M_1M_2M_3$ is an equilateral one, d_0 must be greater than L/6. The numerical calculation yields $d_0 \approx 61$ cm (L/6 = 3 cm) and $d_p \approx 288$ cm. Beam-waist matching is needed to improve the efficiency of the pumping. At a given ω_p, setting $d_0 = f$, the smallest possible beam waist which can be satisfactorily illuminated is $\omega_{0,\,min} \approx f\lambda_p/\pi\,\omega_p$. To do this, ω_p must be positioned at the object focus of lens L.

(d) Let us write $d\bar{I}/\bar{I}\,dz = N_2\hbar\omega\,B_{21}\,n/c$ or again

$$\left[\bar{I}(d) - \bar{I}(0)\right]/\,\bar{I}(0) = N_2\,d\,\hbar\,\omega\,B_{21}\,n/c$$

The rate equation of level 2 is $\dot{N}_2 = N_3 W_p - N_2 A_2 - N_2 B_{21}\bar{\rho}$ and at steady state is $N_{2S} \approx N_t W_p/(A_2 + B_{21}\bar{\rho})$. Thus

$$\left[\bar{I}(d) - \bar{I}(0)\right]/\,\bar{I}(0) \approx N_t\,d\,W_p\hbar\omega\,B_{21}\,n/c\,(A_2 + B_{21}\bar{\rho})$$

giving $a^2[\bar{I}p, \bar{I}(0)] \approx 1 + 2\varepsilon \approx 1 + N_t\,d\,W_p\hbar\omega\,B_{21}\,n/c\,(A_2 + B_{21}\bar{\rho})$. The field amplification $a[\bar{I}p, \bar{I}(0)]$ is $(1+\varepsilon) \approx 1 + N_t\,d\,\hbar\omega\,B_{21}B_{34}\bar{I}p\,n\alpha/2c\,[A_{21} + B_{21}\,\alpha\,\bar{I}(0)]$

(with $W_p = B_{34}\,\alpha\,\bar{I}p$ and $\bar{\rho} = \alpha\,\bar{I}(0)$, where $\alpha = n/c$). In fact, we simplified the pumping sequence and we used an inadequate plane-wave description for the laser emission (see Chapter IV).

(e) We need to incorporate a dispersive element inside the cavity to be able to adjust the wavelength of this tunable laser, and a Fabry-Pérot étalon is needed to have a monomode operating laser (see Chapter IV). We also need an optical diode to eliminate one of the two rotation directions. The self-consistency equation is as follows: $r_2\, a_{min}\, \mathcal{T} \approx 1$ or again $\varepsilon_{min} \approx 1/(r_2\, \mathcal{T}) - 1$. But when emission just starts, we have $\bar{I}(0) \approx 0$ and $\varepsilon \approx N_t\, d\, \hbar\, \omega\, B_{21}\, B_{34}\, n\, \alpha\ \ \bar{I}_p / 2c\, A_{21}$. Thus

$$\bar{I}_{p\,min} \approx 2c\, A_{21}\left[1/\left(r_2\, \mathcal{T}\right) - 1\right] / N_t\, d\, \hbar\, \omega\, B_{21}\, B_{34}\, n\, \alpha$$

The output intensity is $\bar{I}' = \mathcal{T}_2^2\, \bar{I}(d)$ or

$$\bar{I}' \approx 2\, \mathcal{T}_2^2\, c r_2\, \mathcal{T}\, \bar{I}_p\, N_t\, d\, B_{34}\, \hbar\, \omega\, n / \left(1 - r_2\, \mathcal{T}\right)$$

$\bar{I}'$ is proportional to the pump intensity $\bar{I}_p$ and to the number of dye molecules N_t.

A ring cavity makes better use of the amplifying medium which is subjected to a progressive wave, instead of a stationary wave as in the case of linear cavities. This leads to greater emission power.

2.(a) Lens L' focalizes $\bar{I}'$ inside the amplifying medium of the injected laser and it matches the beam waists. If ω_0 represents the size of the beam waist common to both lasers and ω'_0 represents the beam waist after transformation by L', the focal distance f' must be chosen such that

$$\left(\omega'_0 / \omega_0\right)^2 = f'^2 / \left[\left(d_0 - f'\right)^2 + \left(\pi\, \omega_0^2 / \lambda\right)^2\right] \approx 1$$

For instance, we can choose L' such that $f' \approx 50$ cm and $d_0 \approx 1$ m. The distance between the beam waist of the pilot laser and that of the injected laser will be 2 m.

(b) Neglecting spontaneous emission, let us write as we did above: $\overline{\Delta I} / \bar{I} = N_{2S}\, d\, \hbar\, \omega\, n\, B_{21} / c$ with

$$N_{2S} \approx N_t\, W_p / \left(B_{21}\, \bar{\rho} + B_{21\,max}\, \bar{\rho}_{max}\right) \approx N_t\, B_{34}\, \alpha\, \bar{I}'_p \left(B_{21}\, \alpha\, \bar{I} + B_{21\,max}\, \alpha\, \bar{I}_{max}\right)$$

In the same way, $\overline{\Delta I}_{max} / \bar{I}_{max} = N_2\, d\, \hbar\, \omega\, n\, B_{21\,max} / c$.
$a'^2 \approx 1 + \overline{\Delta I} / \bar{I}$ and $a'^2_{max} \approx 1 + \overline{\Delta I}_{max} / \bar{I}_{max}$; $\varepsilon = \overline{\Delta I} / 2\bar{I}$; $\varepsilon_{max} = \overline{\Delta I}_{max} / \bar{I}_{max}$. Looking carefully at these equations, we can see that $\varepsilon / \varepsilon_{max} = \omega\, B_{21} / \omega_{max}\, B_{21\,max}$. At last, we find $\varepsilon = e\, \bar{I}'_p / \left(b\, \bar{I} + b_{max}\, \bar{I}_{max}\right)$ with $e = N_t\, B_{34}\, d\, \hbar\, \omega\, n\, B_{21} / 2c$; $b = B_{21}$; $b_{max} = B_{21\,max}$.

(c) The self-consistency equations take the form $a'_{max}\, r_2 = 1$ (self-sustained oscillation λ_{max}) and $a'\, r_2\, E + \mathcal{T}_2\, E' = E$ (oscillation at the injected wavelength λ). This gives $\varepsilon_{max} = (1/r_2) - 1$ and $\varepsilon = (1 - r_2)\, \omega\, B_{21} / r_2\, \omega_{max}\, B_{21\,max}$.

(d) The equation of section c relative to λ leads to $\bar{I} = \mathcal{T}^2\, \bar{I}' / \left(1 - a'\, r_2\right)^2$. At the injection threshold we have $\bar{I} = \bar{I}_i$ and $\bar{I}_{max} = 0$. Then we have $\varepsilon = e\, \bar{I}'_p / b\, \bar{I}_i$, from which we conclude that $\bar{I}_i = e\, \bar{I}'_p / \varepsilon\, b$. This in turn gives $\bar{I}'_i = \bar{I}_i \left(1 - a'\, r_2\right)^2 / \mathcal{T}_2^2 = e\, \bar{I}'_p \left(1 - a'\, r_2\right)^2 / \varepsilon\, b\, \mathcal{T}_2^2$. And, using a limited power expansion to express r_2, T_2, and a', we end up with $\bar{I}'_i \approx e\, \bar{I}'_p \left(\beta - \varepsilon\right)^2 / 2\, \varepsilon\, \beta$.

Explanation of the name "injected": When $\bar{I}' > \bar{I}'_i$, the injected wave is amplified by stimulated emission at the cost of the population of level 2. The gain of the self-sustained oscillation at λ_{max} decreases and eventually this oscillation vanishes, leaving only the injected wave in the cavity.

(e) Remembering that $\varepsilon / \beta \approx B_{21} / B_{21\,max}$, we find that for $\bar{I}' < \bar{I}'_i$ we have $\bar{I} \approx 2\,\bar{I}' / \beta\left[1 - \left(B_{21} / B_{21\,max}\right)\right]^2$. To find $\bar{I}_{max}$, we can use this expression for $\bar{I}$ in the expression for ε, which gives

$$\bar{I}_{max} \approx \left(B_{21\,max}^2 / \beta\right)\left[\left(e\,\bar{I}'_p / B_{21}^2\right) - 2\,\bar{I}' / \left(B_{21\,max} - B_{21}\right)^2\right]$$

When $\bar{I}' > \bar{I}'_i$, we have $\bar{I}_{max} = 0$ and $\varepsilon = e\,\bar{I}'_p / b\,\bar{I}$.

On the other hand, $\bar{I} = 2\,\beta\,\bar{I}' / (\beta - \varepsilon)^2$. Replacing ε by this value leads to the following solutions:

$$\bar{I} = \left[\bar{I}' + \left(e\,\bar{I}'_p / b\right)\right]\left[1 \pm \sqrt{1 - 1 / \left(1 + b\,\bar{I}' / e\,\bar{I}'_p\right)^2}\,\right]$$

Thus, we obtain two stable injected oscillating states, which means there is a possibility of bi-stability.

CHAPTER VI

Time Structure of a Laser Wave and Its Consequences

VI.1 Time structure and coherence

VI.1.1 Generalities

The standard laser description presented in Chapter IV makes the assumption of a *steady-state* condition. The populations of the various electronic levels, as well as the laser intensity, are assumed to be constant. The variable time seems absent from the description. As a matter of fact, the description presents a very idealized view, and this for two main reasons:

— No continuous-wave laser can deliver an intensity which stays perfectly constant with time. In all cases, *fluctuations* appear. We shall come back later to the cause of these fluctuations.

— In practice, a great number of actually used lasers operate in a pulsed mode. In order to obtain a really large gain from the amplifying medium at a specific instant, outside energy can be injected by means of pulses, instead of continuously as would be required with the steady-state assumption. With pulses, the gain of the amplifying medium can reach very high values during brief times, thus begetting extremely short light pulses. Three operating regimes can be defined: *relaxed, Q-switched,* and *mode-locked regimes.* As said earlier, these can be obtained by *modulating the gain of the amplifier.* But in the description of the laser, we showed (see Section IV.1) that the gain of the amplifier and the losses of the resonator are closely related.

It is therefore easy to see that a pulsed operating mode could be obtained by starting from a continuously pumped laser and then *modulating the cavity losses.* Both these methods, based on extremely different techniques, are commonly used to produce pulses having a repetition rate varying from the single pulse (at demand) to about a hundred megahertz. These will be described later.

What pulse durations can be obtained? Well, one of the most remarkable properties of the laser lies in the extraordinary scope of possible pulses. In fact, the laser offers the choice between an almost infinite pulse duration (continuous wave lasers) and a time width of a few femtoseconds ($1\ \text{fs} = 10^{-15}\ \text{s}$). This property has two consequences which we wish to stress:

— First, at constant energy, shortening the pulse leads to an increase of the intensity $\bar{I}$ of the wave, itself proportional to the square of the amplitude of the electromagnetic field **E** carried by the wave (see Eq. IV.2.1). The large values of the electric field opened the way to *nonlinear optics*, which in turn led to spectacular developments in *laser manipulation of materials* and to the soaring expansion of *opto-electronics.*

— The remarkable diversity of usable pulse durations, including extremely short ones, gave rise to a new type of spectroscopy, namely *time-resolved spectroscopy.* Its applications in fundamental research have been especially fertile during these last years.

Time thus appears as a fundamental parameter in laser physics, and therefore this chapter is entirely devoted to it. Let us remember also that Fourier analysis, which was developed in the 19th century, relates Δt, the time width at half-maximum of a pulse (here it is a light pulse), to its frequency width $\Delta\nu$ in the following way:

$$\Delta t\ \Delta\nu \approx 1 \qquad \text{(VI.1.1)}$$

Emission as a function of time is related to its spectrum profile by *Fourier transform.* Let us recall a few definitions and properties which will be used many times in the following pages of this book.

When two independent measures of Δt and of $\Delta\nu$ lead to a product $\Delta t\ \Delta\nu$ with a value close to unity, the corresponding pulses are said to be *Fourier-transform-limited.*

Before going into the details of the various operating modes of lasers, we want to have a good description of the ideal four-level laser already mentioned in Section IV.2, and which we shall be using presently.

Let N be the population of level 3 (see Figure IV.2.6).

Setting $N_3 = N$, the second equation of IV.2.31 becomes

$$\dot{N} = W_{43}\, N_4 - W_{32}\, N - A_3\, N \qquad \text{(VI.1.2)}$$

Since W_{43} is infinite, we have $W_{43} \approx W_4$, which, using the first equation of IV.2.32, yields

$$W_{43} N_4 = W_p N_1 = W_p (N_t - N) \qquad \text{(VI.1.3)}$$

And Eq. VI.1.2 now becomes

$$\dot{N} = W_p (N_t - N) - N B_{32} \overline{\rho}_{(\nu_{32})} - N A_3 \qquad \text{(VI.1.4)}$$

Defining an overall fluorescence lifetime τ_F of level (3) by

$$\tau_F = A_3^{-1} \qquad \text{(VI.1.5)}$$

we have

$$\dot{N} = W_p (N_t - N) - N B_{32} \overline{\rho}_{(\nu_{32})} - N / \tau_F \qquad \text{(VI.1.6)}$$

Let us define a photon volume-density N_{ph} at frequency ν_{32}, by

---

Inset VI.1 ***Frequency-time Fourier transforms and convolution products***

Definition:

If $g(\nu)$ is a complex or real function of frequency ν, then its Fourier transform, if it exists, is symbolized by ~ and is equal, within a proportionality constant, to the function of time $\tilde{g}(t)$ defined by

$$\tilde{g}(t) = \int_{-\infty}^{+\infty} g(\nu) \exp(-i 2 \pi \nu t)\, d\nu$$

On the condition that it converges, the inverse transform is defined by

$$g(\nu) = \int_{-\infty}^{+\infty} \tilde{g}(t) \exp(+i 2 \pi \nu t)\, dt$$

By definition, the convolution product of two functions f (x) and f'(x) (where x may represent either ν or t) is symbolized by $\otimes$ and is equal to

$$f(x) \otimes f'(x) \int_{-\infty}^{+\infty} f(x') f(x - x')\, dx'$$

Two very important properties:

1. If $g(\nu)$ has a frequency width $\Delta\nu$, then $\tilde{g}(t)$ has a time width Δt such that $\Delta t\ \Delta\nu \approx 1$.

2. $$\overbrace{g(\nu)\times g'(\nu)} = \tilde{g}(t)\otimes\tilde{g'}(t)$$
$$\overbrace{\tilde{g}(t)\otimes\tilde{g'}(t)} = g(\nu)\times g'(\nu)$$

Three Fourier transforms which will be used often throughout this book are illustrated in Figure VI.1.1.

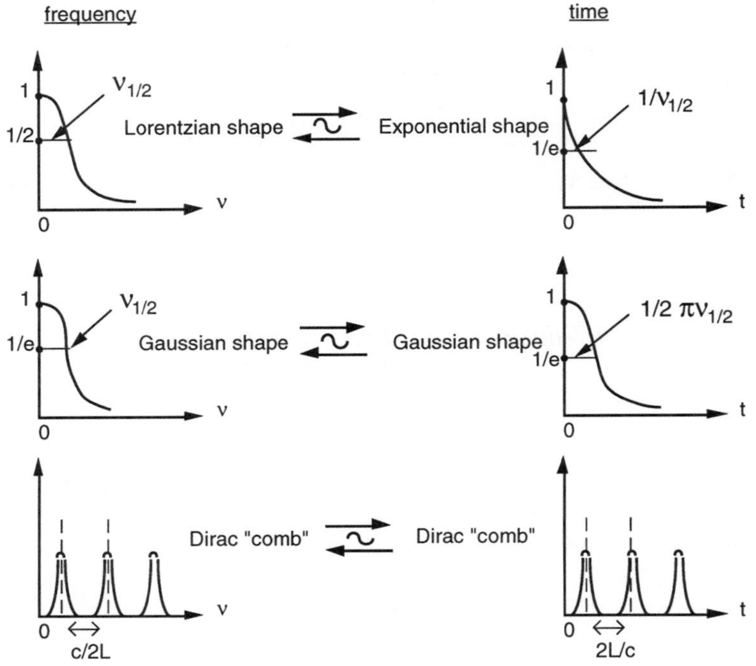

Figure VI.2.1: Illustration of three Fourier transforms. The scales on the coordinate axes are arbitrary.

$$N_{ph} = \bar{\rho}_{(\nu_{32})} / h\nu_{32} \tag{VI.1.7}$$

Starting from the gain $G(\bar{I}_z)$ given by Eq. IV.2.42

$$G(\bar{I}_z) = \sigma\,(N_3 - N_2) = \sigma\,N \tag{VI.1.8}$$

we can define a cross section σ as

$$\sigma = B_{32}\,\bar{\rho}_{(\nu_{32})}\;n / N_{ph}\,c \tag{VI.1.9}$$

so that Eq. VI.1.6 can now be written as

$$\dot{N} = - N c \ \sigma \ N_{ph} / n + W_p(N_t - N) - N / \tau_F \qquad \text{(VI.1.10)}$$

The three terms on the right-hand side of the equation correspond, respectively, to (1) the shrinking of the population of level 3 due to stimulated emission, (2) the increase of population due to pumping, and (3) the decrease due to spontaneous emission. We now wish to write the evolution of N_{ph} with time. Let us remember that every time an atom "falls" from level 3 to level 2, it emits a stimulated photon, so that the differential equation governing the kinetics of N_{ph} must contain the term $Nnc\sigma N_{ph}$. This term is strictly equal, except for the sign which is now positive, to the first term on the right-hand side of Eq. VI.1.10. But we must not forget that the optical cavity plays the part of a *photon trap*. Thus, each photon resulting from stimulated emission remains trapped inside the cavity during a characteristic interval of time τ_{ph} called the *photon confinement time* of the cavity. More details are given in Inset VI.2.

Inset VI.2 ***Two essential time-intervals in laser physics***

Transit time of the optical cavity τ_t:

This is the time it takes light to travel back and forth once through the cavity. It is equal to

$$\tau_t = \sum_i (n'_i \, l_i / c) \approx 2L/c$$

where n'_i and l_i are, respectively, the real part of the refractive index and the length of the ith optical element of the cavity encountered by the wave. τ_t is very close to 2L/c, where 2L is either the total length of a ring cavity, or double the length between the mirrors of a linear cavity. The inverse of the transit time, which is also equal to the frequency interval between two successive longitudinal modes, is called the *free spectral range.*

Photon confinement time of the optical cavity τ_{ph}:

Let us introduce this time interval very simply by starting with the evolution of a progressive wave in the optical cavity shown in Figure VI.1.2. Let the cavity be filled with a medium whose linear field gain g_0 is positive or negative, depending on whether the medium amplifies or absorbs. Let us consider a progressive wave, with angular frequency ω and wavelength λ, propagating from left to right. We assume it is a plane wave. It carries an electric field $\mathcal{E} = \mathcal{E}_0 \exp[-i(\omega t - kz)]$ with $k = 2\pi (n' + in'') / \lambda = -ig_0 + \beta$. Here, $(n' + in'')$ is the complex refractive index of the medium at wavelength λ. Parameter β stands for $\omega n'/c$. Each time the wave reaches mirror M_1, there is reflection of $r_1\mathcal{E}$, and transmission of $\mathcal{T}_1\mathcal{E}$. In the same way, at mirror M_2, $r_2\mathcal{E}$ is reflected and $\mathcal{T}_2\mathcal{E}$ transmitted. The resulting transmitted field is obtained by adding all transmitted fields

$$\mathcal{E}_0 \exp[-i(\omega t - kL)]\,[\mathcal{T}_1 + \cdots + (r_1 r_2)^n\, \mathcal{T}_1 \exp(ink2L) + \cdots]$$

or again

$$\mathcal{E}_0 \exp[-i(\omega t - kL)]\,\mathcal{T}_1\,[1 + \cdots + r_1 r_2 \exp(ik2L)]^n + \cdots]$$

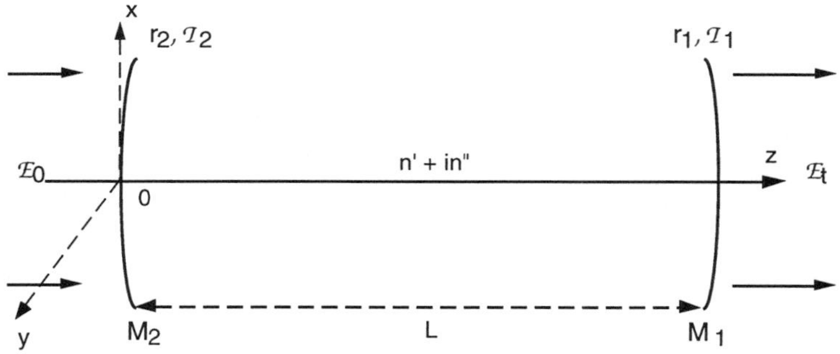

Figure VI.1.2: Geometric and optical parameters of the cavity.

The expression written in square brackets is the sum of a geometric progression of ratio $r_1 r_2 \exp(ik2L)$. It is equal to $1/[1 - r_1 r_2 \exp(ikL)]$, which gives a transmitted field $\mathcal{E}$ of

$$\mathcal{T}_1\, \mathcal{E}_0 \exp[-i(\omega t - kL)] \,/\, [1 - r_1 r_2 \exp(ik2L)]$$

Let us now look at the frequency distribution of the transmitted intensity. Terms $r_1 r_2$ and $\exp(ik2L)$ have values which stay close to one. When squaring the above expression to find the intensity, the only term varying rapidly with frequency gives rise to the following term:

$$1/|1 - r_1 r_2 \exp(ik2L)|^2 \approx 1/\{[1 - r_1 r_2 \exp(2g_0L)]^2 + 4\, r_1 r_2 \exp(2g_0L) \sin^2 \beta L\}$$

Resonance occurs for values which are such that $\beta L = m\pi$ (with m a whole number), yielding

$$\nu = mc / 2\, n'L$$

The half-spectral width at half-maximum corresponds to a $\beta_{1/2}$ equal to

$$\beta_{1/2} \approx [1 - r_1 r_2 \exp(2g_0L)] / 2L\sqrt{r_1 r_2 \exp(2g_0L)}$$

corresponding to frequency $\nu_{1/2}$:

$$\nu_{1/2} \approx (c / 4\pi n'L)[1 - r_1 r_2 \exp(2g_0L)] / \sqrt{r_1 r_2 \exp(2g_0L)}$$

These calculations illustrate the following points:

1. Of course, we rediscover the definitions (introduced earlier) of "free spectral range" ($c/2n'L$) and of "transit time" ($\tau_t = 2n'L/c$).

2. For an empty cavity, $g_0 = 0$. In this case, the *bandwidth* of the cavity is

$$\Delta\nu = 2\nu_{1/2} = c\,(1 - r_1 r_2)\,/\,2\pi n'L\,\sqrt{r_1 r_2}$$

The ratio $1 / \tau_t \Delta\nu$ is called the *finesse* $\mathcal{F}$ of the cavity:

$$\mathcal{F} = \pi \sqrt{r_1 r_2}\,/\,(1 - r_1 r_2) \qquad \text{(VI.1.11)}$$

3. A very important result is obtained when the gain corresponding to one transit time compensates the losses due to the transmission of mirror M_1. In this case, $r_1 r_2 \exp(2g_0L) = 1$ and $\Delta\nu = 0$. Thus, at the oscillation threshold of the laser, we confirm the *a priori* obvious result that a nonzero transmitted wave exists, even for $\mathcal{E}_0 = 0$. The cavity oscillates in a self-sustained mode. Moreover, since the theoretical bandwidth of the emission vanishes, the laser should emit a monochromatic wave lasting forever. At the end of this chapter, we shall show why this is an extremely idealized situation.

At last, using simplified reasoning, we were able to introduce the photon-confinement time τ_{ph}. We showed that the field $\mathcal{E}$ yielded by the cavity was equivalent to an infinite sum of contributions, each term corresponding to an ever-increasing number of back-and-forth trips through the Fabry-Pérot étalon. After n such trips, lasting a time $t = n\,(2n'L/c)$, the amplitude of the initial contribution is multiplied by $(r_1r_2)^n$, which can be written as

$$\exp(n \ln r_1 r_2) \approx \exp[-n(1 - r_1 r_2)] \approx \exp[-(1 - r_1 r_2)\,ct\,/\,2n'L] \approx \exp[-(t/\tau_{ph})]$$

This is clearly an exponential damping, decreasing with the following characteristic time:

$$\tau_{ph} \approx 2n'L\,/\,c\,(1 - r) \qquad \text{(VI.1.12)}$$

where $r = r_1r_2$ is the global field reflectivity of the cavity.

τ_{ph} is called *the confinement time of the cavity.*

Please note that if $r = 1$, τ_{ph} is infinite and no photon can escape from the cavity. We would also like to point out that, as indicated in the beginning of this chapter, the time aspect (τ_{ph}) and the frequency ($\Delta\nu$) aspects are related through

$$\tau_{ph}\,\Delta\nu \approx 1$$

Neglecting spontaneous emission, we can now write the rate of change of the number of photons N_{ph} as

$$\dot{N}_{ph} = Nc\,\sigma\,N_{ph}\,/\,n - N_{ph}\,/\,\tau_{ph} \qquad \text{(VI.1.13)}$$

Equations VI.1.10 and VI.1.13 define the rate equations of the excited atom populations and those of the photon populations. These will be used throughout this chapter.

VI.1.2 Different modes of laser operation

VI.1.2.1 Continuous-wave lasers

In this case, $\dot{N} = \dot{N}_{ph} = 0$ (see Eqs. VI.1.10 and VI.1.13). At steady state, these equations lead to

$$N_S \approx n / c\,\sigma\,\tau_{ph} \tag{VI.1.14}$$

$$N_{ph,\,S} \approx W_p\,N_{tot}\,\tau_{ph} \tag{VI.1.15}$$

Spontaneous emission is neglected here, and we assumed $N_S << N_{tot}$, which is a reasonable assumption for this type of laser.

(a) Longitudinal multimode operation

We showed (see Section IV.2.3) that when the laser operates in its most general mode, it oscillates simultaneously at all the resonance frequencies for which the nonsaturated gain is greater than the cavity losses. Thus the multimode output of a laser is expressed as

$$\mathbf{E}_{00}(t) = \sum_q \mathbf{E}_{00q}(t) \tag{VI.1.16}$$

$\mathbf{E}_{00q}(t)$ is the electric field carried by the TEM_{00q} mode. From now on, we shall simply refer to it as $\mathbf{E}_q(t)$ since all transverse structures considered hereafter are TEM_{00} modes. $\mathbf{E}_q(t)$ is complex and can be written more explicitly:

$$\mathbf{E}_q(t) = \mathcal{E}_q \exp\{i\,[2\pi\nu_q t + \varphi_q(t)]\} \tag{VI.1.17}$$

where $\mathcal{E}_q$ is given by (see Eq. IV.2.65)

$$\mathcal{E}_q = \mathcal{E}_{0q}\,[\omega(0)/\omega(z)]\exp - \Big\{i\Big[\big(k_q z - \phi(z)\big) + k_q\left(x^2 + y^2\right)/2R(z)\Big] + \left(x^2 + y^2\right)/\omega^2(z)\Big\} \tag{VI.1.18}$$

The time distribution of the intensity of the laser critically depends on the phase relations [parameter $\varphi_q(t)$] between the various longitudinal modes. Now phase relations are extremely difficult to evaluate. Very generally, we can say that a close look at a multimode intensity by means of a very fast detector (like, for instance, a "picosecond" photodiode associated with a 1-GHz oscilloscope) shows an intensity which is practically constant in time, but is afflicted with periodic fluctuations at frequency c/2L—i.e., at the inverse of the *transit time* of the cavity—and its multiples. The reader can easily convince himself of this result by noticing that the amplitudes $\mathbf{E}_{0q}$ of the various longitudinal modes are distributed according to a function, $f(\nu_0 \pm qc/2L)$, which depends on the gain and on the saturation of the amplifying medium at frequencies $(\nu_0 \pm qc/2L)$. Assuming that all longitudinal modes have the same spatial dependency and the same polarization, we can write $\mathcal{E}_q = \mathcal{E}_{0q}\, h\,(x, y, z)$, where h is a function which, practically, does not depend on q. The electric field is then expressed as

$$E_{00}(t) = h(x, y, z) \sum_{q=0}^{\infty} f\left(\nu_0 \pm qc/2L\right) \exp i\left[2\pi\left(\nu_0 \pm qc/2L\right) t + \varphi_q(t)\right] \qquad \text{(VI.1.19)}$$

and therefore the intensity is written as

$$I_{00}(t) \sim |E_{00}(t)|^2 \sim \sum_{q=0}^{\infty} \sum_{q'=0}^{\infty} f\left(\nu_0 \pm cq/2L\right) f\left(\nu_0 \pm cq'/2L\right) \exp\left[\pm i2\pi (q - q')\, ct / 2L\right]$$
$$\times \exp\left\{\pm i\left[\varphi_q(t) - \varphi_{q'}(t)\right]\right\} \qquad \text{(VI.1.20)}$$

If the phase fluctuations, described by the term $\exp\left\{\pm i\left[\varphi_q(t) - \varphi_{q'}(t)\right]\right\}$, are slow and random, a very fast detector will be able to detect a beat of frequency c/2L and its multiples, described by the term $\exp\left[\pm i2\pi (q - q')\, ct / 2L\right]$. For most commonly used lasers, these beats have values of about a hundred megahertz. They can be a nuisance. Of course, if a single longitudinal operating mode is forced on the laser, giving an output wave which is close to the ideal monochromatic wave, these beats vanish.

(b) Longitudinal single-mode operation

In single-mode operation, the laser can oscillate only in one of the above-described longitudinal modes. These modes are separated by a frequency interval c/2L. To obtain such a single mode, we could take a very short cavity, so that $c/2L > \Delta\nu_0$ (where $\Delta\nu_0$ is the width of the gain as shown in Figure IV.2.15), but this is technically very difficult. Therefore, it is more common to insert an auxiliary Fabry-Pérot (F-P) interferometer inside the cavity. This F-P interferometer consists of a simple parallel-faced plate with "optical" quality surfaces, whose thickness is much smaller than the cavity length. The free spectral range (see Inset VI.2) and the finesse of this mode selector are chosen in such a way that they produce a modulation of the cavity's loss parameter which is compatible with just one longitudinal mode, of well-defined frequency.

As will be shown in Section VI.3, this procedure greatly improves the time coherence of the emission.

VI.1.2.2 Pulsed lasers

Pulsed lasers emit pulses whose repetition rate can be either fixed or variable. These repetition rates range from single pulses to about a hundred megahertz. The length of the pulse can vary between a millisecond and a few femtoseconds. With pulsed emission, one can obtain considerable instantaneous power (on the order of the gigawatt or even of the terawatt). Very generally, these pulses can be obtained in two entirely different ways:

— By modulating the gain of the amplifying medium in a pulsed way; the medium is illuminated by light pulses provided by a pulsed discharge lamp, or by a second laser playing the part of an optical "pump".

— By modulating the losses of the cavity of a laser whose amplifying medium is pumped continuously: The cavity can have an active "optical gate"—i.e., controlled by the experimenter—or a passive "optical gate", whose opening allows laser oscillation only during a brief period of time.

These two means are used indiscriminately for the three operating regimes of pulsed lasers which will be described presently. However, we shall not go into the details of the technical aspects since there are too many possibilities and, moreover, they are subject to rapid change.

(a) The relaxed regime

In the relaxed regime, a pump pulse causing population inversion is sent onto the amplifying medium during a few hundreds of microseconds. The resulting laser emission simply depends on the shape of the pump pulse and on the response of the medium. In this regime, Eqs. VI.1.10 and VI.1.13 can be used to find the time dependencies of N and of N_{ph}, shown in Figure VI.1.3.

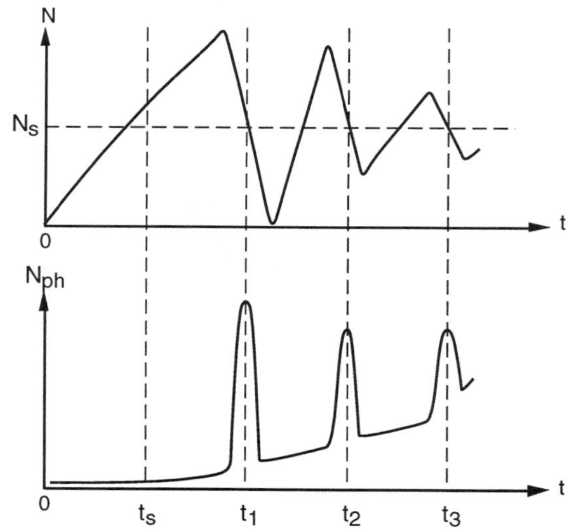

Figure VI.1.3: Time characteristics of relaxed emission.

At instant $t = 0$, a pump pulse is applied to the amplifying medium. No emission takes place until time $t_1 > t_S$ (where t_S is the time required to establish population inversion under constant pumping conditions starting at time $t = 0$). Here we need to stress that the lasers we are now describing operate in a completely different way from continuous-wave lasers. We can see that in this regime and for $t < t_1$, the pump term W_p is much greater than $1 / \tau_F$ and that $N_{ph} = 0$, so that we can write

$$\dot{N} \approx W_p (N_t - N) \tag{VI.1.21}$$

Integrating this equation yields

$$N \approx N_t [1 - \exp(- W_p t)] \tag{VI.1.22}$$

N increases exponentially with t (with time constant W_p^{-1}). At time t_1, N, the population of the highest state, is so great that a flash of light occurs. This extremely rapid increase of N_{ph} of course corresponds to a brutal drop of N and a break-off of the emission. The pumping process starts again, and new laser flashes are emitted at times t_2, t_3 . The population undergoes *relaxation oscillations*, and laser emission occurs in pulses, each lasting about a microsecond and emitted randomly. Multimode relaxed emission can be described by an equation similar to Eq. VI.1.19, provided that one fits in a function of time f(t) which varies slowly as compared to the optical oscillations. This function gives the time envelope of the emission. We can write

$$E_{00}(t) = f'(t) \sum_{q=0}^{\infty} f\left(\nu_0 \pm qc/2L\right) \exp\left\{i\left[2\pi\left(\nu_0 \pm qc/2L\right) t + \varphi_q(t)\right]\right\} \tag{VI.1.23}$$

Now we wish to know whether this time representation is compatible with the frequency representation given by Eq. IV.2.3. The frequency spectrum of this representation is

$$\widetilde{E}_{00}(\nu) = \mathcal{T}.\mathcal{F}\left[\sum_{q=0}^{\infty} f\left(\nu_0 \pm qc/2L\right) \exp\left\{i\left[2\pi\left(\nu_0 \pm qc/2L\right) t + \varphi_q\right]\right\}\right] * \tilde{f}'(\nu) \tag{VI.1.24}$$

We neglected the slow-phase fluctuations $\varphi_q(t)$ by setting $\varphi_q(t) = \varphi_q$, a constant. It is now easy to show that

$$\mathcal{T}.\mathcal{F}\left[\sum_{q=0}^{\infty} f\left(\nu_0 \pm qc/2L\right) \exp\left\{i\left[2\pi\left(\nu_0 \pm qc/2L\right) t + \varphi_q\right]\right\}\right]$$

$$= \sum_{q=0}^{\infty} f(\nu)\, \delta\left[\nu - \left(\nu_0 \pm qc/2L\right)\right] \exp(j\varphi_q) \tag{VI.1.25}$$

where δ is the Dirac function. Equation VI.1.24 then becomes

$$\widetilde{E}_{00}(\nu) = \sum_{q=0}^{\infty} f(\nu)\,\delta\left[\nu - \left(\nu_0 \pm qc/2L\right)\right] \exp(j\varphi_q) * \tilde{f}'(\nu) \qquad \text{(VI.1.26)}$$

and the convolution product becomes

$$\widetilde{E}_{00}(\nu) = \sum_{q=0}^{\infty} f\left(\nu_0 \pm qc/2L\right) \tilde{f}'\left[\nu - \left(\nu_0 \pm qc/2L\right)\right] \exp(j\varphi_q) \qquad \text{(VI.1.27)}$$

Thus that $\widetilde{E}_{00}(\nu)$ shows up as the product of the saturated gain of the amplifier f(ν) and the function

$$\gamma(\nu) = \sum_{q=0}^{\infty} \tilde{f}'\left[\nu - \left(\nu_0 \pm qc/2L\right)\right] \exp(j\varphi_q)$$

γ(ν) represents the Dirac "comb" made up by the longitudinal modes of the resonator in *laser-oscillating conditions* (Fabry-Pérot + amplifier). The spectral width of each "tooth" of $\tilde{f}'(\nu)$ is inversely proportional to the pulse time of the relaxed emission f '(t). These relations are shown in Figure VI.1.4 which is a more realistic version of Fig. IV.2.15, where the resonator and the amplifier were assumed to be completely separated. The reader will have noticed that the longer the pulses last, the narrower the width of the "teeth". Looked at from the point of view of this model, a continuous-wave laser is a laser oscillating on longitudinal modes with zero spectral width (see Inset VI.2). Obviously, this is a simplistic view. Yet, extremely narrow spectral widths—on the order of a few hundred of hertz—have been obtained (see Section VI.3).

Typical powers obtained in relaxed operating conditions are *on the order of the kilowatt*; relaxed operation is often used in applied sciences. However, the randomness of the emission makes this kind of laser unfit for most of fundamental research.

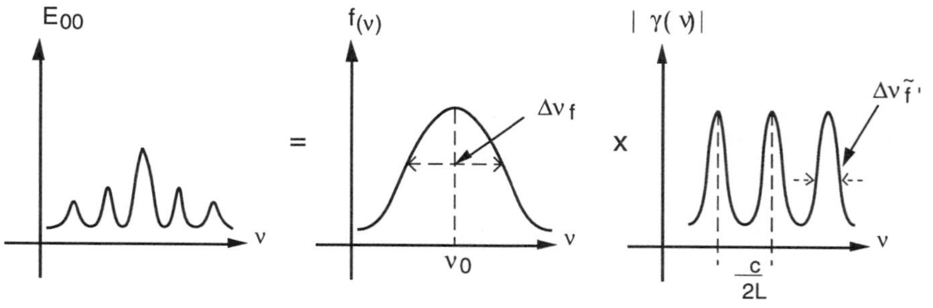

Figure VI.1.4: Frequency structure of relaxed multimode emission.

(b) The Q-switched regime

The fundamental idea behind this kind of operating regime is simple: The experimenter wishes to obtain greater optical power and shorter pulse durations. He will therefore try to avoid all emission from the highest excited state by stimulated emission until the greatest possible population inversion is established. To do this, he will artificially increase the losses of the cavity, i.e. reduce its Q-factor, during part of the pumping process, as indicated in Figure VI.1.5.

The relative loss factor of the cavity, called ξ_r (see definition in Inset VI.3), is written as follows:

$$\xi_r(t) = -\ln r_1^2 + \xi'_r(t) \qquad \text{(VI.1.28)}$$

where ξ'_r (t) represents a discontinuous step function with maximum amplitude $\xi'_{r,\,max}$, with the step located at time t_i. The field reflection coefficient of the output mirror is r_1. The second mirror is assumed to have a reflection coefficient of $r_2 = 1$.

So the value of $\xi_r(t)$ oscillates between $\xi_{r,\,max} = -\ln r_1^2 + \xi'_{r,\,max}$ and $\xi_{r,\,min} = -\ln r_1^2$. The population of the highest level increases from N_f to N_i, while the photon flux, which is nonexistent until time t_i (i for initial), quickly reaches a maximum, to vanish again at time t_f (f for final). During the light pulse, the population of the upper level decreases from N_i to N_f.

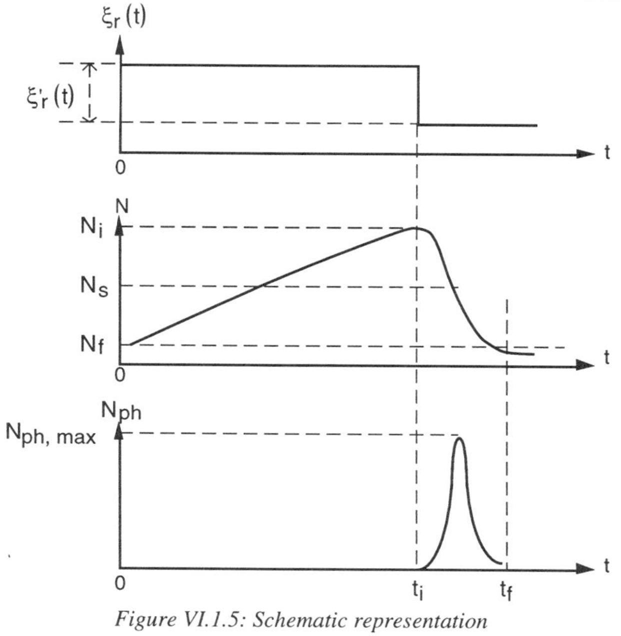

Figure VI.1.5: Schematic representation of the Q-switched regime.

Inset VI.3 ***Relative loss factor of a laser cavity***

Let P_0 be the power of a laser wave at a given point of the optical cavity. The cavity has a length L, and its losses are assumed to be due exclusively to the nonzero transmission of mirror M_1. After traveling back and forth through the cavity once, the power is measured again at the same point and called P_{2L}.

By definition, the relative loss factor of the cavity for one complete transit is the dimensionless parameter ξ_r which is such that $P_{2L} = P_0 \exp(-\xi_r)$.

Now since we know that $P_{2L} = P_0 r_1^2$ we conclude that

$$\xi_r = -\ln r_1^2$$

A simplified mathematical description fitting this triggered pulse is easily found starting with Eqs. VI.1.10 and VI.1.11, in which we want to keep only those terms dealing with stimulated emission. This is appropriate since in this type of regime dealing with intense photon fluxes, stimulated emission is largely predominant. Equation VI.1.10 becomes

$$\dot{N} \approx -N c \sigma N_{ph} / n \qquad \text{(VI.1.29)}$$

Dividing each side of Eq. VI.1.13 by the corresponding side of Eq. VI.1.10 gives

$$dN_{ph} / dN \approx -1 + n / N c \sigma \tau_{ph} \qquad \text{(VI.1.30)}$$

which upon integration gives

$$N_{ph}(t) \approx N_s \ln[N(t) / N_i] + N_i - N(t) \qquad \text{(VI.1.31)}$$

where we inserted the "steady-state" population N_s as defined by Eq. VI.1.14. The reader may check that $N_{ph,\,max}$ is obtained when $N = N_s$ (see Figure VI.1.5) and that the volume density of the maximum peak power transmitted by the output mirror can be written as

$$P_{max} \approx \left[h\nu c\left(1 - r_1^2\right) / 2L\right]\left[N_s \ln(N_s / N_i) + N_i - N_s\right] \qquad \text{(VI.1.32)}$$

This equation contains a lot of information; it shows us that Q-switched lasers need, in general, to be short to give the most power (since P_{max} is inversely proportional to the length of the cavity). Moreover, estimating the order of magnitude of parameters N_i and N_s makes it possible to estimate the order of magnitude of the power and of the duration of the emitted pulse. This kind of regime leads to *powers on the order of the megawatt* for pulses lasting *tens of nanoseconds*.

Inset VI.4 ***How can one block, or inhibit, laser oscillation?***

Manually controlled active blocking:

In this kind of setup, we are dealing with electro-optical polarization blocking, as described in Figure VI.1.6. The active element is a Pockels cell which rotates the linear polarization. It consists of a uniaxial crystal with optical axes 0x and 0y. It is made quarter-wave at the laser wavelength by applying a potential difference V. The polarization of the incident field $\mathbf{E}_1$ is turned by 90° after passing twice through the Pockels cell and undergoing one reflection at the back mirror of the laser. Changing the polarization effectively stops any oscillation. When the potential difference V is suddenly cut off, the Pockels cell is turned off and causes no change in the polarization of the wave. Field $\mathbf{E}_4$ is then parallel to field $\mathbf{E}_1$ and the laser pulse can build up. The experimenter decides of the instant the pulse is triggered by controlling the potential difference V. This setup is an example of an *active optical gate.*

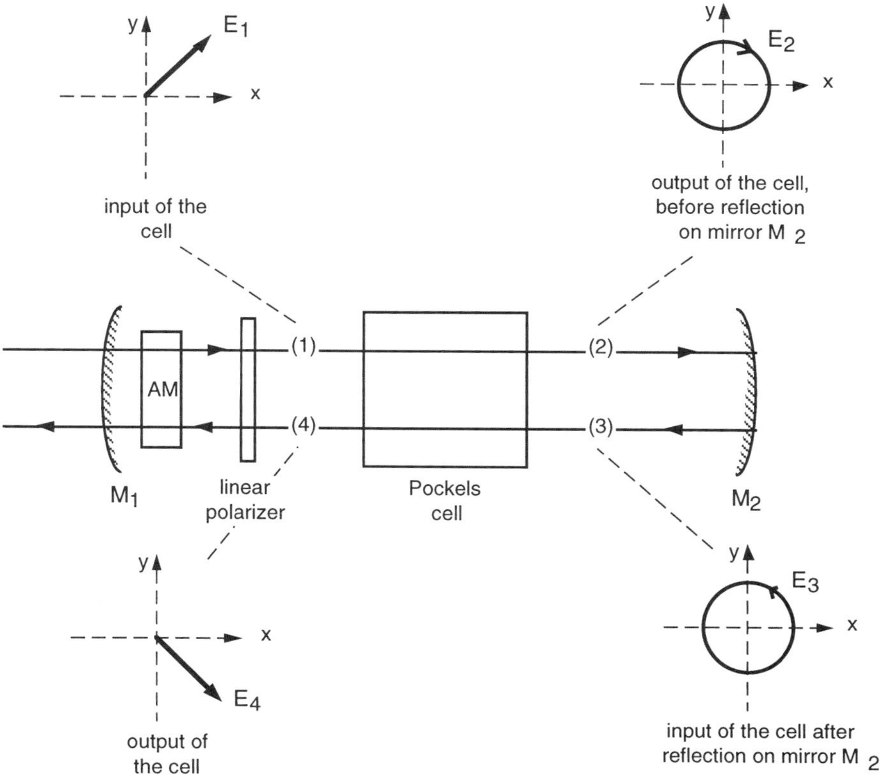

FigureVI.1.6: Active triggering of a laser. Together, the Pockels cell and mirror M_2 act like a half-wave element.

Passive blocking, without the experimenter's control:

Passive blocking is obtained by inserting an absorbing medium inside the laser cavity. The absorbing medium, of thickness d, must have an absorption coefficient which saturates as the intra-cavity light intensity increases. This easily saturated absorber, called *saturable absorber*, often consists of a liquid dye solution in an organic solvent. Figure VI.1.7 shows how the transmission $\mathcal{T}$ of such a cell varies as a function of the light intensity I.

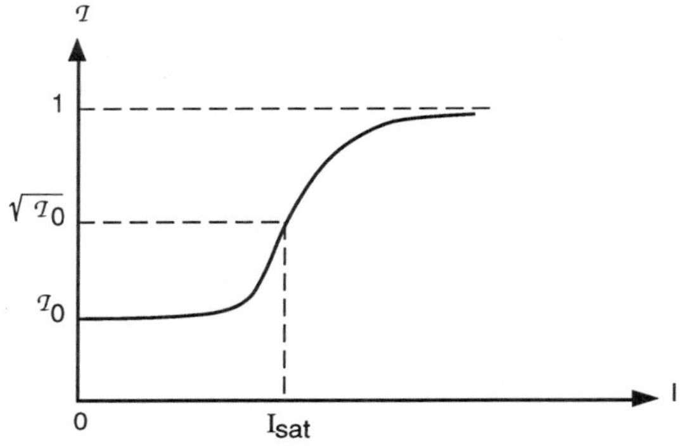

Figure VI.1.7: Transmission of a saturable absorber as a function of intensity. The scales are arbitrary.

I_{sat} is called the *saturation intensity* and is characteristic of the sudden change of the optical transmission. I_{sat} is defined with respect to the absorption coefficient $\alpha = -(1/d)\ \log\ \mathcal{T}$ by

$$\alpha = \alpha_0 / [1 + (I / I_{sat})]$$

α_0 is the absorption coefficient at very low intensity. When $I = I_{sat}$, we have $\alpha = \alpha_0/2$ and $\mathcal{T} = \sqrt{\mathcal{T}_0}$. Calculations show that for most commonly used absorbers, I_{sat} has a value of a few megawatts per square centimeter. When, by means of intense optical pumping of the amplifying medium, the stimulated emission inside the cavity reaches the value of I_{sat}, the *passive optical gate* suddenly opens and the laser emits a pulse. The experimenter cannot control the instant at which the pulse appears.

The Q-switched regime can also be obtained with continuous wave lasers, for instance by inserting a periodically activated Pockels cell, but usually losses are modulated by making use of the acousto-optical effect, described in more detail in the Inset VI.5.

It is very interesting to compare the maximum power P_{max} obtained in the Q-switched regime (and given by Eq. VI.1.32) to the power obtained by the same laser operating in the continuous wave regime. The continuous-wave power, P_c, is obtained by multiplying Eq. VI.1.15 by $h\nu/\tau_{ph}$, which gives

$$P_c = W_p N_t h\nu \qquad (VI.1.33)$$

Now it is easy to show that $W_p N_t = N_\infty/\tau_F$, where N_∞ represents the maximum population obtained *in the absence* of laser effect. Indeed, at very long times, we have $\dot{N}_\infty = 0$, and since for continuous-wave lasers $W_p << 1/\tau_F$, Eq. VI.1.10 leads

Inset VI.5 ***What is an acousto-optical modulator ?***

Figure VI.1.8 describes acousto-optical modulation under Bragg diffraction conditions, which is most frequently seen in laser technology. An acoustical wave (of the megahertz domain) of wavelength Δ is applied to a solid medium having an optical interaction length d. The laser wavelength is λ. Length d is chosen so as to have $\lambda d >> \Delta^2$. In this case, a large part of the incident intensity I_0 is diffracted by the phase grating created inside the medium by the acoustic wave. The diffracted wave makes an angle 2θ with the incident wave, where $\sin\theta = \lambda/2\Delta$. Angle θ is always very small. The diffracted intensity is given by

$$I_1 / I_0 = \sin^2\left(\Delta\phi / 2\right)$$

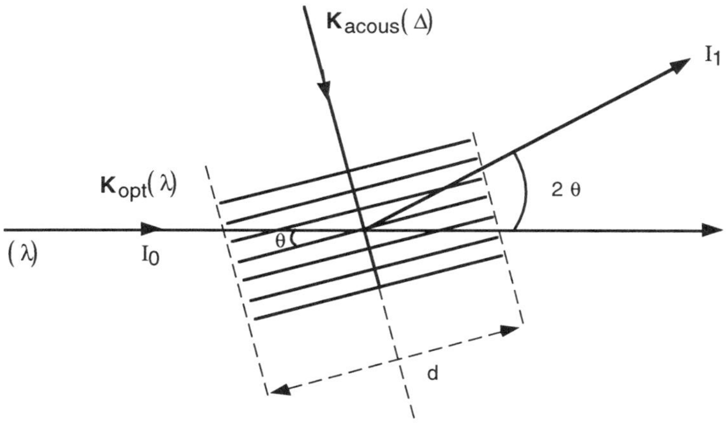

Figure VI.1.8: Accoustic Bragg diffraction.
Angle θ is greatly exaggerated on this figure.

The phase difference $\Delta\phi$ is proportional to the index modulation induced by the applied acoustic wave. As a result, the loss factor of the cavity is also modulated. This can be used to obtain various operating regimes of the laser. In case of a very small phase difference $\Delta\phi$, I_1 / I_0 is proportional to $\Delta\phi^2$, so that an acoustic wave of frequency ν_a induces a modulation of the optical intensity at frequency $2\nu_a$ (see Problem VI.6).

directly to

$$N_\infty \approx W_p N_t \tau_F \qquad \text{(VI.1.34)}$$

yielding

$$P_{max} / P_c \approx (\tau_F / \tau_{ph}) (N_s / N_\infty) [\ln (N_s / N_i) + (N_i - N_s) / N_\infty] \qquad \text{(VI.1.35)}$$

Usually, $\tau_F >> \tau_{ph}$ (for a YAG laser for instance, $\tau_F \approx 0.1$ to 1 ms, and yet $\tau_{ph} = 2L/c(1 - r)$ lies within the 0.1-to 1-μs domain). The ratio τ_F/τ_{ph} has a value of about 1000, while the expression in square brackets is not very much smaller than one. The gain in peak power is spectacular. Q-switched continuous-wave lasers giving intensities on the order of the kilowatt per square centimeter, are widely used in applied science.

(c) Phase-locked longitudinal modes (called the mode-locked regime)

In the free multimode operating regime, described in Section 2.2.1, the modes have random phases $\varphi_q(t)$ (see Eq. VI.1.17) so that the resulting mean intensity is approximately equal to the sum of the intensities of each mode:

$$\bar{I}_{00}(t) = | f'(t)^2| \sum_{q=0}^{\infty} \left| f\left(\nu_0 \pm qc/2L\right)\right|^2 = \sum_{q=0}^{\infty} I_{00q} \qquad \text{(VI.1.36)}$$

The sum of the differential terms between the various electric fields E_{00q} has a practically vanishing mean value, and the laser intensity is almost constant in time (except for the fluctuations already described). However, a more or less random coupling occurs between the phases. This coupling is brought about by intermode competition within the amplifying medium, consisting of more or less incoherent microparticles, as described in Chapter IV.

What would happen if one could organize these mode competitions in such a way that all modes have the same phase? The interference terms would no longer vanish. The Fourier transform gives us a powerful tool to help us imagine the time-dependent shape of the wave thus obtained. Let us start our calculations with Eq. VI.1.26, which describes multimode emission in frequency space.

$$\tilde{E}_{00}(\nu) = \sum_{q=0}^{\infty} f(\nu)\, \delta\left[\nu - \left(\nu_0 \pm qc/2L\right)\right] \exp\left[i\varphi_q * \tilde{f}'(\nu)\right]$$

For simplicity's sake, let $\varphi_q = 0$ for all q. This is a way to force the phases of all longitudinal modes to be equal. We obtain

$$\begin{aligned} E_{00}(t) &= \mathcal{T.F}\left\{ \sum_{q=0}^{\infty} f(\nu)\, \delta\left[\nu - \left(\nu_0 \pm qc/2L\right)\right] \right\} \times f'(t) \\ &= \tilde{f}(t) * \mathcal{T.F}\left\{ \sum_{q=0}^{\infty} \delta\left[\nu - \left(\nu_0 \pm qc/2L\right)\right] \right\} \times f'(t) \end{aligned} \qquad \text{(VI.1.37)}$$

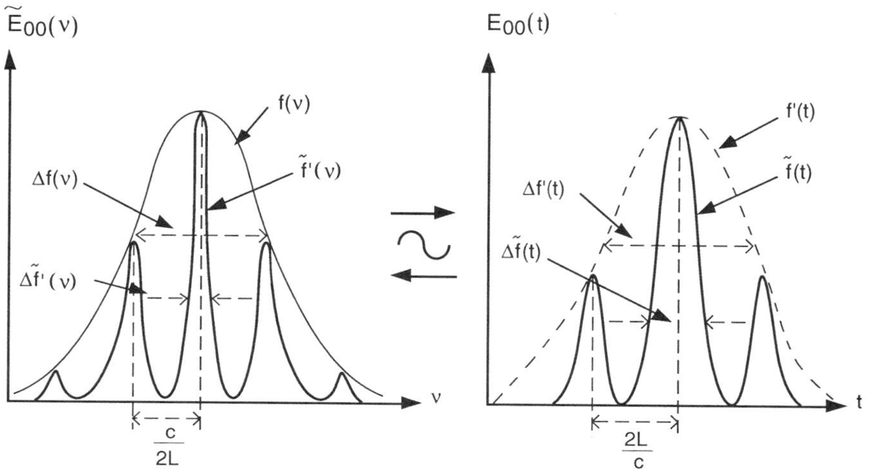

Figure VI.1.9: Frequency and times aspects of pulsed emission with locked phases (mode-locked regime). The continuous wave is obtained by enlarging width $\Delta f'(t)$.

But we know that the Fourier transform of a Dirac "comb" in frequency space is a Dirac "comb" in time space (see Inset VI.1). Therefore

$$E_{00}(t) = \left[\tilde{f}(t) * \sum_{q=0}^{\infty} \delta\left(t \pm q2L/c\right) \right] \times f'(t) \qquad \text{(VI.1.38)}$$

These results are illustrated in Figure VI.1.9. Light emission occurs in trains of pulses, and these pulses *are separated by a time interval equal to the transit time* of the cavity, $\tau_t = 2L/c$. The time width of the emission curve $\tilde{f}(t)$ is *inversely proportional to the frequency width of curve* $f(\nu)$.

By this method, very short pulses, on the order of a few femtoseconds, can be obtained. Naturally, when $f'(t) = 1$, the above equation describes the phase-locked regime of continuous-wave lasers shown in Figure VI.1.10. There is a mathematical demonstration showing that, if there is a point in the cavity where all the modes of the cavity interfere constructively, then they must interfere destructively at all other points of the cavity. This means that a phase-locked laser wave with a time extension of τ has a limited linear extension of length $c\tau$ (which gives 3 microns for a wave of 10 fs inside the cavity). This finite portion of sine curve travels back and forth through the cavity, letting out a short pulse of light to the outside every 2L/c seconds; each time it is partially reflected by the output mirror.

It is clear that the shortest pulses are obtained with amplifiers having a large spectral width, such as dye amplifiers, or the recent Ti/sapphire amplifiers. There also exists a method called the method of *time-compressed* pulses which can lead to pulses lasting just a few femtoseconds and carrying an enormous peak power, on the order of the gigawatt and more.

VI.1.3 Time coherence

Time coherence is defined by the *time autocorrelation* function defined by

$$\gamma(\Delta t, 0) = \overline{E(t, x)\, E^*(t + \Delta t, x)} \;/\; \sqrt{\overline{|E(t, x)|^2 \, |E(t + \Delta t, x)|^2}} \qquad \text{(VI.1.39)}$$

Δt is the time interval separating two observations of the state of the field at point x. Of course, for an ideal continuous-wave laser, we must have $|\gamma(\Delta t, 0)| = 1$ for all x and for all Δt. The reader will have noticed the similarity between Eqs. V.1.4 and VI.1.39. This last equation defines the *characteristic coherence time* Δt_c.

Inset VI.6 ***How can one obtain phase-locking in the longitudinal modes of a laser?***

If one could modulate the amplitude of the qth laser mode at frequency c/2L, the resulting modulated mode would take the form

$$E_{00q}(t) = E_{00q}\,[1 + M \cos(\pi ct/L)]\, \cos 2\pi \nu_q t$$

which can be written equivalently, as the reader may check

$$E_{00q}(t) = E_{00q} \cos 2\pi\nu_q t + (E_{00q} M/2) \cos 2\pi(\nu_q + c/2L)\, t + (E_{00q} M/2) \cos 2\pi(\nu_q - c/2L)\, t$$

To modulate the qth mode in amplitude is equivalent to creating two lateral phase-correlated frequencies which oscillate at the same frequencies as the (q + 1)th and as the (q – 1)th mode of the laser. The competition between the lateral frequencies and the modes within the amplifying medium creates a forced oscillation, forcing the two adjacent modes to oscillate in phase with the lateral frequencies by which they are controlled. The forcing spreads from mode to adjacent mode so that finally there is a global locking of the phases of all modes.

How can one modulate the amplitude of the longitudinal modes of a laser at frequency c/2L ?

Figure VI.1.11 illustrates a few possibilities.

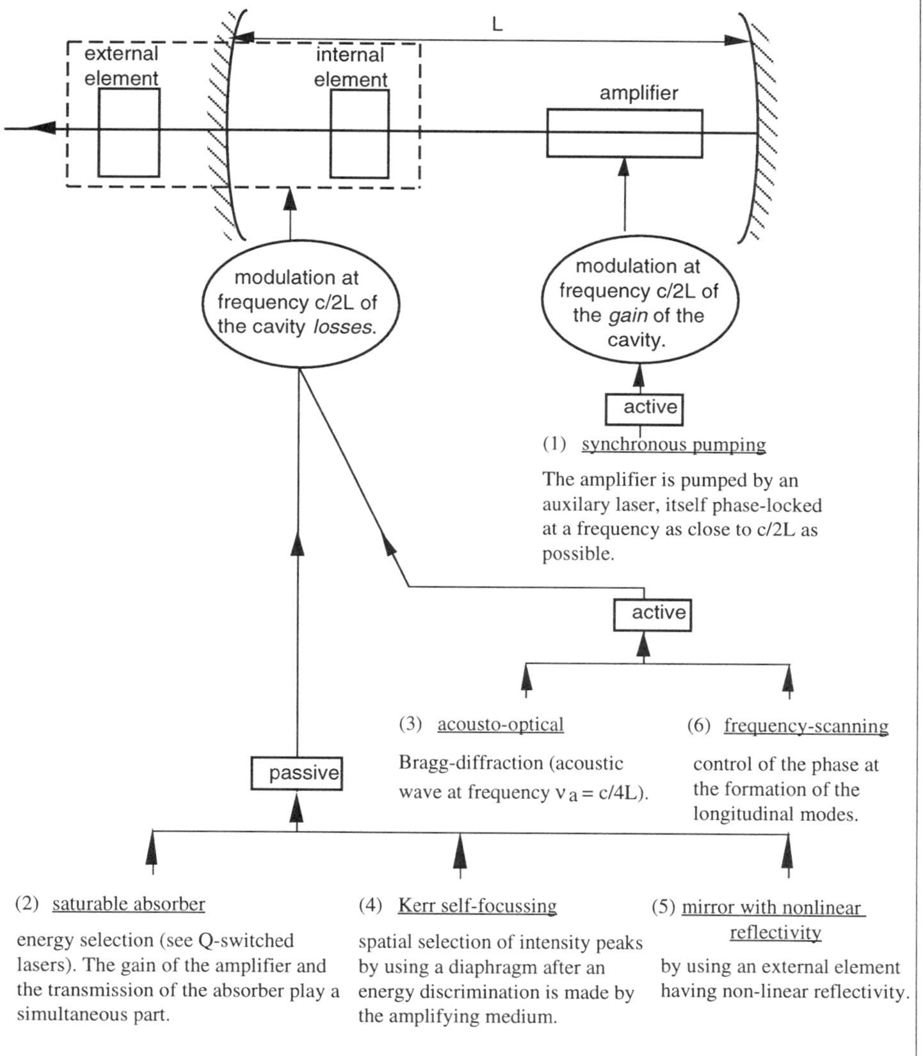

Figure VI.1.11: A few possibilities to obtain phase-locking of the longitudinal modes of the laser.

Procedures (1), (2), and (3) are often used, while procedures (4), (5), and (6) are very recent and apply only to the Ti/sapphire laser (see Chapter IV).

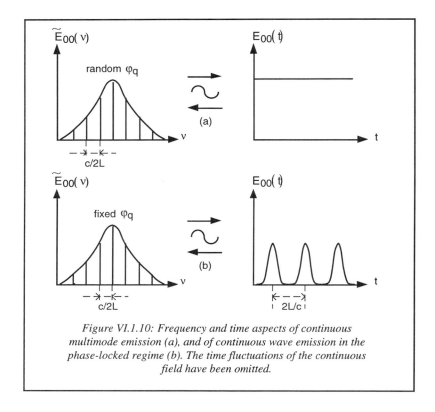

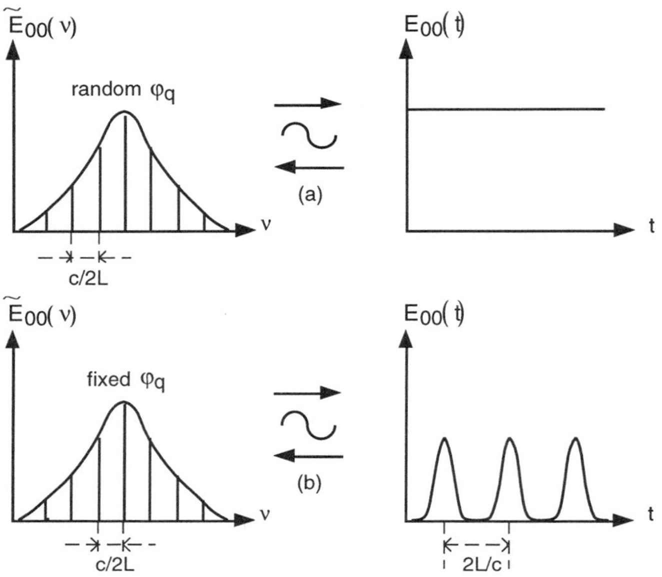

Figure VI.1.10: Frequency and time aspects of continuous multimode emission (a), and of continuous wave emission in the phase-locked regime (b). The time fluctuations of the continuous field have been omitted.

VI.2 Consequences of the time coherence of a laser

VI.2.1 Frequency concentration of the energy

The time coherence is directly related to the spectral distribution of the laser. The theorem of Wiener-Khintchine gives the relation between *spectral density* (mean dissipated power in a unit frequency band) and *the autocorrelation function* $\gamma(\Delta t, 0)$; i.e., they are each other's Fourier transform. This is why a perfectly time-coherent wave (having an infinite characteristic time Δt_c) is also a perfectly *monochromatic* wave.

Quite usually, lasers deliver an energy of a few tens of watts, concentrated in a narrow frequency domain with a spectral width measuring less than a megahertz. In this case, Δt_c is on the order of the microsecond. For frequency-stabilized single-mode lasers, this characteristic time can reach values as large as a few hundreds of milliseconds. Better still, in sophisticated optical metrological measuring setups, widths on the order of a hertz ($\Delta t_c \approx 1$ s) can be obtained during more or less limited time intervals. If one remembers that traditional low-pressure sources have characteristic times Δt_c on the order of a few tens of nanoseconds, the improvement is spectacular.

VI.2.2 Time concentration of the energy

The time coherence of the laser makes it possible to obtain high energy concentrations in time. We just showed that, in order to have very short pulses, one needs to have a next-to-perfect coupling between longitudinal modes. This can only be brought about if the different phases do not vary randomly with respect to each other, i.e. only if the modes possess a certain amount of time coherence.

VI.3 Comparison of the spectral and time properties of classical and laser light sources

This comparison is summarized in Table VI.3.1, which completes Table V.2.1. To conclude, we can say that the coherence property definitely classifies lasers as extremely interesting and original light sources, as compared to classical sources, their performances being incomparably better from almost all points of view. However, one should be aware that in our portrayal of the laser, the deliberately clear separation between space and time coherence is in fact artificial. In reality, the properties of the laser derive from the joint space-time coherence, expressed by the *general coherence* function $\gamma(\Delta t, \Delta x)$. The reader can easily find this function by combining Eqs. V.1.4 and VI.1.39. The four essential properties of the laser—namely, surface, angular, frequency, and time energy concentration—essentially depend on the general coherence function. The time-space dichotomy was introduced only for clarity's sake. Yet, this didactic method does yield results which are excellent first approximations.

Last, we need to state the fundamental limits to laser coherence, mostly due to the technical construction of the laser and to spontaneous emission:

— The quality of the time coherence essentially depends on the stability of the oscillation frequencies ν_q of the longitudinal modes. Fluctuations of the length and of the refractive index of the cavity, as well as thermal agitation within its components, lead to fluctuations of the frequencies and are the cause of a loss of time coherence.

TABLE VI.3.1 — **Comparison between a classical and a laser light source.**

Classical source	Laser source
Frequency concentration	
not less than a few hundreds of megahertz	up to a few tens of hertz
Time concentration	
of the order of the nanosecond	a few femtoseconds

— A decrease of the spatial coherence with respect to theory is due to the fact that since the spatial dimensions of the elements making up the cavity are necessarily finite, a pure TEM_{00q} mode, extending indefinitely, can never exist within a laser. The actual mode existing in the laser is subject to the fluctuations of the refractive index mentioned above. Thus, spatial coherence also has a limit of technological origin.

— Last, even perfect technology cannot do away with the destructive part played by spontaneous emission which is the cause of non-zero line-widths (a few hundreds of hertz), and thus of imperfect time coherence. Moreover, it pollutes the transverse distribution of the TEM_{00q} mode, thereby lowering spatial coherence.

VI.4 Problems and outlined solutions

VI.4.1 Problem VI.1: Fabry-Pérot analysis of laser emission

We wish to analyze a laser emission spectrum with a Fabry-Pérot interferometer whose width e can be adjusted by means of a piezoelectric element.

1. The laser beam can be described by a monochromatic plane wave of intensity I_0 and of frequency ν_0. The beam enters the interferometer at normal incidence (i = 0). The interferometer consists of two identical plane mirrors whose field reflection coefficient r_0 is real. These mirrors are assumed to have no losses $\left(r_0^2 + \mathcal{T}_0^2 = 1\right)$ and their external surfaces have received an anti-reflection coating so that their reflection can be neglected. The space between the two mirrors, of width e, is filled with air (refractive index $n_0 \approx 1$). One of the two mirrors is mounted on a holder which is connected to a piezoelectric element (PZ) used to adjust the position of the mirror. The mirror is adjusted in a direction parallel to its axis by changing the voltage V applied to PZ. Width e is assumed to vary linearly with V in the following manner: $e = e_0 + aV$.

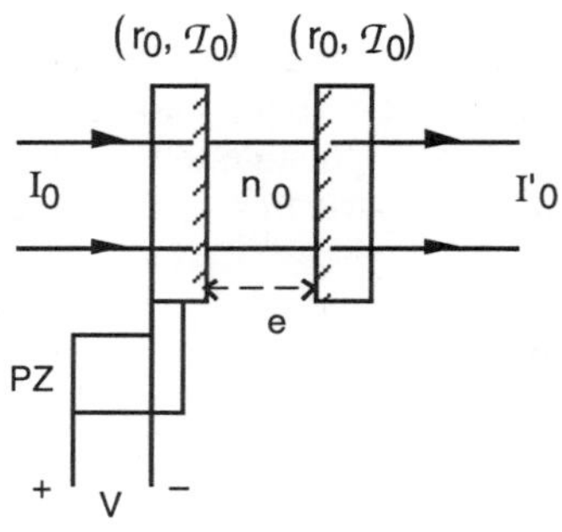

(a) Find the values of the voltages V_m for which the incident wave I_0 is transmitted by the Fabry-Pérot without attenuation. Let $\Delta V = V_{m+1} - V_m$ be the voltage difference between two successive transmission maxima. Calculate how the intensity I'_0 changes as V increases from $(V_{m-1} + \Delta V/2)$ to $(V_{m+1} + \Delta V/2)$. Give the value of the contrast C of the transmission peaks, C being defined as the ratio between the minimum and the maximum value of I'_0. Calculate the finesse $\mathcal{F}$ of the peaks. ($\mathcal{F}$ is defined as the ratio between the interval separating two peaks and the full-width at half-maximum of a peak.) Numerical calculation: $e_0 = 1$ cm, $r_0 = 0.99$, $n_0 \cong 1$, $a = 10^{-9}$ m V^{-1}, $\lambda_0 = 0.5$ µm. Find ΔV, c, and $\mathcal{F}$.

(b) Intensity I'_0 is focused on a photodiode, and the potential difference thus obtained is sent into the y-entrance (vertical entrance) of an oscilloscope. The voltage V applied to the piezoelectric element PZ is the sawtooth scanning voltage provided by the horizontal amplifier of this same oscilloscope. This voltage is adjusted so that it increases linearly from $(V_{m-1} + \Delta V/2)$ to $(V_{m+1} + \Delta V/2)$, then falls back almost instantaneously to $(V_{m-1} + \Delta V/2)$ and starts increasing again. The period of the sequence is the same as the scanning period of the oscilloscope. What does one see on the oscilloscope screen? Next, keeping the same period, the voltage is forced to very great amplitude swings, beyond the original limits of $(V_{m-1} + \Delta V/2)$ and $(V_{m+1} + \Delta V/2)$. What is now seen on the screen?

2. The laser beam now consists of (a) a superposition of two incoherent plane waves having the same intensity I_0 and (b) frequencies (ν_0) and $(\nu_0 + \delta\nu)$, i.e. two longitudinal laser modes with a frequency difference of $\delta\nu << \nu_0$. Find the total transmitted intensity I'_0 as a function of ν (for the numerical calculation, use the same values as those given in question 1). Show that when $\delta\nu \approx 150$ MHz, the setup separates the two modes. What does one see on the screen of the oscilloscope? Which parameters of the Fabry-Pérot can be adjusted in order to see the two modes more clearly? Discuss the practical side of these adjustments.

3. The laser is an Ar^+ laser and its cavity consists of a plane mirror and a spherical mirror with radius of curvature R, placed at a distance L from the plane mirror.

(a) What relation must L and R satisfy for the cavity to be stable? What is the distance $\delta\nu$ between adjacent longitudinal laser modes? Numerical calculation: R = 3 m, L = 1 m, $\lambda_0 = 0.5$ µm. Check whether this is a stable cavity. Calculate $\delta\nu$.

(b) The graph of the gain g (ν) of the amplifying medium (which is assumed not to perturb the modes of the empty cavity) has a total width of $\Delta\nu$. Find N, the number of modes which can self-oscillate simultaneously. Numerical calculation: $\Delta\nu = 8$ GHz.

(c) The multimode laser beam is analyzed by the interferometer of question 1. Describe what one observes on the oscilloscope if the parameters of the interferometer are the same as those given in 1. What happens if one uses other widths e_0, either much larger or much smaller than the width e given above?

(d) Suggest a method leading to a single mode operation of the Ar^+ laser. How can one check that the laser indeed operated in single mode?

4. Actually, it is very difficult to analyze a multimode laser beam (which is in fact a Gaussian wave) by means of an interferometer consisting of two plane mirrors. It is therefore common to use, instead of the preceding interferometer, a resonant cavity consisting of two spherical mirrors having the same radius of curvature R_0 and placed at a distance R_0 from other (confocal setup). This resonator is also easier to line up with the beam.

(a) The beam of the single-mode operating Ar^+ laser enters the confocal Fabry-Pérot without any special precautions. There, it excites all the transverse modes of the cavity. Describe the behavior of the voltage difference $\Delta V' = V'_{m+1} - V'_m$ separating adjacent transmission maxima when the value of e (the distance between the spherical mirrors) changes slightly around R_0 (by adjusting the piezoelectric element).

(b) If we wish to excite exclusively the TEM_{00q} mode of the cavity, the incident beam must have the same transverse distribution as this resonant mode. Indicate qualitatively how this condition may be obtained. Is the voltage difference between adjacent maxima, ΔV", the same as in a)?

1. (a) The interferometer is resonant for $2n_0 e_m = 2n_0 (e_0 + aV_m) = m\lambda_0$, yielding $V_m = m\lambda_0 / 2n_0 a - e_0 / a$ and $\Delta V = \lambda_0 / 2n_0 a$. The transmitted intensity takes the form

$$I'_0 = I_0 / \left[1 + 4 r_0^2 \sin^2(2 n_0 e \pi / \lambda_0) / \left(1 - r_0^2\right)^2\right]$$

(see Chapter VI) with $e = e_0 + aV$. The contrast is

$$C = I_{max} / I_{min} = \left[\left(1 + r_0^2\right) / \left(1 - r_0^2\right) \right]^2$$

The finesse is given by $\mathcal{F} = \Delta e / 2 e_{1/2} = \pi r_0 / \left(1 - r_0^2\right)$ (where $e_{1/2}$ represents the half-width at half-maximum). Numerical calculations yield $C \approx 10^4$, $\mathcal{F} \approx 157$, and $\Delta V \approx 250$ V.

(b) At scanning frequency, the screen of the oscilloscope shows the graph representing $I'_0 = f(V)$ for $(V_{m-1} + \Delta V/2) \leq V \leq (V_{m+1} + \Delta V/2)$. One can see two peaks, separated by $\Delta V = \lambda_0 / 2n_0 a$, each having a width $\delta V = \lambda_0 \left(1 - r_0^2\right) / 2 \pi n_0 r_0 a$. Changing the scanning amplitude changes the number of peaks observed on the screen.

2. When two modes oscillate simultaneously, the transmitted intensity is the sum of the contribution described above with λ_0, and of a second contribution expressed by the same formula, except that λ_0 is replaced by $(\lambda_0 + \delta\lambda)$. The oscilloscope shows four peaks, in two groups of two. Writing the finesse $\mathcal{F}$ leads to the expression of the minimum frequency separation width: $\Delta\nu / \mathcal{F} \approx c / 2n_0 e_0 \mathcal{F} \approx 100$ MHz. Therefore two modes with a frequency difference of 150 MHz can be separated by the interferometer. To improve this separation, one can try to increase the free spectral range $c / 2n_0 e$, for instance by increasing width e_0. One can also try to increase $\mathcal{F}$ by increasing the reflectivity r_0 of the mirrors.

3. (a) We have $h_1 = 1$ and $h_2 = 1 - L/R$ (see Eq. IV.2.67) and the double condition $0 < h_1 h_2 < 1$ yields $L < R$. This condition is satisfied by the numerical data. $\delta\nu = c/2d \approx 150$ MHz.

(b) $N = \Delta\nu / \delta\nu \approx 53$.

(c) The oscilloscope shows two well-separated "combs" (see Figure IV.2.15 c) of 53 peaks, having a total width of 8 GHz each. When e_0 is decreased, the peaks are no longer well separated (the oscilloscope will show the envelope curve). When e_0 is increased, the free spectral range decreases and the two combs start to overlap. In both cases, the image becomes more difficult to interpret.

Single-mode operation is obtained by inserting, and orienting, an auxiliary Fabry-Pérot resonator inside the laser cavity. The free spectral range of this resonator must be smaller than $\delta\nu$.

4. (a) In the case of a laser which does not operate in transverse single mode, there exists an equation giving the frequency of mode m, n, q [see bibliographical references of Chapter IV, Siegman (A) 1986] in a symmetrical resonator: $\nu_{mnq} = c\left[q + 1 + (m + n + 1)\arccos(1 - L/R)/\pi\right]/2L$ (the reader is invited to check that this relation leads to Eq. IV.2.74 when m = n = 0 and when $h_1 = h_2$). In the case of a confocal resonator, we have L = R and $\nu_{mnq} = c\left[q + 1 + (m + n + 1) \times (1/2 \pm k)\right]/2L$, where k is a whole number. In this case, the frequency gap between adjacent TEM_{00q} modes is halved, so that $\Delta V' = \Delta V/2$.

(b) If we wish to excite only the TEM_{00q} modes of the confocal analyzing cavity, we must focus the laser in such a way as to obtain a beam waist identical in *position and in diameter* to the beam waist of the analyzing cavity. In this case, the frequency gap between modes is $\Delta V'' = 2\Delta V' = \Delta V$.

VI.4.2 Problem VI.2: Frequency stabilization of a ring laser

1. Let us consider the laser represented on the diagram below. This laser consists of a ring cavity (with mirrors M_1, M_2, M_3), an amplifying medium AM, and of a set of optical elements F which impose single-mode operation on the laser, with a progressive wave traveling in the direction of the arrows.

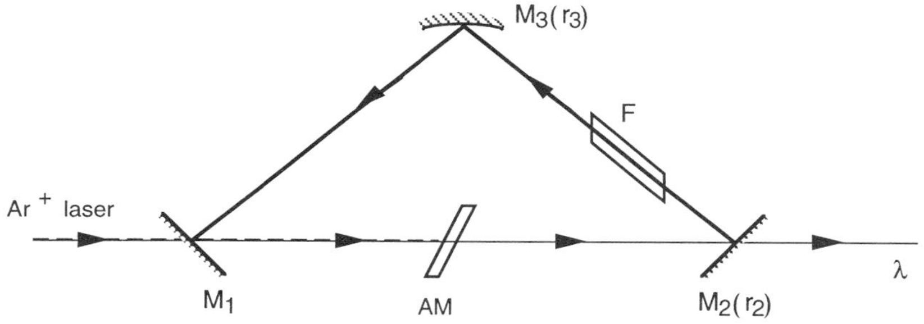

The AM is pumped by an Ar^+ laser through mirror M_1. M_1 is transparent for the pump wavelength λ_p, and its reflection coefficient is equal to unity for the generated wavelength λ.

(a) First, we wish to study the properties of the empty cavity consisting only of two plane mirrors M_1 and M_2, and of a spherical mirror M_3 with radius of curvature R_3. Let L be the total length of the ring. Find the transfer matrix of this cavity. Find the equivalent two-mirror cavity.

(b) Determine the stability domain of the cavity. Give the characteristics of its fundamental Gaussian mode: position and size ω_0 of the beam waist, resonance frequencies.

(c) We now want to pump the AM, which is placed at the beam waist, with a 1-W pump beam. For the AM, the most appropriate pump intensity on the axis of the beam would be of 100 kW cm^{-2}; what would then be the value of the beam waist? Knowing that $R_3 = 10$ cm and $\lambda = 600$ nm, calculate the optical ring-length L required to obtain this value of the beam waist. What can you say about the stability of this cavity?

2. The amplifying medium has a pumping sequence roughly described as follows: The Ar^+

laser lifts the molecules from the ground state (3) to an excited electronic level (4).

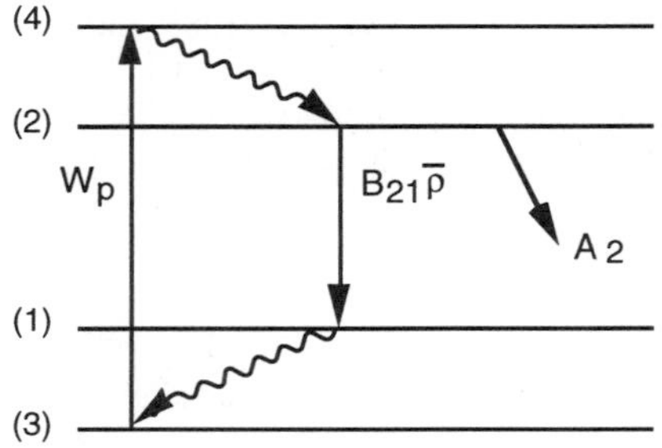

The molecule then relaxes very quickly (in a time assumed to be very fast compared to the processes under consideration) to a level (2) of the same electronic state. Once in level (2), the molecule has a probability A_2 per unit time to relax to level (1) by means of stimulated emission [with Einstein's coefficient B_{21} for transition (2) $\rightarrow$ (1)]. State (1) then relaxes to level (3) so fast that its population N_1 is not affected by the pumping sequence. At the temperature used in this experiment, N_1 is assumed to stay vanishingly small at all times. Let N_t stand for the total density of the molecules. Assume that the number of molecules participating in the pumping sequence is very small compared to N_t .

(a) Calculate N'_{2S}, the population of level (2) at steady state in the absence of laser effect $(\bar{\rho} = 0)$. Then calculate N_{2S}, the same but with laser effect taking place $(\bar{\rho} \neq 0)$. Assume a uniform density W_p for the pump wave.

(b) Let "a" be the field amplification of the AM; "a" is assumed to be a real number. The AM has a width d and has a refractive index n at ω, the frequency of the generated wave. We have $a = E(d) / E(0) = 1 + \varepsilon$ with $\varepsilon << 1$. Calculate ε. Remember that $a^2 = \left[\bar{I}(d) / \bar{I}(0)\right]$. Simplify the expression of ε for the case when $\bar{I}(0)$ is large enough so that spontaneous emission from level (2) is negligible compared to stimulated emission. From now on, this is assumed to be the case throughout the remainder of this problem.

(c) At λ, mirrors M_1 and M_3 have a reflection coefficient equal to unity. The output mirror M_2, which is a mirror without losses, has a real field reflection coefficient $r_2 = 1 - \beta$ with $\beta << 1$. Filter F has a real transmission coefficient $\mathcal{T} = 1 - \gamma$ with $\gamma << 1$. Assume the laser wave is a plane wave and write the steady state condition for the field in the cavity. From this calculate $\bar{I}(d)$, the intensity of the self-oscillating wave as it comes out of the AM. Calculate $\bar{I}'$, the output intensity of the laser.

(d) A servo-element S is used to stabilize the frequency of the selected q mode around a given frequency ν_0. S delivers a potential difference V which is proportional to the difference between ν_q and ν_0, the frequency imposed by the servo-system: $V = a(\nu_q - \nu_0)$. This voltage V is converted into a modification ΔL_1 of the optical length of the cavity by means of a piezoelectrical element PZ controlling the position of mirror M_1: $\Delta L_1 = bV$.

When there is no perturbation, $L = L_0$, and so $\nu_q = \nu_0$. Let ΔL_0 be a statistical perturbation of L; calculate first the frequency change $\Delta\nu_q$ of the laser without servo-system, then its frequency change $\Delta\nu'_q$ in presence of the servo-system. Numerical calculation: $L_0 = 1$ m, $\lambda_0 = 0.6\ \mu$m,

$\Delta L_0 = 100$ nm. Calculate $\Delta\nu_q$. What must the gain of the servo-system be in order to have $\Delta\nu'_q < 50$ kHz?

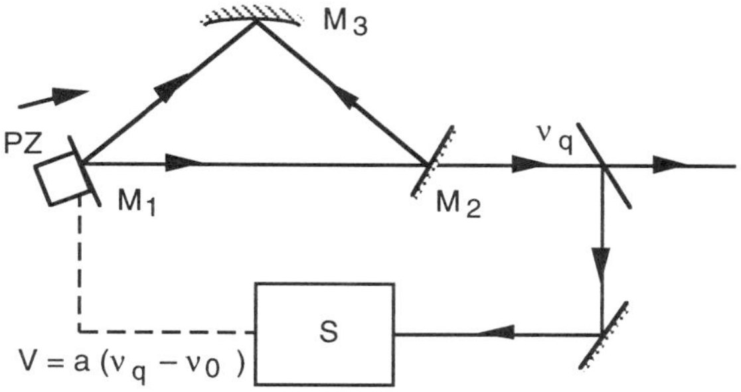

1.(a) As in the case of Problem V.1, a first approximation to the transfer matrix (neglecting the effects of elements F and of the AM) is $\begin{pmatrix} 1 & L \\ -2/R_3 & 1-2L/R_3 \end{pmatrix}$. The equivalent two-mirror cavity is a symmetrical cavity consisting of two spherical mirrors with radius of curvature R_3 and separated by a distance L.

(b) The stability condition is expressed by $L < 2R_3$. The beam waist has radius $\omega_0^2 = \lambda\sqrt{L(2R_3 - L)}\,/\,2\pi$ and is found at an optical distance L/2 from mirror M_3, in the middle of segment M_1M_2. The resonance frequencies are $\nu_q \approx c\,(q+1)\,/\,L$.

(c) The intensity is expressed by $I_p = I_0 \exp\left[-2r^2/\omega_0^2\right]$. The power corresponding to this intensity is $P_p = \int_0^\infty I_p\, 2\pi r\, dr = \pi\,\omega_0^2\, I_0\,/\,2$. The size of the beam waist is $\omega_0^2 = 2\,P_p\,/\,\pi\, I_0$. The numerical calculation yields $\omega_0 \approx 25$ μm, which in turn yields for length L: $L \approx 19.98$ cm. This cavity is near the edge of its stability domain and will be difficult to use.

2. (a) In the absence of stimulated emission, the rate equation writes: $\dot{N}'_2 \approx W_p N_3 - A_2 N'_2$. Therefore $N'_{2S} \approx N_t W_p / A_2$. In the presence of stimulated emission, we find $N_{2S} \approx N_t W_p / \left(A_2 + B_{21}\bar{\rho}\right)$.

(b) Looking back at Problems IV.2 and IV.3, we see that we can write $a^2 \approx 1 + 2\varepsilon$ with $\varepsilon = N_t\, d\, \hbar\omega\, B_{21} B_{34} \bar{I}_p\, n\alpha \,/\, 2c\left[A_{21} + B_{21}\alpha\,\bar{I}_{(0)}\right]$ (where we set $W_p = B_{34}\,\alpha\,\bar{I}_p$ and $\bar{\rho} = \alpha\, I(0)$ with $\alpha = n/c$). In the case of $A_{21} << B_{21}\,\alpha\,\bar{I}_{(0)}$ calculations yield $\varepsilon = N_t\, d\,\hbar\omega B_{34}\,\bar{I}_p\, n \,/\, 2c\,\bar{I}_{(0)}$.

(c) Let us write $r_2\mathcal{T}a = 1$; this is equivalent to $(1-\beta)(1-\gamma)(1+\varepsilon) = 1$, yielding $\varepsilon = \beta + \gamma$. Therefore, $\bar{I}_{(d)} = N_t\,\hbar\omega\, nd\, B_{34} I_p / 2c\left(\beta+\gamma\right)$ and $\bar{I}' = 2\beta\,\bar{I}_{(d)} = N_t\,\hbar\omega\, nd\, B_{34} I_p\,\beta \,/\, c\left(\beta+\gamma\right)$.

(d) We have $\nu_q \approx c\,(q+1)/L$ so $d\nu_q = -\nu q\, dL/L$. Without servo-loop, this gives rise to $\Delta\nu_q = -\nu_0\,\Delta L_0 / L_0$. And when the servo-loop is installed we obtain

$$\Delta\nu'_q = -\nu_0\,\Delta L / L_0 \text{ with } \Delta L = \Delta L_0 + \Delta L_1 = \Delta L_0 + a\,b\,\Delta\nu'_q$$

or again $\Delta\nu'_q = \Delta\nu_q / (1 + ab\nu_0 / L_0)$. The numerical calculations yield $\Delta\nu_q \approx -50$ MHz. At $\Delta\nu'_q = 50$ kHz, the gain of the servo-loop is found to be approximately $ab \approx 2 \times 10^{-12}$ mHz^{-1}.

VI.4.3 Problem VI.3: Study of a relaxed laser

The laser shown on the diagram below has an amplifying medium consisting of a liquid dye solution in a solvent. It is pumped by a laser delivering very powerful pulses. The amplification which takes place in the dye-containing cell gives rise to a laser effect in the cavity pictured in the diagram. The beam of this dye laser (in the fundamental TEM_{00} mode) can be assimilated to a Gaussian beam. Its beam waist of radius ω_0 is located inside the dye cell.

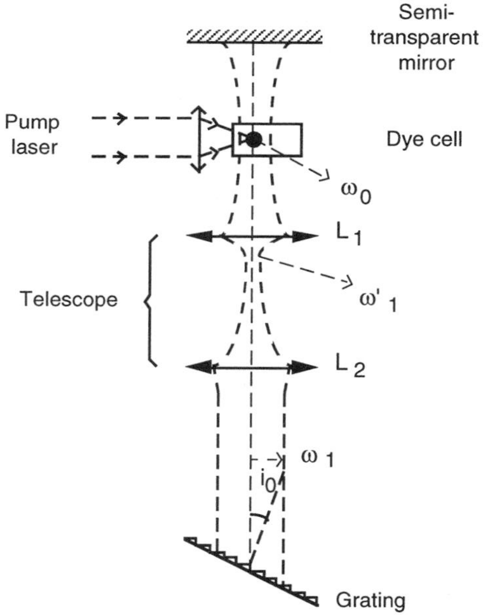

1. Let us first try to understand why a telescope has been placed between the dye cell and the diffraction grating.

(a) Knowing that the beam waist, which is inside the dye cell, measures 80 μm and that the laser oscillates at wavelength $\lambda_1 = 0.6$ μm, calculate the divergence θ_0 of the Gaussian beam between the cell and the telescope. What does this suggest?

(b) As the beam comes out of the dye cell, it travels through a set of two lenses L_1 and L_2, of focal lengths f_1 and f_2, placed at an afocal distance (i.e. such that the intermediary beam waist ω'_1 lies at the object focus of lens L_2). The set of lenses acts as an inverted telescope. Explain the part played by this telescope and its importance with respect to the laser cavity. Calculate the radius of beam waist ω_1 and the divergence θ_1 of the Gaussian beam as it emerges from lens L_2, knowing that the distance between beam waist ω_0 and lens L_1 is d_0. Numerical calculation: $d_0 = 80$ mm, $f_1 = 10$ mm, $f_2 = 180$ mm.

2. Let us now study the part played by the grating. First, we assume that, with respect to the grating, the incident wave behaves like a plane wave whose wave vector makes an angle i_0 with the normal vector of the grating (i.e. the Gaussian nature of the incident wave is not taken into account). The grating has the profile shown in the diagram accompanying (an échelette grating) and is used in reflection. The reflecting face of each groove of the grating makes an angle i_0 with the plane of the grating, so that the incident plane wave is perpendicular to each little facet of the grating. Let "d" be the distance between two successive grooves of the grating.

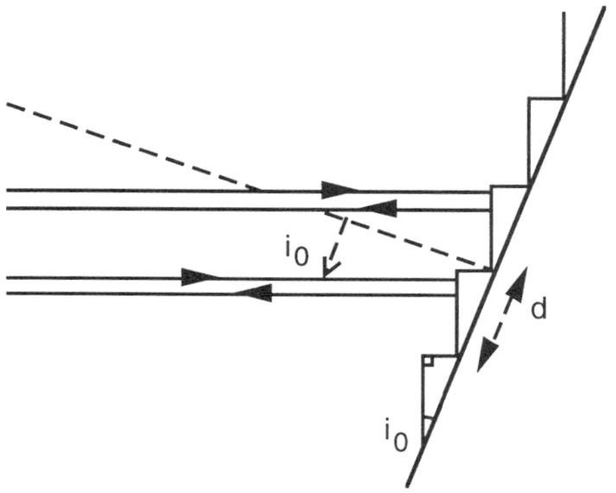

(a) Find the phase difference between the beams diffracted by two adjacent grooves of the grating in the direction directly opposite the incident direction. Find the wavelengths λ_m corresponding to a maximum of constructive interference in this direction. Write them as a function of d and of i_0.

Practical application: We want the grating to reflect wavelength $\lambda_1 = 0.6$ µm in the first order of the grating and in the direction of the incident wave, with i_0 equal to 30°. Calculate the number of grooves per millimeter the grating must have to satisfy this condition. What are the other wavelengths reflected in this direction? Comment on these results.

(b) Show that, for wavelength λ_1, the diffraction function of each groove presents a principal maximum in the direction we are interested in. Show that the other directions of the diffracted beam, which are close to i_0 and which present constructive interference at λ_1, correspond to a zero value of this diffraction function. Comment on this result.

(c) Actually, wavelength λ_1 will not be the only wavelength to come back through the amplifying medium after reflection on the grating. Assume that a band of width $\Delta\lambda$ of wavelengths around λ_1 are reflected by the grating in a direction whose angle with the cavity axis is smaller than θ_1, the divergence of the beam, and re-enter the amplifying medium where they give rise to laser oscillation. Knowing that the pump pulses are so short that no cavity mode has time to build up (this allows you to forget about the cavity-resonance conditions), find the bandwidth of this laser as a function of λ_1, i_0, and θ_1.

(d) In order to change the central operating wavelength λ_1 of the laser, the incident angle i_0 is changed by tilting the grating. Find the limits of the domain which i_0 must cover in order to sweep the whole visible domain (0.4 µm < λ_1 < 0.7 µm). What can you say about this tilting of the grating?

3. To narrow the linewidth of this laser, a Fabry-Pérot is inserted inside the laser cavity. This Fabry-Pérot has width e, refractive index n, and finesse $\mathcal{F}$, and is positioned so that its normal makes a small angle α with the cavity axis. The wave traveling through the Fabry-Pérot is assumed to be a plane wave. Find the difference in wavelength between two successive transmission maxima of the Fabry-Pérot. What is the bandwidth of these transmission maxima? With these values, calculate the linewidth of the laser for the wavelength reflected by the grating and imposed by this new filter. How can one ensure that the Fabry-Pérot continue to play its part

as a filter normally when the grating is tilted to let wavelength λ_1 sweep over a range of wavelengths? Numerical calculation: $\alpha \cong 10$ mrd, n = 1.5, e = 6 mm, $\mathcal{F} = 20$.

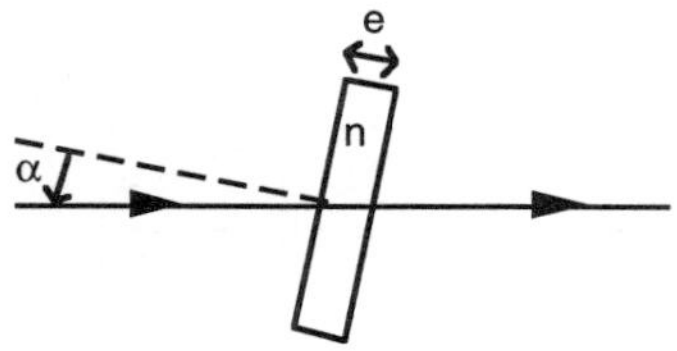

1. (a) The divergence is $\theta_0 = \lambda_1/\pi\omega_0$. The numerical calculations yield $\theta_0 \approx 2.4$ mrd. This is a rather large divergence which will make it difficult to use the grating—which requires waves as close to plane waves as possible—at optimal conditions. To reduce this divergence, it is best to increase the size of the beam waist, for instance by using the inverted telescope.

(b) Applying Eq. V.2.7, we arrive at $\theta_1 = \lambda_1 / \pi\omega_1$ with $\omega_1^2 = \omega_0^2\left[1 + \lambda_1^2 (d_0 - f_1)^2 / \pi^2 \omega_0^4\right] \times (f_2 / f_1)^2$. The part played by the inverted telescope is explained by the term f_2 / f_1, the ratio of the focal distances of the lenses. The numerical calculations yield $\omega_1 \approx 3.3$ mm and $\theta_1 \approx 0.05$ mrd.

2. (a) The phase difference is $\Delta\varphi = 2d \sin i_0$. Wavelengths λ_m are such that $\Delta\varphi = m\lambda_m$ (with m a whole number). $\lambda_m = 2d \sin i_0 / m$. At first order, m = 1 and $\lambda_1 = 2d \sin i_0$. We find that $d \approx 0.6$ μm, yielding 1666 grooves per millimeter. The other reflected wavelengths are those corresponding to higher orders of the grating, with m = 2, 3 .... For m = 2, we calculate $\lambda_2 = \lambda_1 / 2 = 0.3$ μm. This wavelength belongs to the ultraviolet domain, and therefore it does not interact with the amplifying medium.

(b) The diffraction of the grating is the same as that of a rectangular slit of width $d' = d \cos i_0$. The diffraction function can be expressed by $I / I_0 = \sin^2(\pi d' \sin i' / \lambda_1) / (\pi d' \sin i' / \lambda_1)^2$ with $i' = i - i_0$ (the diffraction angle is always assumed to be small). The diffracted intensity is greatest ($I = I_0$) for $i = 0$. The constructive interference directions are those defined by $d[\sin i_0 + \sin i] = (k + 1)\lambda_1$, leading to $i' \approx k\lambda_1/d'$. In other words, constructive interference occurs at the zeros of the diffraction function. A very large part of the intensity diffracted at wavelength λ_1 is therefore concentrated in the first order of the grating.

(c) For the first order of the grating, we can write $d(\sin i_0 + \sin i) = \lambda$ (so, if $i = i_0$, we have $\lambda = \lambda_1$). If we take the derivative of this equation, we obtain $d \cos i_0\, di \approx d\lambda$ which in turn gives $di/d\lambda \approx 2 \operatorname{tg} i_0/\lambda_1$. So if $di_{max} = 2\theta_1$, we get $\Delta\lambda \approx \theta_1\lambda_1/\operatorname{tg} i_0$.

(d) We find the limits of i_0 to be $19° < i_0 < 36°$. When the grating is rotated, the condition stipulated in 2 (b) is no longer perfectly satisfied. As a result, the grating becomes less efficient.

3. The condition for constructive interference is written as $m\lambda_m \approx 2$ ne cos (α/n) (where m is a whole number). The difference in wavelength is $\Delta\lambda = \lambda_m - \lambda_{m+1} \approx 2$ ne cos $(\alpha/n)/m^2$. These two equalities put together yield $\Delta\lambda \approx \lambda^2/2$ ne. The bandwidth is $\delta\lambda = \Delta\lambda/\mathcal{F} \approx \lambda^2/2$ ne $\mathcal{F}$ and is also equal to the linewidth of the laser emission. The parameters of the Fabry-Pérot (geometrical: e, and optical: n) as well as its tilting angle α cannot be chosen at random. The Fabry-Pérot must be resonant at the laser wavelength λ_l: $m\lambda_l = 2$ ne cos (α/n). As the grating is tilted, the Fabry-Pérot must be tilted at the same angle for the above condition to stay true.

VI.4.4 Problem VI.4: Study of a Q-switched laser

The pumping sequence of the Nd^{3+} energy in the YAG crystal matrix (of thickness d and of section S) can be roughly described as follows: The pumping photons (of energy density W_p and of frequency ν_p) lift the Nd^{3+} ions from their ground state called (3) (of energy E_0) to a set of excited states called (4) (of energy E_4). Then these ions relax by nonradiative processes to level (2) (of energy E_2) in a time which is very short compared to the other processes considered here. Once in level (2), the ion has a probability A_2 per unit time to relax to a lower level [the lifetime of level (2) is $\tau = 1/A_2$]. In particular, it can relax to level (1) by emitting a photon: $E_2 - E_1 = h\nu_1$. Level (1) relaxes very rapidly to level (3) so that its population can always be assumed to be the population of thermal equilibrium with respect to level (3). The laser wave is treated like a plane wave of frequency ν and of energy density $\bar{\rho}$. It oscillates between two plane mirrors placed at a distance L from each other and of which one has a reflection coefficient of 1, and the other an intensity transmission coefficient $\mathcal{T}^2$. The refractive index of the medium is assumed to be homogeneous and equal to n.

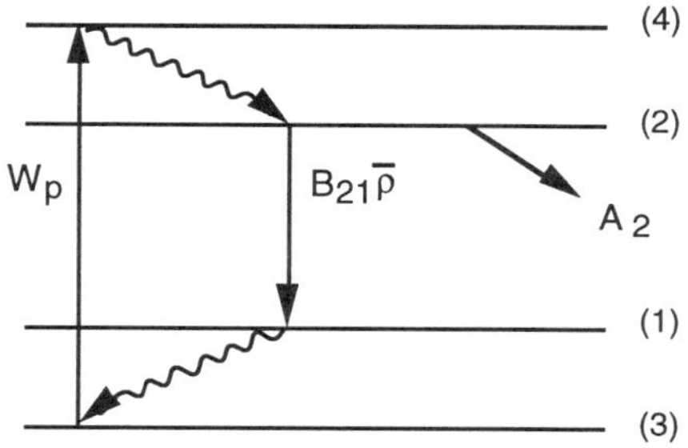

1. Assume that W_p = constant, that $\bar{\rho} = 0$, and that the population participating in the pumping sequence is negligible compared to N_3. Find the population difference $(N_{2S} - N_{1S})$ at steady state, assuming that the medium operates at an equilibrium temperature T which is homogeneous throughout the medium. Show that there is population inversion when W_p is greater than a minimum value W_p^0. Calculate W_p^0.

2. This amplifying medium is placed inside a resonator (W_p = constant). Let us now consider the laser operating in a steady state regime, above the threshold ($\bar{\rho}$ = constant $\neq 0$). Let B_{21} be the Einstein coefficient relative to the (2) → (1) transition. Calculate the population difference $N_{2S} - N_{1S} \cong N_{2S}$ in the presence of a nonvanishing $\bar{\rho}$ [the thermal population of (1) is negligible: $N_{1S} \approx 0$]. Assume $\mathcal{T}^2 \ll 1$ (the gain increases linearly with z). Calculate the intensity $\bar{I}$ inside the laser; the intensity $\bar{I}'$ at the output of the laser; and the threshold power W_p^S, defined as the minimum power at which the laser can actually operate.

3. In this section, we wish to study the laser in a Q-switched regime, where it is pumped by a light flash: W_p = constant for all t such that $-t_p < t < 0$. During the pumping time lasting t_p,

an electro-optical device placed in the cavity masks one of the mirrors, thereby realizing the condition $\bar{\rho} = 0$ until $t = 0$. At $t = 0$, W_p vanishes and the mirror is uncovered. Now, at each plane of coordinate z of the amplifying medium, $\bar{\rho}$ increases, goes through a maximum, decreases, and finally vanishes. During pulsed operation, spontaneous emission from level (2) is assumed to be negligible ($A_2 = 0$).

(a) Find N_2 at $t = 0$, always assuming $N_2 << N_3$.

(b) We wish to study as a function of time the evolution of the total number of photons N_{ph} present in that part of the beam located between the two mirrors. We assume that during one back-and-forth travel of a pulse, i.e. during an interval of time $\Delta t = 2L/c$, the active population N_2 is the same at all points ($N_1 \cong 0$). The relative gain $(\Delta N_{ph} / N_{ph})_g$ on this back-and-forth trip through the amplifying medium, and the relative losses $(\Delta N_{ph} / N_{ph})_l$ due to the transmission of the mirror, take the same form as $\left(\Delta\bar{I} / \bar{I}\right)_g$ and $\left(\Delta\bar{I} / \bar{I}\right)_l$ respectively, of the laser intensity $\bar{I}$ in the continuous wave operating regime, except that the condition "gain" = "losses" is not satisfied in the pulsed regime. Find the total relative change in number of photons $(\Delta N_{ph} / N_{ph})_t$ for a time interval Δt. Assimilate $(\Delta N_{ph} / \Delta t)_t$ to dN_{ph}/dt and show that the equation describing the evolution of N_{ph} can be written as $dN_{ph} / dt = K\, N_{ph}\left[(n_2 / n_S) - 1\right]$, where $n_2 = N_2 LS$ is the total number of ions in level (2) at time t in a volume SL of the crystal. Calculate constants K and n_S.

(c) Remembering the conservation of energy during the wave-ion interaction, find dn_2 / dt. Find the differential equation satisfied by N_{ph} (n_2) by eliminating the factor time. Integrate this equation. For what value of n_2 does N_{ph} show a maximum? Calculate $N_{ph,\, max}$. For what value of n_2 does N_{ph} vanish? How could one go about to calculate the pulse duration?

1. The rate equation of level (2) is $\dot{N}_2 = W_p N_3 - A_2 N_2$. At steady state $N_{2S} = W_p N_{3S} / A_2$. The Maxwell-Boltzmann statistics yield $N_{1S} = N_{3S} \exp(-E_1/kT)$ so that at last we obtain

$$N_{2S} - N_{1S} = W_p N_{3S} / A_2 - N_{3S} \exp(-E_1/kT)$$

Population inversion is reached for $W_p > W_p^0 = A_2 \exp\left(-E_1 / kT\right)$.

2. In the presence of stimulated emission at frequency ν, the above equations take the form $N_{2S} - N_{1S} \approx N_{2S} = N_{3S} W_p / (A_2 + B_{21} \bar{\rho})$. As in Problem V.1, we can write $\Delta\bar{I} / \bar{I} = \mathcal{T}^2 = N_{3S} W_p B_{21} \hbar \omega n\, 2d / c \left(A_2 + B_{21} \bar{\rho}\right)$ yielding

$$\bar{I}' = \mathcal{T}^2 \bar{I} = \left(N_{3S} W_p B_{21} \hbar \omega n\, 2d - A_2 c\, \mathcal{T}^2\right) / n B_{21}$$

At the threshold $\bar{I} = 0$ and $W_p^S = A_2 c\, \mathcal{T}^2 / N_{3S} B_{21} \hbar \omega n\, 2d$.

3.(a) During pumping, the rate of change of the population of level (2) is $\dot{N}_2 = W_p N_3$. The population increases linearly according to $N_2(t) = W_p N_3 (t + t_p)$. At time $t = 0$ we obtain $N_2(0) = W_p N_3 t_p$.

(b) Let us write $\left(\Delta N_{ph} / N_{ph}\right)_g = N_2 B_{21} \hbar \omega n\, 2L/c$ and $\left(\Delta N_{ph} / N_{ph}\right)_l = \mathcal{T}^2$.

So $(\Delta N_{ph})_t = (\Delta N_{ph})_g - (\Delta N_{ph})_l = \left[N_2\, B_{21}\, \hbar\, \omega\, n\, 2L/c - \mathcal{T}^2\right](N_{ph})_t$. The time interval Δt associated to $(\Delta N_{ph})_t$ is the time transit 2L/c. We find $dN_{ph}/dt \approx (\Delta N_{ph} / \Delta t)_t = (\mathcal{T}^2\, c\, N_{ph} / 2L)\left[(2N_2\, B_{21}\, \hbar\, \omega\, n\, LS / c\, S\, \mathcal{T}^2) - 1\right]$. This can then be written as $dN_{ph} / dt = KN_{ph}\left[(n_2 / n_S) - 1\right]$ with $n_2 = N_2 L\, S$; $K = \mathcal{T}^2\, c/2L$ and $n_S = c\, S\, \mathcal{T}^2 / 2\, B_{21}\, \hbar\, \omega\, n$.

(c) When N_{ph} increases by one unit, n_2 decreases by one unit (stimulated emission). $dn_2/dt = -\, kN_{ph}\, n_2 / n_S$; $dN_{ph} / dn_2 = (dN_{ph} / dt)\,(dt / dn_2) = (n_S / n_2) - 1$. Integrating this equation leads to $N_{ph} = n_S \ln\left(n_2 / n_2^{(i)}\right) - \left(n_2 - n_2^{(i)}\right)$. $n_2^{(i)}$ is the initial value of n_2 at the time the triggered pulse is about to start. N_{ph} is largest for $n_2 = n_S$ (set $dN_{ph}/dn_2 = 0$). $N_{ph,\,max} = n_S \ln\left(n_S / n_2^{(i)}\right) - \left(n_S - n_2^{(i)}\right)$. N_{ph} vanishes (zero is its final value) when $n_S \ln\left(n_2^{(f)} / n_2^{(i)}\right) = \left(n_2^{(f)} - n_2^{(i)}\right) n_2^{(f)}$. If n_S were known, $n_2^{(f)}$ could be found graphically by plotting the straight line representing the variation of $\left(n_2 - n_2^{(i)}\right)$ and the graph of $n_S \ln\left(n_2 / n_2^{(i)}\right)$ as a function of n_2. The x coordinate of the intersection would give the value of $n_2^{(f)}$ corresponding to the end of the pulse. To find the pulse duration, one must integrate the following equation:

$$dn_2 / dt = -\left(kn_2 / n_S\right)\left[n_S \ln\left(n_2 / n_2^{(i)}\right) - \left(n_2 - n_2^{(i)}\right)\right]$$

Then let n_2 take on the value for $n_2^{(f)}$ determined graphically just now.

VI.4.5 Problem VI.5: Study of a mode-locked laser: the Ti/sapphire laser

Please refer to Problem IV.5, Questions 1, 2(a) and 2(b) where this kind of laser was already partly investigated. The crystal has a thickness d. The beam waist of the pump laser is located at the front face of the crystal. We wish to study the propagation of the Gaussian beam inside the crystal.

The refractive index n of the medium depends on the intensity of the wave (nonlinear medium). As a first approximation, the refractive index of the medium depends on r, the distance from the point under consideration to the axis of the beam: $n(r) = n_0\,(1 - \beta^2 r^2 / 2)$, where β is a parameter which increases with the central beam intensity I_d and which vanishes when $I_d = 0$.

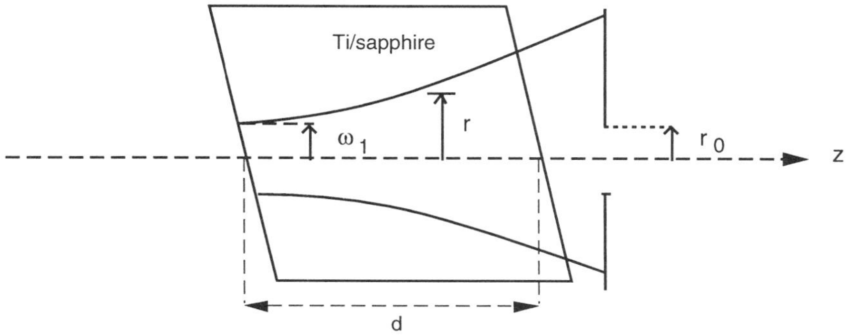

1. Using the transfer matrix to express the complex radial beam parameter q of the Gaussian beam [see Chapter IV Section 2.2.2 (a) and Inset IV.2], calculate the complex beam parameter q_d of the beam at the back face of the crystal. The beam waist of size ω_1 is located on the front face of the crystal. In order to calculate the transfer matrix, use the propagation equation of a light ray

in geometrical optics: $d^2 r/dz^2 = dn(r)/n(r)\,dr$. Find the radius $\omega_S(d)$ of the Gaussian beam at the back face of the crystal.

2. Now a circular diaphragm (pinhole) of radius r_0 is placed behind the crystal. Calculate the transmission of the diaphragm. Show that it is a function of I_d and that the diaphragm has the same effect as a saturable absorber, provided that d is chosen correctly. Explain qualitatively why the laser now operates in a mode-locked regime.

1. Referring to Problem IV.8, we write $d^2 r / dz^2 = dn(r) / n(r)\,dr$. Now we know that $dn/dr = -n_0 \beta^2 r$, $d^2 r/dz^2 = -\beta^2 r$, and $r = A \cos \beta z + B \sin \beta z$. The transfer matrix takes the form

$$\begin{pmatrix} \cos(\beta z) & \sin(\beta z)/\beta \\ -\beta \sin(\beta z) & \cos(\beta z) \end{pmatrix}$$

yielding

$$(qd)^{-1} = \left[-i\pi\omega_1^2 \beta \sin(\beta d)/\lambda + \cos\beta d\right]\left[-i\pi\omega_1^2 \beta\cos(\beta d)/\lambda + \sin(\beta d)/\beta\right] / \left[\sin^2(\beta d)/\beta^2 + \pi^2\omega_1^4 \cos^2(\beta d)/\lambda^2\right]$$

and identifying the imaginary part of $(qd)^{-1}$ with $-\lambda/\pi\omega_S^2(d)$, we find

$$\omega_S(d) = \omega_1 \sqrt{\cos^2(\beta d) + \lambda^2 \sin^2(\beta d)/\beta^2\pi^2\omega_1^4}.$$

2. The power transmitted by the diaphragm can be written

$P_t = \int_0^{r_0} \exp\left[-2r^2/\omega_S^2(d)\right] 2\pi r\,dr$. The total output power P_i can be expressed by a similar integral, but with an infinite upper integration limit. The transmission of the diaphragm is written as $\mathcal{T}^2 = P_t / P_i = 1 - \exp\left[-2r_0^2/\omega_S^2(d)\right]$. If we choose d, the thickness of the amplifying medium, such that $d \ll 1/\beta$, then we have

$$\omega_S(d) \approx \omega_1 \sqrt{1 - (\beta l)^2/2 + \lambda^2 d^2/\pi^2\omega_1^4}$$

$\omega_S(d)$ decreases when β increases. But β increases with intensity. Therefore $\omega_S(d)$ decreases as the intensity I increases. The diaphragm plays the part of a saturable absorber: The hole lets through beams of high intensity, but cuts off beams of weak intensity. Thus we succeeded in making a passive-mode blocking device. In this case, the nonlinearity of the refractive index has an electronic origin. It is extremely fast ($\sim 10^{-15}$ s).

VI.4.6 Problem VI.6: Mode-locking of the longitudinal modes of a laser by acousto-optical effect

Let us consider a transparent plate of thickness e and of width d. An acoustic progressive wave of angular frequency ω_a travels at speed v (the speed of sound) through this plate along the x-axis. The plate is then illuminated by an optical monochromatic plane wave having an angular frequency in vacuum of ω_0 (with $\omega_0 >> \omega_a$), a wavelength λ_0, and carrying an electric field E_0. The amplitude of this electric field is A_0. The refractive index of the plate through which travels the acoustic wave can be written in the form

$$n_p(x, t) = n_0 + \Delta n \cos \omega_a (t - x/v)$$

n_0 is the refractive index in the absence of acoustic wave, Δn is the amplitude of the induced modulation, and t represents time.

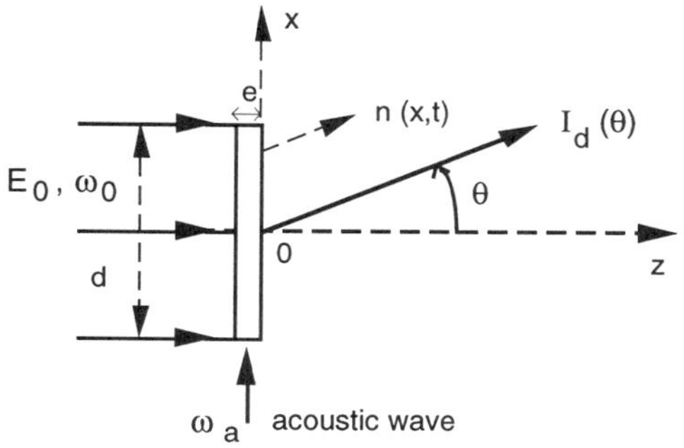

1. Find the phase difference $\Delta\phi(x, t)$ induced by the plate in the optical wave. Infer from this the electric field E(x, t) (a complex number) in the xOy plane at the back face of the plate.

2. Assuming that $k_0 \Delta n e << 1$, show that, in this approximation, E(x, t) is the sum of three fields, E_0, E_1, E_{-1}, which oscillate at three different frequencies ω_0, $\omega_0 + \omega_a$, $\omega_0 - \omega_a$. Give the exact form of these three fields and compare their amplitudes.

3. We now wish to analyze the diffraction pattern $I_d^{(1)}(\theta)$ [the superscript (1) refers to first order diffraction] of the pupil function defined by E(x, t). The width of the plate, d, is assumed to be infinite. How could one actually observe this diffraction pattern? Describe the pattern by specifying the mathematical form of $I_d^{(1)}(\theta)$.

4. Let us now consider a plate identical to the one describe above, but in which there is a stationary acoustic wave of frequency ω_a. In this case, the refractive index takes the form

$$n_S(x, t) = n_0 + \Delta n \cos(\omega_a t) \cos(\omega_a x / v)$$

Assuming the same approximations as for the progressive wave, and proceeding in a similar way, show that a plane wave traveling along the z-axis and passing through the plate has time-modulated losses at infinity in the z-direction. Calculate the frequency of these modulations as a function of ω_a.

5. Let us consider the multimode operating laser represented by the accompanying diagram. Let L be the length of the optical path between the two mirrors. What is the frequency difference δv between two adjacent modes of the laser? One method to lock together the phases of all the modes consists in inserting losses into the cavity at frequency δv. To this purpose, the plate described in the first part of this problem is inserted next to the amplifying medium (AM). What value must be chosen for frequency ω_a in order to obtain a correct mode-locked operation?

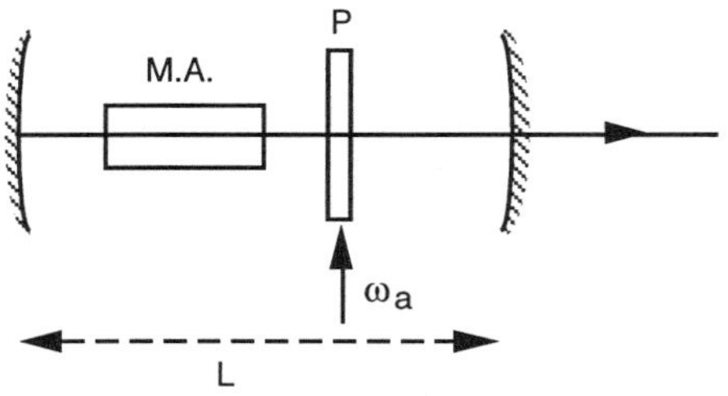

1. The phase difference $\Delta\phi$ can be expressed as $\Delta\phi = 2\pi (n - 1) e / \lambda_0$. Replacing n by $n_p(x, t)$ yields $\Delta\phi (x, t) = 2\pi (n_0 - 1) e/\lambda_0 + 2\pi \Delta n e \cos \omega_a (t - x/v) / \lambda_0$. The electric field is equal to $E (x, t) = A_0 \exp \{- i [\omega_0 t - \Delta\phi (x, t)]\}$ and if $\Delta\phi(x, t)$ is written out in full, we obtain

$$E(x, t) = A_0 \exp \{- i [\omega_0 t - (n_0 - 1) e k_0]\} \exp \{ik_0 \Delta n e \cos \omega_a (t - x/v)\}$$

(we let $2\pi/\lambda_0 = k_0$).

2. By writing the electric field as

$$\exp (ik_0) \Delta n e \cos [\omega_a (t - x/v)] \approx 1 + (ik_0 \Delta n e / 2) \times \{\exp [i\omega_a (t - x/v)] + \exp [- i\omega_a (t - x/v)]\}$$

we clearly show that E (x, t) is the sum of three fields E_0, E_1, and E_{-1} such that

$$E_0 = A_0 \exp\{- i [\omega_0 t - (n_0 - 1) e k_0] \}$$

$$E_1 = (iA_0 k_0 \Delta n e / 2) \exp[i (n_0 - 1) e k_0] \exp[- i\omega_a x/v] \exp[- i (\omega_0 - \omega_a) t]$$

$$E_{-1} = (iA_0 k_0 \Delta n e / 2) \exp[i (n_0 - 1) e k_0] \exp[i\omega_a x/v] \exp[- i (\omega_0 + \omega_a) t]$$

The three amplitudes have the following values:

$$|E_0| = A_0$$

$$|E_1| = |E_{-1}| = A_0 k_0 \Delta n e / 2$$

It is obvious that the amplitudes of fields E_1 and E_{-1} are smaller than that of E_0.

3. To go from a description in terms of x to a description in terms of angle θ, we make a Fourier transform by writing $q = k_0 \sin \theta$ (see reference 5.5 of Chapter IV); for the mathematics to be correct, we must assume d to be infinite. We find

$$E_d^{(1)} (q, t) \sim \int_{-\infty}^{+\infty} E (x, t) \exp[- i q x] dx$$

We beget Dirac delta functions. We finally have

$$\begin{aligned} E_d^{(1)} (q, t) \sim\ & A_0 \exp\{- i [\omega_0 t - (n_0 - 1) e k_0] \} \delta (q) \\ & + (iA_0 k_0 \Delta n e / 2) \exp[- i (\omega_0 - \omega_a) t] \exp[i (n_0 - 1) e k_0] \delta [q + (\omega_a / v)] \\ & + (iA_0 k_0 \Delta n e / 2) \exp[- i (\omega_0 + \omega_a) t] \exp[i (n_0 - 1) e k_0] \delta [q - (\omega_a / v)] \end{aligned}$$

There is diffraction at infinity. This can be observed at the image focal point of an appropriate converging optical device. The three Dirac functions are seen as three spots.

The most intense spot is at q = 0, i.e. when $\theta = 0$ (transmission through the plate without diffraction); the two others, much less intense, are localized at $q = \pm(\omega_a / v)$, i.e. at angles θ such that $\sin\theta = \pm(\omega_a / vk_0) = \pm(k_a / k_0)$. Since $\lambda_a >> \lambda_0$, these spots appear at angles $\theta \approx \pm(\lambda_0 / \lambda_a)$. These are the two diffraction spots induced by the progressive acoustic wave.

We wish to remind the reader that Inset VI.5 of this chapter indicates experimental conditions (Bragg conditions) under which a much larger part of the intensity is diffracted.

4. Only fields E_1 and E_{-1} are modified. One simply needs to replace $n_p(x, t)$ by $n_S(x, t)$. With the same approximations as above, we find

$$E = E_0 + \left(i\, k_0 A_0 \Delta n\, e / 2\right) \exp\left\{-i\left[\, \omega_0 t - k_0\left(n_0 - 1\right) e\right]\right\} \exp\left(i\, \omega_a x / v\right) \cos \omega_a t$$
$$+ \left(i\, k_0 A_0 \Delta n\, e / 2\right) \exp\left\{-i\left[\, \omega_0 t - k_0\left(n_0 - 1\right) e\right]\right\} \exp\left(-i\, \omega_a x / v\right) \cos \omega_a t$$

The diffraction pattern shows three diffraction spots, but a frequency modulation in time appears at angular frequency $2\omega_a$ (giving a frequency of ω_a / π), described by the terms in $\cos^2 \omega_a t$.

5. The frequency difference is $\delta\nu = c / 2L$. Therefore we must choose a frequency $f = \omega_a / \pi = \delta\nu = c / 2L$ or $\omega_a = c\, \pi / 2L$. This way, we create a modulation of the cavity losses at a frequency which is equal to the free spectral range of the resonator, thereby ensuring the locking of the phases of the modes.

PART THREE

Fundamentals of the Laser-Molecule Interaction

The knowledge acquired in the first two parts of this book—though these parts were probably more of a memory refresher with some additional information for most readers—will now be put to use in order to establish the basis of *laser-molecule interaction*, in other words, of *molecular nonlinear optics.*

We did not intend to present an exhaustive work on the subject. A relatively large number of excellent books exist dealing with nonlinear optics. Some of these are used extensively in our laboratories and are listed in the bibliographical references at the end of Chapter VIII. We hope to have opened the way to these works by having provided the reader with some indispensable basic knowledge and by having shown him the various possible paths leading to further studies.

Thus we deliberately decided to present two examples which differ completely as well in content as in shape:

— The first example will be dealt with in Chapter VII. It consists in the study of *resonant laser-molecule interaction* by using the *density matrix formalism.* A microscopic two-level system, with eigenfrequencies ω_a and ω_b, will be illuminated by a laser wave of frequency ω_L such that $\omega_L \approx \omega_0 = \omega_b - \omega_a$. This wave will reveal nonlinear optical effects which will be studied at *the same frequency as the laser frequency.* Only the steady state will be taken under consideration. Two practical applications will be presented.

— The second example will be given in Chapter VIII. It consists of using the *state vector formalism* to calculate the optical susceptibility of order n. Unlike the first example, the interaction is a *nonresonant* one in which non-negligible effects will occur at *frequencies different from the exciting laser frequency.* Some of these effects will be shown, in *steady state as well as in non -steady -state interactions.*

But before starting, we want to make the reader aware of the three approximations which will be used all through the section using the semiclassical formalism (see Table IV.3.1) and which are valid at the optical frequencies (10^{14} or 10^{15} Hz) studied here.

— *First approximation: The approximation in which matter is treated quantically while the field is treated classically.* We shall use this semiclassical approximation in which only matter is described by quantum physics. Moreover, we shall use the nonrelativistic Hamiltonian (at 10^{18} Hz, the "speed" of the electrons is still only on the order of c/10).

— *Second approximation: The excited electronic levels do not acquire population thermally (i.e. they are not thermalized) in the absence of laser radiation.* Calculations show that at frequencies greater than 10^{13} Hz, in particular at the optical frequencies (10^{14} or 10^{15} Hz), the electromagnetic field of the thermal radiation is negligible at room temperature ($\approx$ 300 K). Therefore, only the electronic state of lowest energy is populated. This state is called the *ground state.*

The states corresponding to an electronic excitation, called the *excited states*, are assumed to be empty at thermal equilibrium conditions.

— *Third approximation: The electric dipole approximation.*

We showed (see Chapter I) that the part played by the Hamiltonian is fundamental in the description of microscopic systems (as, for example, in the expression of the sixth postulate of evolution when using the interaction representation, and so on...) To construct this operator, we shall start by using the seventh postulate to express the *energy of the microsystem* illuminated by the laser wave. Therefore, we must first express this energy classically. Problem VII.1 shows the details of these calculations, which are based on a fundamental approximation called the *electric dipole approximation* (EDA). In this problem, we showed that the Hamiltonian function of spatial coordinate $\mathbf{r}_j$ (relative to electron j), of its derivative $\dot{\mathbf{r}}_j$, and of time t can be expressed, in the case of an atom with atomic number Z whose nucleus is placed at the origin 0 of a coordinate system, as

$$H\left(\mathbf{r}_j, \dot{\mathbf{r}}_j, t\right) \approx \sum_j \left[\left(m_e \dot{\mathbf{r}}_j^{\,2}/2\right) - \left(e\,\phi\left(\mathbf{r}_j\right)/2\right)\right] + Z\,e\,\phi(0)/2 - \mathbf{p}\,\mathbf{E}(0, t)$$

where m_e and e are the mass and charge of the electron, respectively, $\mathbf{E}$ is the field of the applied laser wave, $\mathbf{p}$ the dipole moment defined by

$$\mathbf{p} = -\,e \sum_j \mathbf{r}_j$$

(see Problem I.4), and $\phi(\mathbf{r})$ is the potential created by the set of electric charges defined by Eq. II.5.1.

The first three terms of the right-hand side of the above equation represent, respectively, the kinetic energy of the electrons, the potential energy of the electrons, and the potential energy of the nucleus of charge Ze, which is placed at the coordinate origin 0. The sum of these terms represents the total energy of the system *in the absence of radiation.*

The last term, $-\,\mathbf{pE}(0,t)$, represents the *laser-atom interaction* in the electric dipole approximation (EDA). Please note that the electric field is taken at the origin of the reference frame, meaning at the level of the nucleus. EDA makes use of the fact that at atomic dimensions (100 pm), the value of the field around the electrons differs very little from the value of the field at the nucleus. This shows that the power expansion of Problem VII.1 was justified. The EDA assimilates the atom to a dipole of moment $\mathbf{p}$ and of energy $-\mathbf{pE}$ placed in the field. The present calculations can be pushed to the next order of the power expansion to reveal electric quadrupole interactions, magnetic dipole interactions, and so on. However, these interactions are at least two orders of magnitude smaller than the dipole interactions. This is why we shall keep only the dipole interactions, symbolized by the term V of Problem I.5 which will be the starting point of our presentation. Now let us proceed to our two selected examples.

CHAPTER VII

Application of the Density Operator Formalism to a Stationary Resonant Laser-Molecule Interaction

---

In this chapter, we wish to examine the interaction between an atomic system and a quasi-monochromatic laser wave whose angular frequency ω_L stays close to one of the resonance frequencies of the atom, namely that between levels $|\, a >$ and $|\, b >$, of energies $E_a = \hbar\omega_a$ and $E_b = \hbar\omega_b$ with $E_b > E_a$. In other words, we have

$$\omega_L \approx \omega_0 = \omega_b - \omega_a$$

In our case, we shall restrict ourselves to the so called *two-level system* to describe the atomic system. In this approximation, the state vector $|\, \psi >$ representing the system can be developed on a two -dimensional basis, consisting of the two eigenvectors $|\, a >$ and $|\, b >$ of Hamiltonian H_0. This approximation means that the system in interaction with the wave is restricted to *just the two atomic levels under consideration, and to the field.* All other levels are included in the *outside environment.* In Chapter I, we showed several examples based on this approximation, and we also showed how the coupling between the system and the outside environment can be described phenomenologically.

Let us now proceed successively to (1) continue the calculations started in Problem I.5, (2) describe the physical consequences of the mathematical results obtained for the optical susceptibility at frequency ω_L and for the refractive index of the system illuminated by a laser, and (3) show three practical applications of these results (namely, saturated absorption spectroscopy, Lamb's semiclassical theory of the laser, and laser manipulation of microscopic systems).

VII.1 Expression of the coherences and of the populations

Let us take the differential equations established in Problem I.5. We shall improve these by taking into account relaxation and population changes phenomenologically. Let us write V explicitly as $-\boldsymbol{p}\mathbf{E}(0,t)$. This yields the following differential equations:

$$\dot{\rho}_{ba} = -(i\omega_a + \gamma_{ab})\,\rho_{ba} + (i/\hbar)\,p_{ab}\,E(0, t)\,(\rho_{aa} - \rho_{bb}) \tag{VII.1.1}$$

$$\dot{\rho}_{aa} = \lambda_a - \gamma_a\,\rho_{aa} + (i/\hbar)\left[p_{ab}\,E(0, t)\,\rho_{ba} - c\,c\right] \tag{VII.1.2}$$

$$\dot{\rho}_{bb} = \lambda_b - \gamma_b\,\rho_{bb} - (i/\hbar)\left[p_{ab}\,E(0, t)\,\rho_{ba} - c\,c\right] \tag{VII.1.3}$$

where cc stands for the conjugated complex term.

In these expressions, from which the formal time dependencies of ρ_{ab}, ρ_{aa}, and ρ_{bb} have been omitted, we used the following notation:

$$p_{ab} = \langle a\,|\,p_E\,|\,b\rangle$$

where p_E represents the projection of the dipole moment on the electric field $\mathbf{E}$. We henceforth assume that $p_{ab} = p^*_{ab} = p_{ba} = p$.

Operator V is of course Hermitian. Let us proceed to a change of variables defined by

$$\begin{aligned} S &= \rho_{aa} + \rho_{bb} \\ D &= \rho_{aa} - \rho_{bb} \end{aligned} \tag{VII.1.4}$$

This yields

$$\dot{\rho}_{ba} = -(i\omega_0 + \gamma_{ab})\,\rho_{ba} + (i/2\hbar)\,p\,E_0\,D\left[\exp(i\omega_L t) + \exp(-i\omega_L t)\right] \tag{VII.1.5}$$

$$\dot{S} + \frac{\gamma_a + \gamma_b}{2}S + \frac{\gamma_a - \gamma_b}{2}D = \lambda_a + \lambda_b \tag{VII.1.6}$$

$$\dot{D} + \frac{\gamma_a + \gamma_b}{2}D + \frac{\gamma_a - \gamma_b}{2}S = \lambda_a - \lambda_b + (i/\hbar)\,p\,E_0\left[\exp(i\omega_L t) + \exp(-i\omega_L t)\right] \times (\rho_{ba} - cc) \tag{VII.1.7}$$

where the laser field is expressed as

$$E(0, t) = (E_0/2)\left(e^{\,i\omega_L t} + e^{-i\omega_L t}\right) \tag{VII.1.8}$$

Next, we wish to find a physically acceptable stationary solution of Eqs. VII.1.5, VII.1.6, and VII.1.7. Now we have seen (see Problem I.5) that for an isolated system we have $\dot{\rho}_{aa} = \dot{\rho}_{bb} = 0$ and $\dot{\rho}_{ba} = -i\omega_0\,\rho_{ba}$. The populations remain constant while the free coherence oscillates at angular frequency ω_0. So in our case, we can try to look for a particular solution in which functions S and D would be constant and respectively equal to S_s and D_s, where the subscript s stands for "stationary".

To find the expression of a particular solution for ρ_{ab}, let us integrate Eq. VII.1.5 after replacing D by D_s and multiplying both members of the equation by $\exp[(i\omega_0 + \gamma_{ab})t]$. We obtain

$$\rho_{ba} \exp\left[(i\omega_0 + \gamma_{ab})\, t\right] = (ip\, E_0\, D_s / 2\hbar) \int_{-\infty}^{t} dt' \exp(\gamma_{ab}\, t')$$

$$\times \left\{\exp\left[i\,(\omega_L + \omega_0)\, t'\right] + \exp\left[-i\,(\omega_L - \omega_0)\, t'\right]\right\} \qquad \text{(VII.1.9)}$$

Please note that if the amplitude of the field E_0 depends on time, D_s is no longer an acceptable solution. Yet even then, $\rho_{ba}(t)$ can be written in the form of an integral:

$$\rho_{ba} \exp\left[(i\omega_0 + \gamma_{ab})\, t\right] = (ip / 2\hbar)$$

$$\times \int_{-\infty}^{t} dt' E_0(t')\, D(t') \exp(\gamma_{ab}\, t') \left\{\exp\left[i\,(\omega_L + \omega_0)\, t'\right] + \exp\left[-i\,(\omega_L - \omega_0)\, t'\right]\right\}$$

If $E_0(t')$ and $D(t')$ vary slowly during times on the order of $1/\gamma_{ab}$, we can take them out of the integral with their value at t' = t, which gives

$$\rho_{ba} \exp\left[(i\omega_0 + \gamma_{ab})\, t\right] = (ip\, E_0(t)\, D(t) / 2\hbar)$$

$$\times \int_{-\infty}^{t} dt' \exp(\gamma_{ab}\, t') \left\{\exp\left[i\,(\omega_L + \omega_0)\, t'\right] + \exp\left[-i\,(\omega_L - \omega_0)\, t'\right]\right\} \qquad \text{(VII.1.9 bis)}$$

This approximation gives a solution of the evolution of $\rho_{ab}(t)$ with $E_0(t)$ and provides a complete solution to the above system of equations, so that D(t) can be calculated. It is known under the name *"rate equations approximation"*. The approached values of the populations obtained by this method are identical to those obtained by Einstein's phenomenological treatment. Let us come back to the simple case of a stationary field $E_0(t) = E_0$. The integral of Eq. VII.9 can be simplified by making an approximation: *the rotating wave approximation* (RWA).

Inset VII.1 ***The rotating wave approximation***

Mathematical justification

The integrand under consideration consists of two terms of very different amplitudes, because, as said earlier, we have

$$\omega_L - \omega_0 \approx 0 \quad \text{and} \quad \omega_L + \omega_0 \approx 2\,\omega_L$$

After integrating, we obtain two terms, proportional respectively to $\gamma_{ab}{}^{-1}$ and to $[\gamma_{ab} - i2\omega_L]^{-1}$. Since $\gamma_{ab}{}^{-1} >> \omega_L^{-1}$, the first term completely dominates the other one, so that we can omit the calculation of the second term because it would bring only a minor contribution to the whole.

This constitutes the rotating wave approximation. It is widely used in theoretical analysis in atomic physics. In fact, it consists of saying that only one of the two rotating components of the laser field, $\exp(i\omega_L t)$ and $\exp(-i\omega_L t)$, contributes to the real field. Why the first of the two components was chosen is explained by the following physical interpretation.

Physical interpretation

The above mathematical reasoning can be illustrated by the example of a particle of spin 1/2 placed inside a superposition of two magnetic fields; one constant field B_0 pointing in the direction of 0z (0z is the quantification axis), and a sinusoidal field $B_0' \cos \omega_L t$ which points in the 0x direction of an orthonormal reference frame. The Hamiltonian of this system is the same as that of a two-level atom placed inside an electromagnetic field. As a consequence, both systems will display similar behavior. However, a classical description of the evolution of spin inside a field says that if the spin is placed in the $\mathbf{B}_0$ field only, it will rotate around 0z at the angular frequency $\omega_0 = \gamma B_0$ (Larmor precession). In the case of the atom, this precession corresponds to the free evolution of the atomic coherence in the absence of an electromagnetic field. The sinusoidal field acts on the spin in the same way as the electromagnetic field acts on the atom. This field, equal to $B_0' \cos \omega_L t$ on the 0x axis, can be decomposed in two fields which turn in opposite directions at angular speed ω_L in the plane perpendicular to the 0z axis (i.e., the x0y plane).

In the spin case, the rotating wave approximation consists in saying that only that field which turns around 0z in the same direction as the spin has a non-negligible influence on its movement. Indeed, seen from the spin, the second field has a relative angular frequency of $(\omega_0 + \omega_L)$. This frequency is so high that it cannot have much influence on the spin.

In the same way, in the case of a two-level atom, only that field which, in the phase diagram, turns in the same direction as the free atomic dipole (i.e., in the absence of any electromagnetic field) needs to be taken into account. All through the remainder of this book, we shall be using the rotating wave approximation.

Correspondence between magnetic and optical parameters

This correspondence is explained in Figure VII.1.1 showing the correspondence between electromagnetic and optical physical quantities. The vectorial treatment of the two-level system placed within an optical wave will be further developed in Problem VII.4.

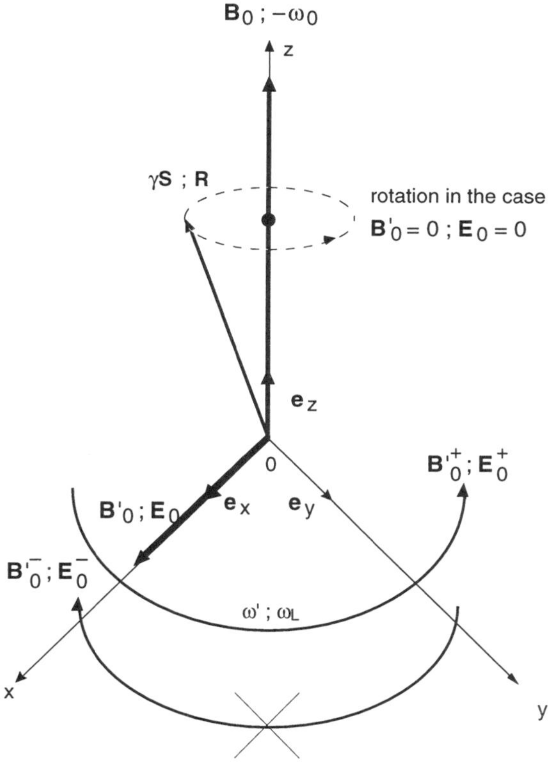

Figure VII.1.1: Illustration of the rotating wave approximation

Spin 1/2 placed inside a superposition of two magnetic fields	Two-level system within an oscillating electromagnetic field:
$\mathbf{B}_0$ constant (0z axis)	$\mathbf{E}_0 \cos \omega_L t$ (0x axis)
$\mathbf{B}'_0 \cos \omega' t$ oscillating (0x axis)	

Correspondences in the absence of any perturbation

Spin 1/2 in constant inductive field $\mathbf{B}_0$	$\leftarrow$ System $\rightarrow$	Isolated two-level system

Rotation at frequency $\omega'_0 = -\gamma B_0$	← Frequency →	Oscillation of the coherence at frequency $\omega_0 = \omega_b - \omega_a$

Origin of the perturbation

Field $\mathbf{B}'_0$ oscillating at frequency ω'	← $\omega' = \omega_L$ →	Linearly polarized electromagnetic field oscillating at frequency ω_L

Correspondences in presence of a perturbation

$H = H_0 + V = -\gamma \mathbf{S}\, \mathbf{B}_0 - \gamma \mathbf{S}\, \mathbf{B}'_0 \cos \omega' t$	← Hamiltonian →	$H = H_0 + V = H_0 - \mathbf{p}\, \mathbf{E}_0 \cos \omega_L t$

Matrix representation of the operators

(where the energy zero is chosen to lie half-way between the eigen energies)

$\begin{pmatrix} -\gamma B_0 \hbar/2 & 0 \\ 0 & \gamma B_0 \hbar/2 \end{pmatrix}$	← (H_0) → $\omega'_0 = \omega_0$	$\begin{pmatrix} \hbar\omega_0/2 & 0 \\ 0 & -\hbar\omega_0/2 \end{pmatrix}$
$-\left(\gamma B'_0 \hbar / 2\right) \cos \omega' t \begin{pmatrix} 0 & 1 \\ 1 & 0 \end{pmatrix}$	← (V) →	$-p\, E_0 \cos \omega_L t \begin{pmatrix} 0 & 1 \\ 1 & 0 \end{pmatrix}$

$(\gamma \mathbf{B}'_0 / 2) = \Omega$ ($\Omega = p\, E_0 / \hbar$ is the Rabi frequency)

Vectorial analogy	$\mathbf{S} \leftrightarrow \mathbf{R}$	*(in reference frame Oxyz)*
γS_x		$R_x = \rho_{ab} \exp(-i\omega_L t) + cc$
γS_y	← vector components →	$R_y = -i\, \rho_{ab} \exp(-i\omega_L t) + cc$
γS_z		$R_z = -D$ $(\gamma_a = \gamma_b)$
$\dot{\mathbf{S}} = \gamma\, \mathbf{S} \times \mathbf{B} - \left(S_z / T_1\right) \mathbf{e}_z$ $-(S_x\, \mathbf{e}_x + S_y\, \mathbf{e}_y) / T_2$	Bloch's ← kinetic → equations	$\dot{\mathbf{R}} = \mathbf{R} \times \boldsymbol{\mathcal{R}}' - R_z\, \gamma_a\, \mathbf{e}_z$ $-(R_x\, \mathbf{e}_x + R_y\, \mathbf{e}_y)\, \gamma_{ab}$ $\mathcal{R}'_x = \Omega$ $\mathcal{R}'_y = 0$ $\mathcal{R}'_z = -(\omega_0 - \omega_L)$
T_1	← longitudinal relaxation times →	$1 / \gamma_a = 1 / \gamma_b$
T_2	← transverse relaxation times →	$1 / \gamma_{ab}$

Physical interpretation	
A precession of $\gamma\mathbf{S}$ around $\mathbf{B}$ at frequency: $\Omega' = \gamma\lvert \mathbf{B} \rvert$ $= \gamma\sqrt{(B'_0 / 2)^2 + (\omega' - \omega_0)^2 / \gamma^2}$	A precession of $\mathbf{R}$ around $\mathfrak{R}'$ at frequency : $\Omega' = \lvert \mathfrak{R}' \rvert = \sqrt{\Omega^2 + (\omega_L - \omega_0)^2}$ At resonance, the system oscillates at the Rabi frequency, $(p\, E_0 / \hbar)$, between state $\lvert a > (R_z = -1)$ and state $\lvert b > (R_z = +1)$. Far from resonance, $\mathbf{R}$ precesses around vector $\mathfrak{R}'$ in the $\mathbf{e}_x$, $\mathbf{e}_z$ plane. Relaxation decreases the amplitudes of the population and of the coherence.

Thus, only the second term of Eq. VII.1.9, which varies as $\exp[-i(\omega_L - \omega_0)t]$, will contribute significantly to the integral. After integration, we find

$$\rho_{ba}(t) \approx (ip\, E_0\, D_s / 2\hbar) \exp(-i\,\omega_L t) / [-i(\omega_L - \omega_0) + \gamma_{ab}] \tag{VII.1.10}$$

Now, we can calculate S_s and D_s. Let us look more especially at parameter D_s, the population difference. Setting $\dot{S} = 0$ in Eq. VII.1.6 yields

$$S_s = 2(\lambda_a + \lambda_b)/(\gamma_a + \gamma_b) - D_s(\gamma_a - \gamma_b)/(\gamma_a + \gamma_b)$$

Carrying this result into Eq. VII.1.7 with $\dot{D} = 0$, and using the rotating wave approximation, gives

$$\begin{aligned} &D_s\left[(\gamma_a + \gamma_b)/2 - (\gamma_a - \gamma_b)^2 / 2(\gamma_a + \gamma_b)\right] \\ &\approx \lambda_a - \lambda_b - (\lambda_a + \lambda_b)(\gamma_a - \gamma_b)/(\gamma_a + \gamma_b) + (ip\, E_0 / \hbar)(\rho_{ba,\, s} - c\, c) \end{aligned} \tag{VII.1.11}$$

with

$$\rho_{ba,\, s} \approx \rho_{ba}(t) \exp(i\,\omega_L t) \tag{VII.1.12}$$

At last, we can write the expression

$$D_s \approx \frac{D_0}{1 + S_{sat} / \left[1 + (\omega_0 - \omega_L)^2 / \gamma_{ab}^2\right]} \tag{VII.1.13}$$

where S_{sat} is a saturation parameter defined by

$$S_{sat} = \bar{I}_L / I_{sat} \tag{VII.1.14}$$

with $\bar{I}_L = \varepsilon_0\, c n\, E_0^2 / 2$ and

$$I_{sat} = \varepsilon_0\, c n\, \hbar^2\, \gamma_a\, \gamma_b\, \gamma_{ab} / p^2 (\gamma_a + \gamma_b) \tag{VII.1.15}$$

$D_0 = (\lambda_a / \gamma_a) - (\lambda_b / \gamma_b)$ represents the population difference in the absence of electromagnetic wave.

VII.2 Fundamental consequences and physical interpretation

VII.2.1 Coherence

Equation VII.1.10 expresses the coherence. In the absence of an electric laser field, the coherence $\rho_{ab}(t)$ follows its free course, starting from its initial value and falling to zero in damped oscillations as described by the law $\rho_{ab}(t) = \rho_{ab}(0) \exp\left[(-i\omega_0 + \gamma_{ab})\, t\right]$, with *a phase turning at angular frequency* ω_0 in a well-defined direction. But Eq. VII.1.5 shows that the laser field *"forces"* the phase of the coherence to turn in the same direction as the free coherence but *at the angular frequency* ω_L *of the field.* The induced coherence is proportional to the electric field. As a matter of fact, Eq. VII.1.10 only gives the terms of the coherence which oscillate at the frequency of the field ω_L. The terms oscillating at $3\omega_L$, $5\omega_L$ ... have been eliminated by the rotating wave approximation. The coherence amplitude at frequency ω_L is a linear function of E_0, but is saturated by the dependency of D_s on E_0^2. At resonance, $\rho^0_{ab,\,s} = \alpha E_0 / \left(1 + \beta E_0^2\right)$ and expanding $\rho^0_{ab,\,s}$ as a power series of E_0 gives us the terms in E_0, E_0^3 ... which we could have obtained by doing a perturbation calculation.

VII.2.2 Population difference

Equations VII.1.2 and VII.1.3 are very similar in form to an equation the reader has already encountered in the phenomenological description of the laser in the second part of this book (see the rate equations IV.2.2). The terms $(-\gamma_a \rho_{aa})$ and $(-\gamma_b \rho_{bb})$ describe population reductions due to the relaxing of the populations. The terms in $(i/\hbar)\,[pE(0, t)\rho_{ba} - cc]$ describe population variations due to absorption and to stimulated emission. *The advantage of the semiclassical theory is that it specifies the physical meaning and describes the frequency dependency of Einstein's phenomenological constant* B_{21}*, which he introduced in order to describe stimulated emission.*

Moreover, the semiclassical theory makes it possible to calculate the polarization of the medium, by way of the coherence, whereas Einstein phenomenological theory does not describe the response of a medium to a perturbing electromagnetic wave.

Equation VII.1.13 shows that whatever the sign of the population difference D_0 in the absence of the laser field, applying the field *makes this difference smaller.* This is not a surprising result. If D_0 is positive, we have an *absorbing medium*, and the laser field lifts more atoms into state $|\, b >$ by absorption than it sends into state $|\, a >$ by stimulated emission.

The difference therefore becomes smaller. On the other hand, if D_0 is negative, we are dealing with an *amplifying medium* and so the laser field sends more atoms into state | a > by stimulated emission than it carries into state | b > by absorption. Here too, the population difference diminishes. This decrease of the population difference has a *resonant nature:* The denominator $\left[1 + (\omega_0 - \omega_L)^2 / \gamma_{ab}^2\right]$ has the same structure as that of Eq. IV.2.14, $\left[1 + 4(\omega_L - \omega_0)^2 / \Delta\omega_0^2\right]$, which was introduced during the description of the classical microscopic theory developed in Chapter IV. *The further* ω_L *gets away from the resonance frequency* ω_0 *of the illuminated system, the smaller the effect of the wave on the medium.* In what follows, we shall use the notation $(\omega_L - \omega_0) / \gamma_{ab} = \Delta\omega_L$.

We shall presently discuss the very important part played by parameter S_{sat}. Assuming resonance ($\omega_L = \omega_0$), we shall consider three successive limit cases:

(a) $S_{sat} << 1$

In this case, $\bar{I}_L << I_{sat}$ (a laser with very weak intensity, or (and) a medium which is not very sensitive to radiation and is characterized by a weak transition moment p). Then $D_s \approx D_0$: the laser does not modify the population difference between states | a > and state | b >. This is the domain where *Lambert-Beer's law* applies, since it assumes that the populations under consideration stay quasi-constant.

(b) $S_{sat} \approx 1$

Now $\bar{I}_L$ (equal to I_{sat}) represents that laser intensity, inversely proportional to p^2, with which the medium must be illuminated if one wishes to halve the population difference. I_{sat} is called the *saturation intensity of the transition under consideration* (i.e., the transition between states | a > and | b >). Nowadays, this saturation intensity can be obtained for most EDA-allowed (EDA: electric dipole approximation) atomic and molecular transitions.

(c) $S_{sat} >> 1$

In this limit, we have $\bar{I}_L >> I_{sat}$ (a very intense laser beam, or (and) a medium which is very sensitive to radiation and which is characterized by a large transition moment). Now $D_s \approx 0$, and the laser radiation equalizes the populations. Absorption is compensated by stimulated emission. The laser wave behaves as if it were no longer absorbed.

As a first approximation, this mechanism gives the outline of the phenomenon of *saturable absorption* which is met in laser physics (see Inset VI.4). The reader will notice the complete similarity between Eq. VI.1.33 and Eq. VII.1.13 (taken at resonance).

However, we have to say that as the result of the approximations used, Eq. VII.1.13 is only valid for not-too-large values of S_{sat}. For very strong laser intensities, more complex phenomena—which are not reported here—appear.

These phenomena somewhat upset our analysis, but it remains qualitatively valid.

Let us examine the consequences of the interaction in terms of polarization, susceptibility, and refractive index.

VII.2.3 Expression for the polarization

The following expression gives the component of the polarization which is parallel to field **E** in terms of the volume density N of the microsystems, and of the mean quantum value of the induced microscopic dipolar moment **p** (see Eq. IV.2.8 and Section I.1.1.2):

$$|P_E(t)| = N\,|<|\mathbf{p}|>| \qquad \text{(VII.2.1)}$$

On the other hand, we have shown (see Problem I.4)

$$|<|\mathbf{p}|>| = p\,(\rho_{ab} + c\,c) = 2\,p\,\mathcal{R}(\rho_{ba}) \qquad \text{(VII.2.2)}$$

Substituting expression VII.1.10 into VII.2.2, we obtain

$$|P_E(t)| \approx \mathcal{R}\left\{\left[p^2 E_0 N D_0 / \hbar\, \gamma_{ab}\right] \frac{\left(i - \Delta\,\omega_L\right)\exp(-i\omega_L t)}{1 + S_{sat} + \Delta\,\omega_L^2}\right\} \qquad \text{(VII.2.3)}$$

In the same way as in Chapter IV (see Eq. IV.2.11), let us define

$$P_E(t) \approx \mathcal{R}\left[\mathcal{P}_E(\omega_L)\exp(-\,\omega_L\,t)\right]$$

and the polarization $\mathcal{P}_E(\omega_L)$, *induced at frequency* ω_L by the laser in the medium, measured in the direction of the field **E**, takes the form

$$\mathcal{P}_E(\omega_L) = \left[N\,p^2 E_0 D_0 / \hbar\, \gamma_{ab}\right] \frac{i - \Delta\omega_L}{1 + S_{sat} + \Delta\omega_L^2} \qquad \text{(VII.2.4)}$$

Here, we need to make a few comments:

— We already observed (see Section II.3.2.1) that in the expression of polarization $P_E(t)$, *harmonic overtones* of frequency ω_L appear. But the model developed here, based on the rotating wave approximation, limits the expression to just the single component of the *polarization oscillating at frequency* ω_L.

— As is shown when using the classical microscopic model developed in Chapter IV, the polarization is *a complex quantity*.

— It is also a *nonlinear quantity* (S_{sat} depends on I_L). At resonance, and assuming S_{sat} small compared to unity, a power series expansion of expression VII.2.4 shows only *uneven powers of the electric field*. Indeed, the medium is assumed to be isotropic, so that polarization must be uneven with respect to spatial inversion (see Chapter II). Therefore, *all polarizations of even order must vanish.* Let us now define the optical susceptibility of the medium.

VII.2.4 Expression for the susceptibility

The susceptibility is expressed by

$$\chi(\omega_L) = \mathcal{P}_E(\omega_L) / \varepsilon_0 E_0 \qquad \text{(VII.2.5)}$$

yielding

$$\chi(\omega_L) = \left[N\, p^2\, D_0 / \varepsilon_0 \hbar\, \gamma_{ab}\right] \frac{i - \Delta\omega_L}{1 + S_{sat} + \Delta\omega_L^2} \qquad \text{(VII.2.6)}$$

The susceptibility is also a complex quantity, and it is nonlinear by its very nature. A power series expansion reveals only *even powers of the field.* Only uneven order susceptibilities (even ranked tensors) are nonvanishing. As indicated above, this property is a consequence of the spatial isotropy of the medium under consideration.

VII.2.5 Expression for the refractive index

We can write the electric displacement vector **D** (see Eq. II.2.13) as

$$\mathbf{D} = \varepsilon_0\, \varepsilon_r\, \mathbf{E} = \varepsilon_0\, \mathbf{E} + \mathbf{P} = \varepsilon_0\,(1 + \chi)\,\mathbf{E} \qquad \text{(VII.2.7)}$$

and the refractive index as $n^2 \approx \varepsilon_r$.

If we assume $|\chi| << 1$, which is true for diluted media (such as gases for example), then we can write:

$$n = n' + in'' \approx 1 + \chi'/2 + i\,\chi''/2 \qquad \text{(VII.2.8)}$$

As always, the superscripts (') and (") indicate, respectively, the real and the imaginary part of a complex quantity. This equation leads us to define an *index increment* Δn which adds to the real part of the index, as well as an *absorption-amplification coefficient* α, as follows:

$$\begin{aligned} \Delta n &= \chi'/2 \\ \alpha &= -2\pi\chi''/\lambda_L \end{aligned} \qquad \text{(VII.2.9)}$$

λ_L is the wavelength of the laser wave propagating along the 0z axis and α is defined by the relation

$$d\bar{I}(z) / \bar{I}(z) = \alpha(z)\, dz \qquad \text{(VII.2.10)}$$

$\alpha > 0$ corresponds to amplification while $\alpha < 0$ corresponds to absorption. This equation is formally equivalent to Eq. IV.2.37 which was established during the phenomenological study of the laser. However, as we noted earlier, the semiclassical theory gives a physical content to the phenomenological parameter. Identification with Eq. VII.2.6 yields

$$\Delta n \approx -\left(N\, p^2\, D_0 / 2\, \varepsilon_0\, \hbar\, \gamma_{ab}\right) \frac{\Delta\omega_L}{1 + S_{sat} + \Delta\omega_L^2} \qquad \text{(VII.2.11)}$$

$$\alpha \approx -\left(2\, \pi\, N\, p^2\, D_0 / \varepsilon_0\, \hbar\, \lambda_L\, \gamma_{ab}\right) \frac{1}{1 + S_{sat} + \Delta\omega_L^2} \qquad \text{(VII.2.12)}$$

Figure VII.2.1 shows the variations of Δn and of α as a function of $\Delta\omega_L$ for two values of parameter S_{sat}.

The reader will have noticed that the curves are similar to those of Figure IV.2.4 except for the symmetry with respect to axis $\Delta\omega_L$ in the case of $S_{sat} = 0$, i.e. when the laser wave does not perturb the populations of the energy levels. But when S_{sat} increases, the two curves have decreasing amplitudes. Both have a tendency to flatten out as $S_{sat} \rightarrow \infty$. At this limit, the medium is "transparent" for the optical wave (we discussed the meaning of this adjective earlier). On these curves, *the nonlinear nature* of the refractive index is clearly seen.

VII.2.6 Evolution of the laser intensity within the medium

This evolution is described classically by Eq. VII.2.10, where $\alpha(z)$ is given by Eq. VII.2.12. It is important to remember that the quantities Δn and α defined in the last section are in fact functions of the z coordinate, 0z being the propagation direction of the laser wave. Indeed, Δn and α depend on the intensity $\bar{I}_L$ (through the saturation parameter S_{sat}), and of course, $\bar{I}_L$ is a function of z.

In particular, at resonance we have

$$\alpha\,(z) = \alpha_0 \,/\, (1 + S_{sat}) \qquad \text{(VII.2.13)}$$

In this expression α_0 represents the nonsaturated absorption coefficient, which is *independent of z*, obtained by setting $S_{sat} = 0$ in Eq. VII.2.12. Please notice that this expression is similar to the one giving the saturated gain G of the laser (see Chapter IV). Equation VII.2.10 can always be integrated analytically, but integration is especially simple in the two following limit cases:

— $S_{sat} \approx 0$: Integration gives an *exponential evolution* (Lamber-Beer law) for the wave intensity inside the medium (see Eq. IV.2.39).

— $S_{sat} >> 1$: In this case, we obtain a *linear regime*, characteristic of saturation (see Eq. IV.2.40).

Let us proceed to the description of three applications of this first example of a theoretical treatment.

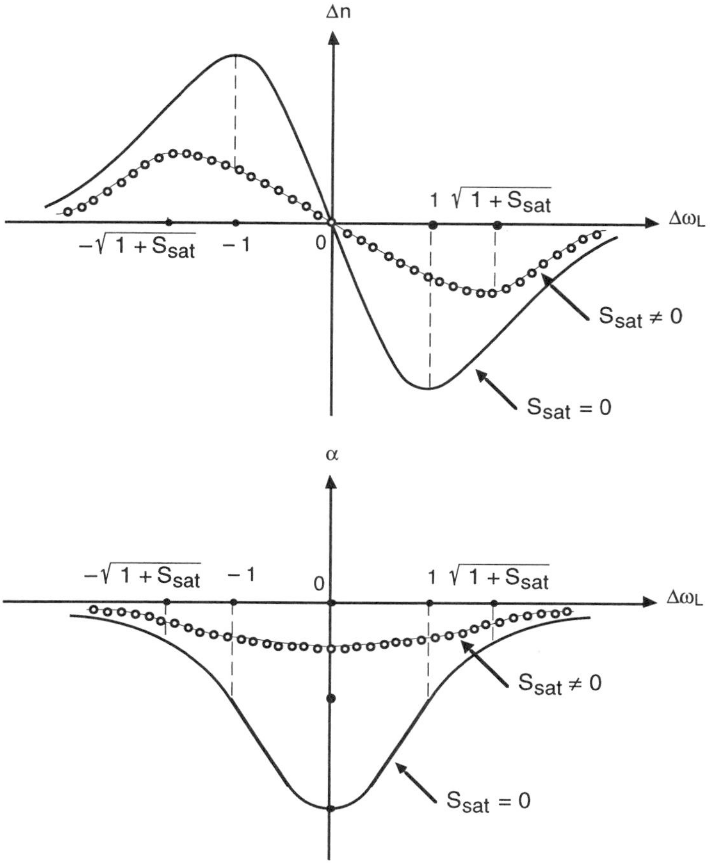

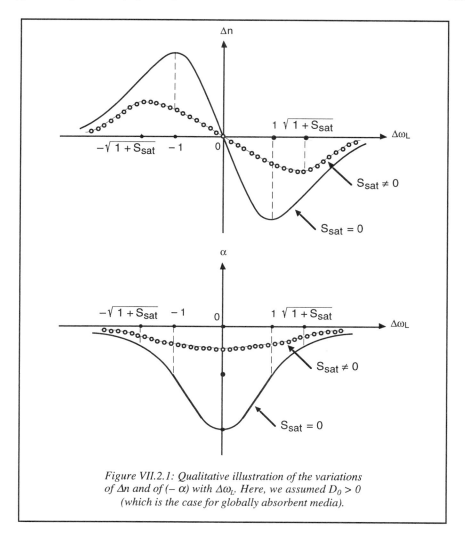

Figure VII.2.1: Qualitative illustration of the variations of Δn and of $(-\alpha)$ with $\Delta\omega_L$. Here, we assumed $D_0 > 0$ (which is the case for globally absorbent media).

VII.3 Some applications

VII.3.1 Lamb's semiclassical theory of the laser (1964)

We shall give only an outline of this theory, which the reader can study more in detail in reference IV.5.3.

As said before, only matter is quantified in this theory.

The thought line followed to calculate the field at each point of the laser is illustrated below:

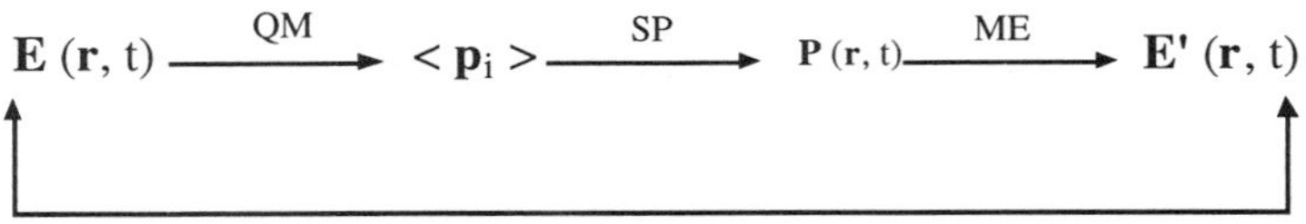

fields **E** and **E'** are identical

The electric field $\mathbf{E}(\mathbf{r}, t)$, assumed to exist within the active medium of the laser, induces *microscopic dipolar moments* $\mathbf{p}_i$. We know how quantum mechanics (QM) goes about to calculate the mean *quantum value* $< \mathbf{p}_i >$ (see Chapter I). Statistical physics (SP) shows how the statistical sum of all these microscopic polarizations leads to a *macroscopic polarization* of the medium $\mathbf{P}(\mathbf{r}, t)$ (see Chapter III). This polarization plays the part of a *source term* in Maxwell's equations (ME) whence it begets a resultant field $\mathbf{E}'(\mathbf{r}, t)$. But since this is in fact the very field that induces the dipolar moments in the first place, it must be identical with $\mathbf{E}(\mathbf{r}, t)$.

Let us suppose that we calculated the macroscopic polarization of the medium (see Section VII.2.3), and let us try to include it in the wave equation so as to find the value of field $\mathbf{E}(\mathbf{r}, t)$. We already wrote the wave equation in vacuum (see Table IV.2.2).

Inside a medium, this equation is written as (see reference II.7.3)

$$\Delta \mathbf{E} - \mu_0 \sigma \, \partial \mathbf{E} / \partial t - (1/c^2)\, \partial^2 \mathbf{E} / \partial t^2 = \mu_0 \, \partial^2 \mathbf{P} / \partial t^2 \qquad \text{(VII.3.1)}$$

μ_0 is the permittivity of vacuum and σ is the electric conductivity (the reader could try to obtain this equation starting with the data of Table II.2.2).

In fact, σ is here considered to be a loss factor concerning losses distributed evenly all through the amplifying medium. This is indeed a rather unrealistic approximation to the Lamb theory. As a matter of fact, in most lasers the major part of the losses are localized on the mirrors and on the other optical elements present in the cavity. Yet, this approximation makes it possible, as we shall see later, to write at each point an equilibrium equation between the losses and the gain, thus avoiding a painstaking integration over the length of the cavity. Within the frame of this theory, applied to a linear resonator (on the 0z axis) and to multimode emission, we shall choose a stationary solution of the form

$$E(z, t) \approx (1/2) \sum_q \left[A_q(t)\, U_q(z) + cc\right] \qquad \text{(VII.3.2)}$$

where $A_q(t)$ represents the time dependency of the qth mode:

$$A_q(t) \approx E_q(t) \exp\left\{-i\left[\omega_q t + \phi_q(t)\right]\right\} \qquad \text{(VII.3.3)}$$

and $U_q(z)$ represents the spatial dependency of the stationary wave, written as

$$U_q(z) = \sin(\omega'_q z/c) \qquad \text{(VII.3.4)}$$

ω'_q and ω_q stand for the angular frequencies of the qth mode, respectively, inside the empty cavity and inside the cavity containing the amplifying medium. It must be added that Lamb's theory also uses the plane-wave approximation, which corresponds to the case where the beam waist of the wave is infinite (see Chapter IV). Let us write the polarization in a form similar to that of the field:

$$P(z, t) = (1/2) \sum_q P_q(t) \exp\left\{-i\left[\omega_q t + \phi_q(t)\right]\right\} U_q(z) + cc \qquad \text{(VII.3.5)}$$

At the cost of a few approximations [that $E_q(t)$, $\phi_q(t)$, $P_q(t)$ vary slowly with time as compared to the optical period ; that we are dealing with diluted active media; etc.] we end up with a differential equation whose solution is field **E**:

$$[i\,\sigma\,\omega_q/\varepsilon_0 + \omega_q^2 - \omega'^2_q + 2\omega_q\,\dot{\phi}_q(t)]\,E_q(t) + 2i\,\omega_q\,\dot{E}_q(t) = -(\omega_q^2/\varepsilon_0)\,P_q(t) \qquad \text{(VII.3.6)}$$

Or again, writing the equality of the real parts and of the imaginary parts of this equation we obtain

$$[\omega_q^2 - \omega'^2_q + 2\omega_q\,\dot{\phi}_q(t)]\,E_q(t) = -(\omega_q^2/\varepsilon_0)\,\mathcal{R}\,[P_q(t)] \qquad \text{(VII.3.7)}$$

$$(\sigma\omega_q/\varepsilon_0)E_q(t) + 2\,\omega_q\,\dot{E}_q(t) = -\left(\omega_q^2/\varepsilon_0\right)\mathcal{J}\left[P_q(t)\right] \qquad \text{(VII.3.8)}$$

We can now define a *quality factor* of mode q, Q_q, such that the ratio $Q_q/2\pi$ is equal to the ratio between the mean volume energy density stored $\left(\varepsilon_0\,|\mathbf{E}|^2/2\right)$ and the mean volume energy density dissipated per period $\left(\pi\,\sigma\,|\mathbf{E}|^2/\omega_p\right)$ by mode q. Then, making the approximation $\omega_q + \omega'_q \approx 2\omega_q$, Eqs. VII.3.7 and VII.3.8 can be written as

$$\omega_q + \dot{\phi}_q(t) = \omega'_q - (\omega_q/2)\,\mathcal{R}\,[P_q(t)/\varepsilon_0\,E_q(t)] \qquad \text{(VII.3.9)}$$

$$\dot{E}_q(t) + \left(\omega_q/Q_q\right)E_q(t) = -\left(\omega_q/2\right)\mathcal{J}\left[P_q(t)/\varepsilon_0\right] \qquad \text{(VII.3.10)}$$

We notice that $P_q(t)\,/\,\varepsilon_0\,E_q(t)$ defines the susceptibility $\chi_q = \chi'_q + i\,\chi''_q$ of the medium (see Eq. VII.2.5). The above differential equations can be written in a condensed form as follows:

$$\omega_q + \dot{\phi}_q(t) - \omega'_q = -(\omega_q/2)\,\chi'_q \qquad \text{(VII.3.11)}$$

$$\dot{E}_q(t)\,/\,E_q(t) + \omega_q\,/\,2Q_p = -(\omega_q/2)\,\chi''_q \qquad \text{(VII.3.12)}$$

Equations VII.3.11 (the equation describing the dispersion of frequency ω_q) and VII.3.12 (equation describing the damping of the field) express the self-consistency condition of the field at each point of the amplifying medium.

If we replace χ'_q and χ''_q by the expressions we established in Section VII.2, we can solve these equations which describe the operating laser.

But let us just point out some of the more important assets of this theory. First, we must analyze the consequence of the presence or the absence of an active medium:

— *If there is no active medium:*

o Then $\chi'_q = \chi''_q = 0$. Equation VII.3.12 shows that $E_q(t)$ relaxes exponentially with a time constant of Q_q/ω'_q which is equal to the confinement time τ_{ph} of the empty cavity (see Eq. VI.1.13). Thus, reffering to Inset VI.2, we have in this case

$$Q = \omega'_q \, \tau_{ph} = 2\pi \,/\, (1 - r) = \omega'_q \,/\, \Delta\, \omega'_q$$

And we can see that the quality factor—or Q factor—of a mode is indeed equal to the inverse of its relative-frequency half-width.

o Equation VII.3.11 shows that $\phi_q(t)$ = constant: There is no time drift of the phase in the empty cavity, and, of course, the oscillation frequencies are those of the empty cavity.

— *If an active medium is present:*

o Then the stationary field inside the cavity is described by Eq. VII.3.12 in which we set $\dot{E}_q(t) = 0$. Clearly χ''_q must be negative, which corresponds indeed to a gain in field of

$$\chi''\,(E_q) = -\,1/Q \qquad \text{(VII.3.13)}$$

At each point of the active medium, the saturated gain is now equal to the distributed losses of the cavity.

o The frequency of the qth mode $\left[\omega_q + \dot{\phi}_q(t)\right]$ is shifted by a quantity $(-\,\omega_q\,\chi'_q\,/\,2)$ with respect to ω'_q, the frequency of the same mode in the empty cavity. For small values of the saturation parameter S_{sat}, defined by Eq. VII.1.14, χ'_q does not depend on the field of the laser. The frequency shift is then at its maximum. This is the so-called *"mode pulling"* phenomenon. At higher laser intensities, S_{sat} increases and the frequency shift becomes smaller. This inverse phenomenon is called *"mode pushing"*.

So this is how Lamb's theory describes monomode and multimode longitudinal laser operation. It expresses the evolution of the oscillation frequencies and that of the laser intensity. It gives the threshold condition and it yields the main characteristics of the steady state operating regime of the laser. It also yields a very satisfying description of mode-to-mode coupling in case of multimode emission and relates it to the saturation of the amplifying medium.

This theory is based on a certain number of *approximations*: Waves are assumed to be plane waves; the medium is viewed in the two-level system approximation; line broadening is assumed to be homogeneous; losses are assumed to be distributed continuously all through the active medium. It is a *semiclassical* theory in which the electric field is treated as if it were a *scalar field*.

While we are on the subject, we would like to point out the important contribution to our understanding of laser-matter interaction made by the *vectorial* semiclassical theory, developed by Le Floch and Le Naour (see reference 5.1), which proved to be especially fertile. This theory reveals the shortcomings of the scalar theory.

VII.3.2 Saturated absorption spectroscopy

VII.3.2.1 Homogeneous broadening of absorption lines in the absence of saturation

Let us suppose that, in a standard absorption experiment using a tunable laser, we are dealing with identical and frozen microsystems (i.e., the microsystems are motionless) and that the laser intensity is weak enough so as to have $S_{sat} << 1$. Then, using the approximation $\alpha(z) \approx \alpha_0$ (see Eq. VII.2.13), Eq. VII.2.10 can be integrated and leads to the following value of the intensity $\bar{I}(d)$ transmitted through a tank of width d (such that $\alpha_0 d << 1$):

$$\bar{I}(d) \approx \bar{I}(0)(1 + \alpha_0 d)$$

Replacing α_0 by its value yields

$$\bar{I}(d) \approx \bar{I}(0)\left[1 - \left(2\pi N p^2 D_0 d / \varepsilon_0 \hbar \lambda_L \gamma_{ab}\right) \frac{1}{1 + \Delta\omega_L^2}\right] \qquad \text{(VII.3.14)}$$

The curve of the transmitted intensity as a function of $\Delta\omega_L$ has a Lorentzian shape with a half-width at half-maximum of unity (yielding a frequency half-width equal to γ_{ab}). This curve is similar to that of coefficient α shown in Figure VII.2.1.

In this case of *homogeneous broadening* (see Inset IV.1), an absorption experiment gives access to the following *three direct statements*:

— The absolute value of the resonance frequency ω_0 (found by varying ω_L),

— The value of the relaxation constant γ_{ab} of the dipolar moment (given by the frequency linewidth),

— The value of p, the transition dipole moment (which is related to the amplitude of the intensity variation induced by varying ω_L).

Moreover, starting from the absorption curve, it is possible to make an *indirect* construction of the dispersion curve by using the Kramers-Kronig relations mentioned earlier (in Chapter IV).

VII.3.2.2 Saturation of homogeneous broadening

If the intensity of the laser increases, the above equation is no longer valid. The curve of $\bar{I}(d)$ stays Lorentzian, but its amplitude decreases as $\bar{I}(0)$ (and therefore S_{sat}) increases. The half-width at half-maximum is $\sqrt{1 + S_{sat}}$, and so it also increases with S_{sat} (see Figure VII.2.1). This widening greatly hinders the determination of the fundamental constants (ω_0, γ_{ab}, and p) of the microsystems.

VII.3.2.3 Inhomogeneous broadening; the case of Doppler broadening

As a matter of fact, the experimentally obtained absorption spectra are very often broadened because of the presence of several classes of molecules, each characterized by a slightly different resonance frequency; this is called *inhomogeneous broadening* (described in Inset IV.1). As an example, we take the Doppler broadening of a gas which is due to the fact that the microsystems under consideration move at different velocities. These velocities, following Maxwell's statistics, are distributed according to the characteristic distribution function $f(v_z)$:

$$f(v_z) = (\sqrt{\pi}\, v_0)^{-1} \exp\left[- v_z^2 / v_0^2\right] \qquad \text{(VII.3.15)}$$

v_z is the projection on the 0z axis and v_0 is the modulus of the most probable velocity at thermal equilibrium, assuming the experiment takes place at thermal equilibrium. But a microsystem moving at a velocity of modulus v "sees" the laser frequency in its own reference frame at frequency ω_L^* such that

$$\omega_L^* \approx \omega_L \left(1 - v \cos\theta / c\right) \approx \omega_L - k_L\, v_z \qquad \text{(VII.3.16)}$$

where θ is the angle between vectors $\mathbf{v}$ and $\mathbf{k}_L$ (the wave vector of the laser wave, assumed to lie on the 0z axis).

The frequency differences kv_z add to the experimental line-broadening, which thus appears as a mixture of homogeneous broadening (the homogeneous line width broadened by saturation) and of inhomogeneous broadening. The experimental signal is a superposition of the absorption of all the velocity classes, weighed by the number of molecules in each class. When $k_L v_z >> \gamma_{ab} \sqrt{1 + S_{sat}}$, which is practically always the case for the media used in experiments in the optical domain, we observe a *Gaussian Doppler envelope* which, partly or completely, masks the desired homogeneous broadening. During the last 10 to 40 years, various techniques have been devised so as to obtain *spectroscopies without Doppler effect*. Some of these techniques are as follows

— The so-called *optical pumping* method (Kastler, 1950), which studies the evolution of excited states (filled with the help of auxiliary sources) undergoing outside perturbations in the form of electromagnetic fields. This evolution cannot be concerned by Doppler effect.

— The so-called *molecular jet* methods, which select only one class of velocities. But these methods are experimentally very difficult and they are marred by the divergence of the jet.

— The so-called *two-photon* methods (see Problem VII.3) where the two absorbed photons have opposite wave vectors. In this case, the resonance condition implies that $(\omega_L - k_L v_z) + (\omega_L + k_L v_z) = \omega_0$, i.e., $2\,\omega_L = \omega_0$, and this does not depend on velocity class. This very clever method makes it possible to study transitions between states of same parity which are forbidden for one-photon transitions by the electric transition dipole. Unfortunately, two-photon absorption is very weak, making this method of limited use. As it is, two-photon spectroscopy seems to be complementary to one-photon spectroscopy, which can only study electric dipole-permitted transitions.

— And, last, the so-called *laser saturation* methods which we shall discuss presently.

VII.3.2.4 The method of saturated absorption spectroscopy

(a) Principle

The principle of this method is described by Figure VII.3.1.

A tunable laser of variable frequency ω_L sends two waves: a pump (pu) wave which is relatively strong and which carries a "label" (for instance, it is labeled by an external low frequency modulation); and a probe (pr) wave of lesser intensity. The two waves travel in opposite directions through a cell containing a medium with eigenfrequency ω_0. The microsystems interacting resonantly with the pump laser are those that move with velocity **v** such that $v_{z,\,pu} = (\omega_L - \omega_0) / k_L$; those interacting with the probe wave have a velocity such that $v_{z,\,pr} = (\omega_0 - \omega_L) / k_L = - v_{z,\,pu}$. Due to these two waves, two "holes" appear in the distribution giving the population difference D_s as a function of v_z (presenting a maximum at $v_z = 0$, see Eq. VII.3.15). These two "holes" have a width of $\gamma_{ab} \sqrt{1 + S_{sat}} / k_L$, and they lie symmetrically with respect to the origin (see Figure VII.3.2).

The width of the "hole" caused by the probe wave is smaller than that of the pump "hole" since $I_{pu} >> I_{pr}$ and therefore $S_{sat,\,pu} >> S_{sat,\,pr}$. This also implies (see Section VII.1.13) that the pump "hole" is much deeper than the probe "hole". What is fundamental, though, is that the two "holes" thus created correspond to different velocity classes: the molecules interacting with the pump wave have speeds such that $v_z > 0$, while those interacting with the probe wave have speeds with $v_z < 0$. *Therefore the detected probe response does not, in general, carry the label of the pump wave.*

Yet, when the experimenter varies ω_L, the two holes move symmetrically with respect to the origin of v_z. When $\omega_L = \omega_0$, i.e. for $v_z = 0$, pump and probe wave interact with the same class of molecules, namely those moving perpendicularly to the 0z axis (see broken line in Figure VII.3.2). The detector, which is programmed to detect only waves carrying the pump label, will therefore detect a signal only when $\omega_L = \omega_0$. This signal, which is not marred by Doppler effect (since in this case **v** and $\mathbf{k}_L$ are perpendicular), is shown in Figure VII.3.3.

The width at half-maximum now leads to a correct determination of γ_{ab}.

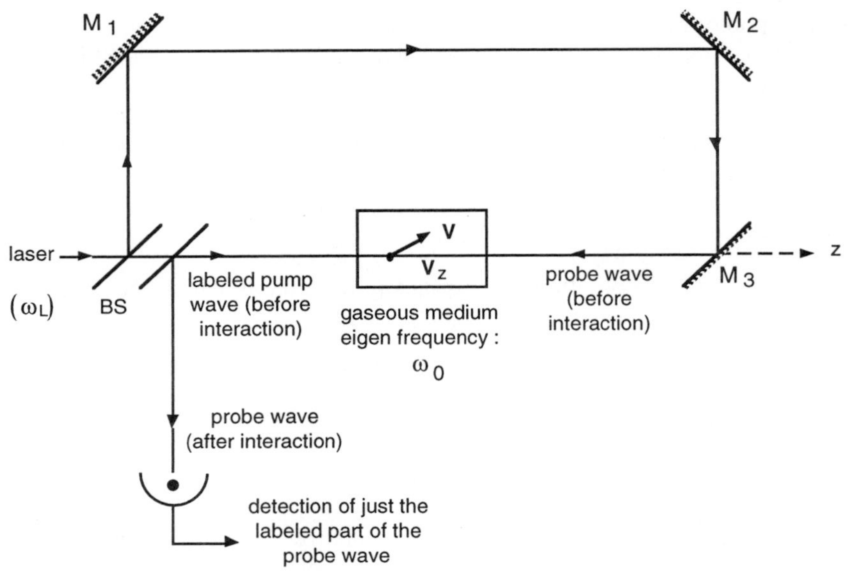

Figure VII.3.1: Principle of saturated absorption spectroscopy; BS is a beam splitter such that $I_{pr} << I_{pu}$.

(b) Limitations and results

The above description is, of course, not a very thorough one. In reality, there are many causes that restrict the saturation techniques. For instance, the quality, the geometry, the spatial inhomogeneities of the laser beams, and the low-frequency modulation imposed on the pump wave make the interpretation of the results very complex. Also, using these saturation techniques in precision measuring experiments revealed the existence of many other causes of broadening other than the Doppler effect which was the only one we spoke about in our analysis [for instance, second-order Doppler effect (proportional to v^2/c^2), molecular recoil upon absorbing a photon, Zeeman effect due the earth's magnetic field].

Yet, notwithstanding these restrictions, saturated absorption has led to many spectacular advances in spectroscopy and especially in the domains of hyperfine structures, of isotopic line displacements, and in the physics of molecular gases (see reference 5.2).

VII.3.3 Laser manipulation of microscopic systems

The most commonly studied interaction between electromagnetic waves and microscopic systems is the one which couples the wave to the internal degrees of freedom of the system. The electromagnetic wave imposes a cyclic behavior onto

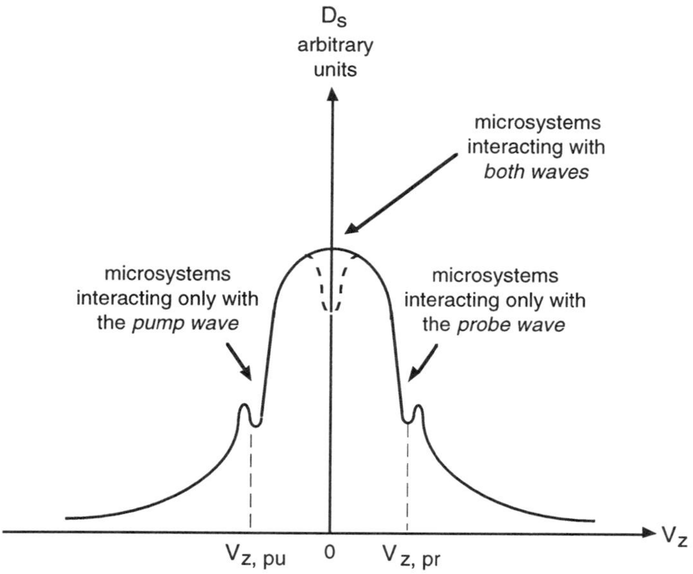

Figure VII.3.2: Distribution of the velocities (case of a gaseous medium and with Doppler effect).

the system: It will lift it from its ground state into an excited state (absorption), or it will induce it to return to a lower energy state by emitting a photon (stimulated emission). However, the wave can also interact with the movements of the system —that is, with its external degrees of freedom. Beginning in the years 1980, this kind of interaction has been used to *levitate microscopic particles* and also *to cool* (i.e. *to slow down*) or *to trap* ions or atoms. In this interaction, there is an exchange of momentum between the wave and the systems. We shall briefly describe the main aspects of this exchange, and then we shall try to show how such exchanges can be used to manipulate systems of sizes ranging from atomic sizes (a few tens of nanometers) to micrometric dimensions (a few micrometers). The interested reader can refer to references 5.3 and 5.4 for further information on this subject.

VII.3.3.1 Coupling between an electromagnetic wave and a microsystem

(a) Radiative scattering forces

Let us first look at the momentum exchange between a wave and an atom during the absorption of a photon by the atom. The law of conservation of the total momentum implies that the momentum of the atom $(m\mathbf{v})_a$ changes by a quantity $\Delta_a(m\mathbf{v})$ which is equal to $\hbar\mathbf{k}$, the momentum of the absorbed photon: $\Delta_a(m\mathbf{v}) = \hbar\mathbf{k}$.

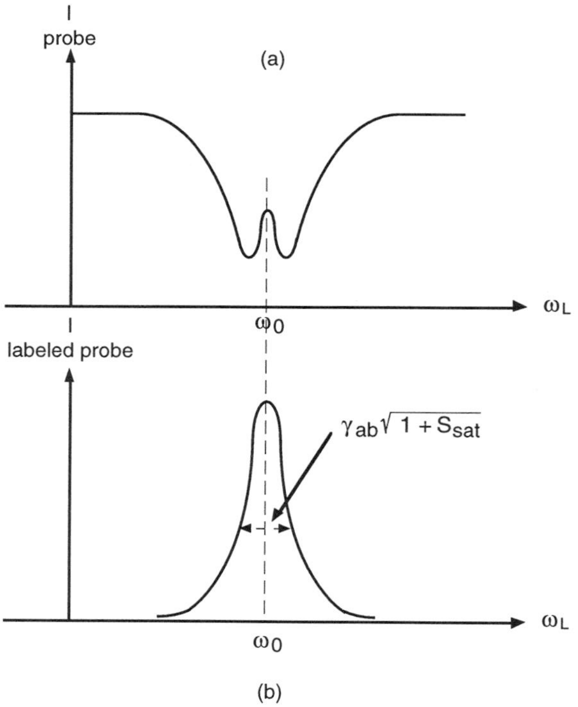

Figure VII.3.3: Detected intensities:
(a) of the probe wave
(b) of the "labeled" part of the probe wave.

If this absorption is followed by the stimulated emission of a photon, and assuming that the exciting wave is a plane wave associated to photons all having the same momentum $\hbar\mathbf{k}$, then the photon emitted by stimulation will in its turn carry away a momentum $\hbar\mathbf{k}$. During the stimulated emission procedure, the atom undergoes a change of momentum $\Delta_{st}(m\mathbf{v}) = -\hbar\mathbf{k}$. After a complete absorption-stimulated emission cycle, the change in momentum of the atom is zero:

$$\Delta_t(m\mathbf{v}) = \Delta_a(m\mathbf{v}) + \Delta_{st}(m\mathbf{v}) = \mathbf{0} \qquad \text{(VII.3.16 bis)}$$

But these absorption-stimulated emission cycles can be interrupted by an absorption-spontaneous emission cycle. During a spontaneous emission, the change in momentum of the atom is equal to $\Delta_{sp}(m\mathbf{v}) = -\hbar\mathbf{k}_{sp}$, where $\mathbf{k}_{sp}$ is the wave vector of the spontaneously emitted photon, which means that $\mathbf{k}_{sp}$ has random direction. After an absorption-spontaneous emission cycle, the total momentum change no longer vanishes:

$$\Delta_t(m\mathbf{v}) = \Delta_a(m\mathbf{v}) + \Delta_{es}(m\mathbf{v}) = \hbar\mathbf{k} - \hbar\mathbf{k}_{sp} \qquad \text{(VII.3.17)}$$

A spontaneous emission occurs with a probability which depends on the probability of the atom being in excited state $|b>$, which is $|b|^2$. It also depends on γ_b, the probability per second for an atom in state $|b>$ to relax spontaneously. Therefore $N = \gamma_b |b|^2$ is the mean number of spontaneous emission photons emitted per second by one atom. It is also equal to the number of absorption-spontaneous emission cycles per atom and per second. $10^8\ s^{-1}$ is a typical value for γ_b. N can be very large if $|b|^2$ is not very much smaller than unity. Now in the case of a laser wave saturating transition $|a> \rightarrow |b>$, where $|a>$ is the ground state of the atom, the atom goes through the absorption-stimulated emission Rabi cycles at an extremely high frequency (see Inset VII.1), so that in average we can consider $|b|^2$ to be equal to 1/2.

To get an idea of the force exerted on the atom, i.e. the change in momentum per second, we can multiply N by the mean change in momentum over one cycle, $< \Delta_t(m\mathbf{v}) >$. This yields

$$< \Delta_t (m\mathbf{v}) > = < \hbar\mathbf{k} > - < \hbar\mathbf{k}_{sp} > \tag{VII.3.18}$$

but because of the isotropy of spontaneous emission, $< \mathbf{k}_{sp} > = \mathbf{0}$, so that the radiative force $\mathbf{F}_{rad}$ is equal to

$$\mathbf{F}_{rad} = N < \Delta_t (m\mathbf{v}) > = |b|^2 \gamma_b \hbar\mathbf{k} \tag{VII.3.19}$$

If one wishes to calculate the radiative force for systems which are not at resonance with the wave, but which scatter it (or reflect it, or refract it in case of objects whose dimensions are greater than the wavelength), one cannot use the same method. To calculate the force exerted on the system, one needs to write the momentum difference between the momentum per second carried by the transmitted, the reflected, or the scattered wave on the one hand, and the momentum of the incident wave on the other.

(b) Dipolar forces

Let us now take an exciting wave which is not a plane wave. Then, the total momentum change after a complete absorption-stimulated emission cycle is no longer equal to zero at all points of the wave since the wave vector of the absorbed photon may be different from the wave vector of the photon coming from stimulated emission. We should therefore try to establish a new momentum balancing equation of the cycles, in which we take into account the intensity distribution of the non-plane wave.

An equivalent way to find the solution to this problem, but one that is much easier to handle, consists in (a) writing the energy balance equation of the system using the intensity gradient of the wave and (b) calculating the force which tends to push the particle toward an energy minimum. Let α be the polarizability of the system. For the time being, we shall assume it to be real and isotropic. Let $\mathbf{p}$ be the dipolar moment of the system induced by the electric field. $\mathbf{p}$ increases with the electric field according to $d\mathbf{p} = \alpha\, d\mathbf{E}$. U, the coupling energy between this dipolar moment and the field which induced it, decreases in the following way:

$$dU = -\mathbf{E}\, d\mathbf{p} = -\alpha\, \mathbf{E}\, d\mathbf{E} \tag{VII.3.20}$$

So in an electric field E_0, the energy of the system is

$$U = -\alpha \mathbf{E}_0^2 / 2 \qquad \text{(VII.3.21)}$$

For an electromagnetic field, we have $\mathbf{E}_0 = \mathcal{E}_0 \cos(\omega_L t - kz)$. In this case, U is a periodic function oscillating at an angular frequency of $2\omega_L$. The frequency of these energy vibrations are much too high to have any influence on the movement of the system, which can feel only the mean value, averaged over time, of energy U:

$$\overline{U} = -\alpha \overline{E_0^2} / 2 = -\alpha \mathcal{E}_0^2 / 4 \qquad \text{(VII.3.22)}$$

If α is positive, or negative, $\overline{U}$ will have a minimum at the maxima, or at the minima, of $\overline{E_0^2}$. So the system will have a tendency to move toward maxima, or minima, of the intensity, depending on the sign of α. The corresponding force $\mathbf{F}_{dip}$ is proportional to the intensity gradient as given by Eq. VII.3.22:

$$\mathbf{F}_{dip} = \alpha \nabla \overline{E_0^2} / 2 \qquad \text{(VII.3.23)}$$

For an atomic system at near-resonance with the wave, force $\mathbf{F}_{dip}$ is obtained by replacing the polarizability α in Eq. VII.3.23 by the macroscopic volume susceptibility χ provided by Eq. VII.2.6.

When calculating $\overline{U}$ (see Eq. VII.3.22), one discovers that only the real part of χ contributes to that part of the value of U which does not vanish when averaged over time. Therefore, in the case of our atomic system, we want to replace α by the real part of χ. Doing this leads to

$$\mathbf{F}_{dip} = -\left[\left(N p^2 D_0 / 2 \varepsilon_0 \hbar \gamma_{ab}\right) \Delta\omega_L / \left(1 + S_{sat} + \Delta\omega_L^2\right) \right] \nabla \overline{E_0^2} \qquad \text{(VII.3.24)}$$

By using a laser frequency greater than or smaller than the resonance frequency (a positive or negative $\Delta\omega_L$) we are able, at will, to attract the atoms toward the axis of the laser beam or to push them away from it.

We would like to point out that for micrometric particles which are not at resonance with the wave, the sign of the polarizability α depends on the difference in refractive index between the particles and their environment. A standard calculation for this case shows that

$$\alpha = n_e^2 \left[(n_p / n_e)^2 - 1 \right] a^3 / \left[(n_p / n_e)^2 + 2 \right] \qquad \text{(VII.3.25)}$$

where n_p, n_e, and a are, respectively, the refractive index of the particle, that of its environment, and the radius of the particle (assumed to be spherical). So, depending whether n_p/n_e is greater than or smaller than unity, the particles are attracted or repelled by the axis of the laser beam.

VII.3.3.2 A few examples of manipulation

(a) Atom cooling

Lately, radiative forces have been used to slow down atoms of atomic jets (and slowing these atoms in vacuum is of course equivalent to cooling them down). A laser wave which is resonant with the atoms of a jet having speed $\mathbf{v}_0$ propagates in the direction opposite to the jet (see diagram). The radiative force exerted on each atom causes it to slow down; the velocity of the atom then decelerates uniformly until its speed is vanishingly small. We can calculate (1) the characteristic parameters of this movement for sodium atoms with velocity $v_0 \approx 10^5$ cm s^{-1} and (2) a laser wave ($\lambda_L \approx 600$ nm) saturating a transition of the sodium atom characterized by a γ_b of 10^8 s^{-1}. We find the following values:

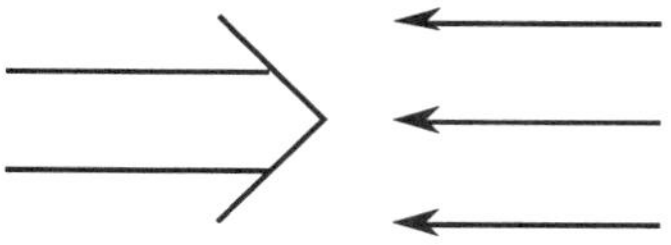

— Change in speed per absorption-spontaneous emission cycle:

$$(\hbar k) / m \approx 3 \text{ cm s}^{-1}$$

— Number of cycles per second:

$$N_c \approx \gamma_b / 2 \approx 50 \text{ MHz}$$

— Mean deceleration:

$$\gamma_b \hbar k / 2m \approx 1{,}5 \times 10^6 \text{ m s}^{-2}$$

— Time constant for the stopping of the atoms:

$$2\, m v_0 / \gamma_b \hbar k \approx 600\ \mu\text{s}$$

— Mean stopping distance of the atoms:

$$v_0^2\, m / \gamma_b \hbar k \approx 30 \text{ cm}$$

The details of these calculations are given by Problem VII.6. But there remains a fundamental question concerning this cooling procedure: The laser wave must stay in resonance with the atoms even as their speed changes. Two solutions exist to maintain this resonance:

— One can apply a magnetic field in the interaction zone, taking care that the field along the axis is such that the Zeeman effect it induces compensates exactly the change in resonance frequency due to the Doppler effect (see Problem VII.6).

— One can try to change the frequency of the laser wave so as to satisfy the resonance condition in spite of the Doppler effect due to the changing speed of the atoms.

These slowed atoms can now be trapped. The first trapping experiments used dipolar forces which tend to push the atoms toward a maximum of the field intensity. Still, the potential well thus obtained is not deep enough to stop residual movement from dispersing the atoms, making the well rather inefficient. Other, more subtle types of interactions, whose descriptions lie outside the scope of this book, are used nowadays to trap atoms in an efficient way. Let us just mention that the advances made, not only in defining efficient experimental configurations, but also in the theoretical descriptions of the interactions, have made it possible to cool atoms down to temperatures on the order of the microkelvin. This temperature is defined on the basis of the mean residual speed of the molecules. Actually, experiments destined to reach the nanokelvin are on their way.

(b) Levitation and trapping of micrometric particles

Lately, a great number of particles have been levitated and trapped. The radiative forces produced by a laser beam pointed directly upward can compensate gravity. The dipolar forces associated to the Gaussian structure of the beam, can, under certain conditions, take care of stabilizing the equilibrium of the levitating particle.

In this domain, other types of interaction have shown up. For instance, experiments have shown recently that the oscillation of the light intensity around a particle, due to interference of the exciting wave with the light scattered from the particle, can compel two particles to pair up very strongly. The explanation is that the dipolar forces associated to the intensity oscillations push the second particle into the potential well created by the first one. This optical bond induced between two particles by the wave was used to create whole networks of micrometric particles. This type of particle manipulation could be of great interest in biology, to trap cells, bacteria, liposomes, and the like. The laser beam acts like an "optical plier" which could, for instance, extract the nucleus of a cell without touching it materially. Recently, it was shown that it is possible to manipulate particles of intermediate size, between atoms and micrometric particles, with laser beams. This has been demonstrated in liquids presenting supramolecular structures, such as inverse micellar phases of microemulsions. In these phases, droplets of water measuring a few nanometers across and surrounded by amphiphilic soap molecules (inverse micelles), form a stable suspension in oil. On the other hand, it has been shown that dipolar forces can be used to move the micelles, and thus to change the phase composition.

These effects are especially interesting at and around a phase transition. For instance, when the liquid is close to the critical point of dissolution between a single micellar phase and a diphasic equilibrium between two micellar phases, its susceptibility diverges. Actual experiments show that the associated optical nonlinearity also diverges. Moreover, manipulating the phase composition by means of lasers can also induce hitherto unknown phase transitions.

Clearly, laser manipulation of microscopic systems, a field which has expanded rapidly since the early 1980s, opens attractive perspectives, in fundamental research as well as in applied research; this also applies to the study of gaseous media as well as that of liquid media showing supramolecular structures.

VII.4. Problems and outlined solutions

VII.4.1 Problem VII.1: The electric dipole approximation

Let us take an atom. It consists of a nucleus of charge +Ze, surrounded by Z electrons (of charge –e and with a rest mass of m_e) identified by the subscript j (j going from 1 to Z). Assume the center of mass of the atom is located at the nucleus and take it as the origin of the spatial reference frame. The atom is illuminated by a laser wave whose electric field (**E**) and magnetic field (**H**) derive from a vector potential **A** as follows:

$$\mathbf{E} = -\dot{\mathbf{A}}, \ \mathbf{H} = (1/\mu_0)\,\nabla \times \mathbf{A}$$

where μ_0 is the magnetic permittivity of vacuum.

The static electric field **F** induced by the electric charges of the atom (see Eqs. II.5.5 and II.5.6) derive from a scalar potential ϕ by $\mathbf{F} = -\nabla\phi$.

1. Give a qualitative explanation for the following expression of the Lagrangian function L:

$$L(\mathbf{r}_j ; \dot{\mathbf{r}}_j ; t) = \sum_j \left[m_e\, \dot{\mathbf{r}}_j^{\,2}/2 - e\, \dot{\mathbf{r}}_j \mathbf{A}(\mathbf{r}_j ; t) + e\,\phi(\mathbf{r}_j)/2\right] - Ze\,\phi(0)/2$$

2. Compare the characteristic dimensions of the atoms with those of the optical wave and show that it is legitimate to use $\mathbf{A}(0; t)$ as an approximation for $\mathbf{A}(\mathbf{r}_j; t)$.

3. Show that the Lagrange equations are not modified by adding the extra term $d\left[e \sum_j \mathbf{r}_j \mathbf{A}(0 ; t)\right]/dt$ to the Lagrange function. Write the expression of the Langrangien thus obtained.

4. Last, find the Hamiltonian function, using $\mathbf{p} = -\sum_j e\,\mathbf{r}_j$.

1. The reader may report to specialized books on analytical mechanics (see references in Section I.6) where the Lagrangian and Hamiltonian functions are defined and analyzed. In order to do the calculation, one must neglect the mass of the electrons with respect to that of the nucleus (which is at the same time the center of mass of the atom). In this case, the kinetic energy of the nucleus vanishes and the expression $\sum_j (m_e\, \dot{\mathbf{r}}_j^{\,2}/2)$ represents the (nonrelativistic) kinetic energy of the electrons. The expressions $\sum_j e\,\phi(\mathbf{r}_j)/2$ and $-Ze\,\phi(0)/2$ correspond respectively to the electronic and to the nuclear part of the potential energy V as they appear in the Lagrangian function (except for a change of sign).

The term $\sum_j e\,\dot{\mathbf{r}}_j \mathbf{A}(\mathbf{r}_j ; t)$ describes a potential energy due to the interaction between the electrons and the electric field of the laser [see first reference of Section I.6, volume II]. Starting with these terms of the Lagrangian, the reader can easily show that the Lagrange equations lead to the Coulomb and the Lorentz forces, whose existence can be shown experimentally. This correspondence between the Lagrange equations and the fundamental dynamic equations explains why we chose this particular form of Lagrangian.

2. Let us write a power expansion of $\mathbf{A}(\mathbf{r}_j;\ t)$ as a function of $\mathbf{r}_j$, around $\mathbf{r}_j = 0 : \mathbf{A}(\mathbf{r}_j\ ;\ t) = \mathbf{A}(0\ ;\ \mathbf{r}_j) + \mathbf{r}_j \left[\nabla_{\mathbf{r}_j} \mathbf{A}(\mathbf{r}_j\ ;\ t)\right]_{\mathbf{r}_j = 0} + \cdots$

If one looks carefully at the second term of this development, one can see that it contains two characteristic lengths: The first is an "atomic" characteristic length on the order of the angstrom $(\mathbf{r}_j \text{ and } \nabla_{\mathbf{r}_j})$, and the second, an "optic" length on the order of the micron, is characteristic of the vector potential **A**. The large difference (three to four orders of magnitude) between these lengths implies that **A** can be considered as constant over atomic lengths, yielding $\mathbf{A}(\mathbf{r}_j, t) \approx \mathbf{A}(0, t)$.

3. The Lagrangian function is not defined uniquely. Without changing its properties, one may add to it the total derivative with respect to time of a function of the spatial coordinate **r**. This leads us to build a second Lagrangian function, L', which is as follows: $L' = L + d\left[e \sum_j \mathbf{r}_j \mathbf{A}(0,\ t)\right] / dt = L + e \sum_j \mathbf{r}_j \dot{\mathbf{A}}(0,\ t) + e \sum_j \dot{\mathbf{r}}_j \mathbf{A}(0,\ t)$.

(We leave it to the reader to check that this Lagrangian function does indeed lead to the same Lagrangian equations as the first one.) So

$$L'(\mathbf{r}_j\ ;\ \dot{\mathbf{r}}_j\ ;\ t) \approx \sum_j \left[m_e \dot{\mathbf{r}}_j^{\,2} / 2 + e\, \mathbf{r}_j \dot{\mathbf{A}}(0,\ t) + e\, \phi(\mathbf{r}_j) / 2\right] - Ze\, \phi(0) / 2$$

4. The Hamiltonian function H can be written as $H = -L' + \sum_j m_e \dot{\mathbf{r}}_j^{\,2}$, yielding

$$H = \left[\sum_j m_e \dot{\mathbf{r}}_j^{\,2} / 2 - e\, \phi(\mathbf{r}_j) / 2 + Ze\, \phi(0) / 2\right] - e \sum_j \mathbf{r}_j \dot{\mathbf{A}}(0,\ t)$$

The expression in square brackets represents the energy of the atom in the absence of radiation. Let us call it H_0. The last expression can be written as

$$-e \sum_j \mathbf{r}_j \dot{\mathbf{A}}(0,\ t) = e \sum_j \mathbf{r}_j \mathbf{E}(0,\ t) = -\mathbf{p}\, \mathbf{E}(0,\ t)$$

and it represents the atom-radiation interaction energy in the EDA (electric dipole approximation).

VII.4.2 Problem VII.2: Introduction to self-induced transparency

Let $\mathbf{E} = \mathbf{E}_0 \cos \omega_L t$ be a linearly polarized laser wave of amplitude E_0 and of angular frequency ω_L, interacting with a two-level system (whose Hamiltonian is H_0 in the absence of radiation). Frequency ω_L is very close to the eigenfrequency $\omega_0 = \omega_b - \omega_a$ of the electric-dipole-permitted transition between two eigenstates $|\, a >$ and $|\, b >$ of H_0. The corresponding eigenvalues are $\hbar\omega_a$ and $\hbar\omega_b$ $(\omega_b > \omega_a)$. Let $\boldsymbol{p}$ be the dipole-moment operator and let $p = <a\,|\,\boldsymbol{p}_z\,|\,b>$. p is assumed to be real.

1. We want to define a nonstationary state $|\psi(t)\rangle = a(t)|a\rangle + b(t)|b\rangle$, representing the evolution in time of the system. The system is assumed to be in state $|a\rangle$ at $t = 0$.

(a) Show that p_z has vanishing diagonal elements in the $|a\rangle$, $|b\rangle$ basis.

(b) First, write the Hamiltonian of the system as $H = H_0 - \boldsymbol{p}\mathbf{E}$, then write the system of differential equations governing the time evolution of a(t) and b(t).

(c) Using the transformation

$$A(t) = a(t) \exp(+ i \omega_a t)$$
$$B(t) = b(t) \exp(+ i \omega_b t)$$

calculate $\dot{A}(t)$ and $\dot{B}(t)$ in the rotating wave approximation. (Set $\Omega = p E_0 / \hbar$; this is the Rabi frequency).

(d) If one solves this system of equations (but we do not ask you to do it), how can its solution help determine the time evolution of the populations of states $|a\rangle$ and $|b\rangle$?

(e) How can this system of equations be modified phenomenologically so as to take into account possible population relaxations from $|a\rangle$ or $|b\rangle$ to other levels which are not mentioned in our model?

2. Neglecting the relaxations mentioned by 1 (e), let us assume resonance (i.e. $\omega_L = \omega_0$). Now we apply a laser pulse of duration t_p to the medium (a pump pulse).

(a) Show that there is a special value of t_p (called $t_{p,\,\pi/2}$), for which the populations of $|a\rangle$ and $|b\rangle$ reach the same value. Calculate $t_{p,\,\pi/2}$.

(b) Show that for a value $t_{p,\,2\pi} = 4\, t_{p,\,\pi/2}$, the system returns into its initial state. Show that for this value of the pulse duration, there is no global energy transfer from the wave to the medium (self-induced transparency). Give a physical explanation.

1. (a) The diagonal elements of the matrix representing operator p_z are written as $\langle a|p_z|a\rangle$, whatever the base used. Here, $|a\rangle$ represents any eigenvector of Hamiltonian H_0 (in the absence of perturbations). If H_0 commutes with the parity operator (see references of Chapter I), state $|a\rangle$ has a well-defined parity ($|a\rangle$ is even, or it is uneven). $\langle a|p_z|a\rangle$ represents the integral taken from $\mathbf{r} = -\infty$ to $\mathbf{r} = +\infty$ of a spatially uneven function (p_z behaves like the z coordinate and is therefore always uneven). This means the integral vanishes for any state $|a\rangle$, and the matrix representing p_z always has vanishing diagonal elements, whatever the basis of the representation.

(b) Let us use postulate P6 of quantum physics (Eq. I.1.10): $i\hbar|\dot{\psi}\rangle = (H_0 + V)|\psi\rangle$. Making two successive left multiplications, the first by $\langle a|$, the second by $\langle b|$, we obtain

$$\dot{a} + i\,\omega_a\, a = (i\, p / \hbar)\, E_0 \cos(\omega_L\, t)\; b$$
$$\dot{b} + i\,\omega_b\, b = (i\, p / \hbar)\, E_0 \cos(\omega_L\, t)\; a$$

(c) Thanks to the rotating wave approximation, the suggested change of variables yields

$$\dot{A} \approx (i\,\Omega / 2) \exp[-\, i\,(\omega_0 - \omega_L)\, t]\, B$$
$$\dot{B} \approx (i\,\Omega / 2) \exp[\; i\,(\omega_0 - \omega_L)\, t]\, A$$

(d) Solving the above differential equations leads to a determination of A(t) and of B(t). The populations (see Chapter I) of states $|\, a >$ and $|\, b >$ are $|\, a(t)|^2 = |\, A(t)|^2$ and $|\, b(t)|^2 = |\, B(t)|^2$.

(e) If we want to take into account the possible relaxation of states $|\, a >$ and $|\, b >$ to other levels, we can introduce phenomenological constants γ_a and γ_b (see Section I.4.2).

2. (a) At resonance, the differential equations can be written as

$$\dot{A} \approx (i\,\Omega / 2)\, B, \quad \dot{B} \approx (i\,\Omega / 2)\, A$$

We can take the derivative of the first equation a second time, which yields $\ddot{A} \approx (i\,\Omega / 2)\, \dot{B} \approx -(\Omega / 2)^2 A$. The general solution of this equation is $A = C_1 \cos(\Omega\, t / 2) + C_2 \sin(\Omega\, t / 2)$, where C_1 and C_2 are constants. So now we have $B \approx i\,[C_1 \sin(\Omega\, t / 2) - C_2 \cos(\Omega\, t / 2)]$ and since B(0) = 0 we find $C_2 = 0$; $A(0) = \pm 1$ leads to $C_1 = \pm 1$. Finally, we obtain

$$A = \pm \cos(\Omega\, t / 2)\, ;\; B = \pm\, i \sin(\Omega\, t / 2)$$

The populations of states $|\, a >$ and $|\, b >$ are, respectively, equal to $\cos^2(\Omega t/2)$ and to $\sin^2(\Omega t/2)$. We can see that the populations oscillate at Ω, the Rabi frequency. The populations are equal for a pulse duration of $t_{\pi/2} = (\pi/2) / \Omega$.

(b) If a pulse has a duration of $t_{p,\, 2\pi} = 4 t_{p,\, \pi/2} = 2\pi/\Omega$, i.e. equal to the Rabi period, the system is back in its initial state $|\, a >$ at the end of the pulse. Globally, no energy has been transferred from the wave to the medium ; the wave comes out of the medium with just as much energy as it went into it. This is called *self-induced transparency*. Of course, the medium has at first absorbed a photon of the incoming wave front of the pulse, thereby jumping to state $|\, b >$, but it has restored this photon to the wave by stimulated emission, thus returning to state $|\, a >$. We must point out, though, that even if there is no exchange of energy, the time shape of the pulse may have been modified.

VII.4.3 Problem VII.3: Optical transitions in one- and two-photon spectroscopies

1. *Single-photon spectroscopy*

In this problem, we want to study a hydrogen jet in which a large number of atoms are in the metastable 2s state. A laser beam, linearly polarized along 0z, is sent onto the atoms of the jet. It is sent perpendicularly to the atomic velocities so that we can neglect the Doppler effect in this experiment. ω_L, the frequency of the laser beam, is close to the resonance frequency ω_0 corresponding to the following atomic transition: $|2s> (2^2 S_{1/2}) \to |3p> (3^2 P_{1/2})$; $E_{3p} - E_{2s} = \hbar(\omega_{3p} - \omega_{2s}) = \hbar\omega_0$. Upon interacting with the electromagnetic wave, the atoms, which were initially in state $|2s>$, are lifted into a nonstationary state: $|\psi(t)> = a(t)|2s> + b(t)|3p>$. The interaction is detected by measuring the intensity of the spontaneous emission from state $|3p>$. This spontaneous emission is proportional to $|b(t)|^2$, and we assume it does not perturb state $|\psi(t)>$.

(a) Show that the electric dipole transition moment between these states does not vanish (the transition is "allowed"). Conclude from this that in the electric dipole approximation the Hamiltonian can be written as $V = -\boldsymbol{p}\,\mathbf{E}$, where $\mathbf{E} = \mathbf{E}_0 \cos(\omega_L t)$ is the electric field of the wave and $\boldsymbol{p}$ is the dipole moment operator of the atom. Show that $\boldsymbol{p}$ has no diagonal terms in basis $|2s>, |3p>$. For simplicity's sake, use the notation $<2s|p_z|3p> = <3p|p_z|2s> = p$.

Compare the equation describing the evolution of $|\psi(t)>$ to that describing a state of spin 1/2 state placed within a constant uniform magnetic field, to which is added an oscillating field of small amplitude and perpendicular to the first field. What do you conclude?

(b) Let us write $A(t) = a(t)\exp(i\omega_{2s} t)$; $B(t) = b(t)\exp(i\omega_{3p} t)$. Find the equations governing the evolution of A and B in the rotating wave approximation. Find particular solutions for B of the form $\exp(i\omega t)$. With the help of these solutions, show that B can be written as $B(t) = \alpha_+ \exp(i\,\omega_+ t) + \alpha_- \exp(i\,\omega_- t)$. Calculate the values of $\omega_\pm$. Take as the origin of time the instant the atoms enter the laser wave and calculate α_+ and α_-. Compare this result to the evolution of spin in a magnetic field. Now assume that the intensity of the laser beam is uniformly distributed over its diameter and that the detector analyzes the light emitted by one point of the interaction zone. What can you say about the detected signal?

2. *Two-photon spectroscopy*

We wish to study the transition $|1s> (1^1 S_{1/2}) \to |2s> (2^2 S_{1/2})$.

(a) Show that this transition is electron-dipole forbidden.

(b) In order to fill state 2s starting from the ground state 1s, a linearly polarized laser beam of amplitude E_0 is sent onto a tank containing atomic hydrogen. The angular frequency of the wave, ω_L, is approximately equal to $(\omega_{2s} - \omega_{1s})/2$. The wave lifts the atoms, which were initially in the $|1s>$ state, into a nonstationary state described by

$$|\psi(t)> = a(t)|1s> + b(t)|2s> + \sum_n n(t)|n>$$

where $|n>$ stands for states of energy $E_n = \hbar\omega_n$. States $|n>$ all have the same parity, which is opposite to that of states $|1s>$ and $|2s>$ (i.e., dipolar transitions are allowed between any state $|n>$ and states $|1s>$ and $|2s>$). Within the frame of the electric dipole approximation (interaction Hamiltonian equal to $V = -\boldsymbol{p}\mathbf{E}$), show, assuming the rotating wave approximation, that b(t) vanishes for a first-order perturbation calculation. Then show that the second-order term of b(t) does not vanish. Calculate its value, assuming the $<n|p_z|2s>$ matrix elements are known. The result can be simplified by noticing that for all states under consideration, the laser frequency is such that $\omega_{2s} - \omega_{1s} - 2\omega_L << \omega_{2s} - \omega_n - \omega_L$. What does this result suggest? What practical comments can you make about the experiment?

1. (a) States $|2s>$ and $|3p>$ do not have the same parity. Indeed, state $|1s>$ has quantum number $\ell = 0$ while for state $|3p>$ $\ell = 1$, therefore $|1s>$ is even with respect to spatial inversion, while $|3p>$ is uneven. Now operator $\boldsymbol{p}$ is uneven with respect to this same spatial inversion, so that $<2s|\boldsymbol{p}|1s>$ is even. This expression stands for the integral over all space with respect to $\mathbf{r}$ (i.e., from $-\infty$ to $+\infty$) of the product of the bra $<2s|$ by the ket $\boldsymbol{p}|3p>$. The fact that the expression to be integrated (the integrand) is even means that the integral cannot vanish. Therefore we can say $<2s|\boldsymbol{p}|3p> \neq 0$, meaning that the transition $|2s> \rightarrow |3p>$ is allowed in the electric dipole approximation. The interaction Hamiltonian is $V = -\boldsymbol{p}\mathbf{E}$. Its matrix representations are nondiagonal (i.e., have vanishing diagonal elements (see Problems I.5 and VII.1). The comparison of the evolution of $|\psi(t)>$ with that of a state of spin 1/2 is given in detail in Inset VII.1.

(b) The evolution equations are the same as those obtained in Problem VII.2 (where we define $V = V_{2s,\,3p}$): $\dot{A} = (-i V/\hbar)\exp(-i\,\omega_0\,t)B$; $\dot{B} = (-i V/\hbar)\exp(i\,\omega_0\,t)A$ with $\omega_0 = \omega_{3p} - \omega_{2s}$. Applying the rotating wave approximation yields

$$\dot{A} \approx (i\,p\,E_0/2\hbar)\exp[-i(\omega_0 - \omega_L)t]B \; ; \; \dot{B} \approx (i\,p\,E_0/2\hbar)\exp[-i(\omega_0 - \omega_L)t]A.$$

We are looking for a solution of the form $B = \exp(i\omega t)$. Identifying the constants gives a second-degree equation in ω: $\omega^2 + (\omega_L - \omega_0)\omega - \Omega^2/4 = 0$ where Ω is the Rabi frequency ($\Omega = pE_0/\hbar$). This equation has two roots, ω_+ and ω_-, which are $\omega_\pm = [(\omega_0 - \omega_L)/2] \pm (1/2)\sqrt{(\omega_L - \omega_0)^2 + \Omega^2}$. Therefore we can write $B = \alpha_+\exp(i\,\omega_+\,t) + \alpha_-\exp(i\,\omega_-\,t)$. At $t = 0$, the atoms are all in state $|2s>$, so $A(0) = \pm 1$ and $B(0) = 0$. Writing down the initial condition $B(0) = 0$ yields $\alpha_+ + \alpha_- = 0$. Similarly, writing $A(0) = \pm 1$ yields $\alpha_+ = \Omega / 2\sqrt{(\omega_L - \omega_0)^2 + \Omega^2}$ so that at last we obtain

$$B \approx \left[i\Omega \;/\; \sqrt{(\omega_L - \omega_0)^2 + \Omega^2} \right] \exp[i(\omega_0 - \omega_L)t/2]$$

$$\times \sin\left(\sqrt{(\omega_L - \omega_0)^2 + \Omega^2}\; t/2\right)$$

The detected signal (of spontaneous emission from state $|3p>$) is proportional to the population of $|3p>$, that is, to $|B|^2$ and

$$|B|^2 \approx \Omega^2 \sin^2\left(\sqrt{(\omega_L - \omega_0)^2 + \Omega^2}\; t/2\right) / [(\omega_L - \omega_0)^2 + \Omega^2]$$

At resonance, the reader may check that the equation leads to the same value for $|B|^2$, as that found in Problem VII.2, namely, $\sin^2(\Omega\,t/2)$.

The characteristic time of the interaction is $t' = \Delta x / v_0$ (Δx is the distance between the detection point and the point corresponding to the origin of time, $t = 0$. v_0 is the velocity of the atoms). Clearly, the detected signal, which is proportional to $|B|^2$, depends on the point of interaction observed. The fact that there is a distribution of the velocities of the atoms implies that a mean signal is observed whose amplitude is greatest when $\omega_L = \omega_0$. If the velocities are very dispersed, the observed mean signal will practically not depend on time.

2. (a) For the same reasons as those mentioned in 1. (a), since states $|1s>$ and $|2s>$ are of the same parity, the integrand of integral $<1s|\boldsymbol{p}|2s>$ is uneven so that the integral vanishes. That means that transition $|1s> \rightarrow |2s>$ is forbidden in the electric dipole approximation.

(b) The evolution of $|\psi(t)>$ is governed by the following equation (see Eq. I.1.10):

$$i\hbar|\dot{\psi}> = [(H_0 + V)]|\psi> \text{ with } V = -\boldsymbol{p}\,\mathbf{E}.$$

Left-multiplying first by $<1s|$, then by $<2s|$, and last by $<n|$ yields

$$i\hbar\dot{a} = \hbar\omega_{1s}\,a + \sum_n n\,V_{1s,\,n}\,, \qquad i\hbar\dot{b} = \hbar\omega_{2s}\,b + \sum_n n\,V_{2s,\,n}$$

$$i\hbar\dot{n} = \hbar\omega_n\,n + a\,V_{n,\,1s} + b\,V_{n,\,2s}$$

Here, we took into account the fact that Vab = Vnn = 0 (all transitions between states of same parity are forbidden within the EDA).

Let us change the variables as follows:

$$a = A\exp(-i\omega_{1s}t),\quad b = B\exp(-i\omega_{2s}t),\quad n = N\exp(-i\omega_n t)$$

Then we find

$$\dot{A} = (i\,E_0\cos\omega_L t/\hbar)\sum_n N\exp[i(\omega_{1s}-\omega_n)t] < 1s|p_z|n >$$

$$\dot{B} = (i\,E_0\cos\omega_L t/\hbar)\sum_n N\exp[i(\omega_{2s}-\omega_n)t] < 2s|p_z|n >$$

$$\dot{N} = (i\,A\,E_0\cos\omega_L t/\hbar)\exp[i(\omega_n-\omega_{1s})t] < n|p_z|1s > + (i\,B\,E_0\cos\omega_L t/\hbar)\exp[i(\omega_n-\omega_{2s})t] < n|p_z|2s >$$

From this system of coupled equations we can conclude that since $A(0) \approx 1$ and since $B(0) \approx$ and $N(0) \approx 0$ at the zero order of field E_0, then only N will be different from zero at the first perturbation order. This in turn will induce nonzero terms for A and B in the second order. Let us calculate $N^{(1)}$, the first order term of N in our perturbation calculation, within the rotating wave approximation:

$$\dot{N}^{(1)} \approx (i\,E_0\cos\omega_L t/\hbar)\exp[i(\omega_n-\omega_{1s})t] < n|p_z|1s >$$
$$\approx (i\,E_0/2\hbar) < n|p_z|1s > \exp[i(\omega_n-\omega_{1s}-\omega_L)t]$$

Integration gives

$$N^{(1)} \approx [E_0 < n|p_z|1s > / 2\hbar(\omega_n-\omega_{1s}-\omega_L)][\exp[i(\omega_n-\omega_{1s}-\omega_L)t] - 1]$$

And now we can calculate $B^{(2)}$, taking account of the approximation mentioned and of the initial condition $B^{(2)}(0) = 0$:

$$B^{(2)} \approx \left[E_0^2 / 4\hbar^2 \left(\omega_{2s} - \omega_{1s} - 2\omega_L\right)\right] \sum_n \left[< 2s \,|\, p_z \,|\, n > \; < n \,|\, p_z \,|\, 1s > \right] / \left(\omega_n - \omega_{1s} - \omega_L\right)$$
$$\times \left\{ \exp\left[i\left(\omega_{2s} - \omega_{1s} - 2\omega_L\right)t\right] - 1\right\}$$

Again, spontaneous emission from state $|\, 2s >$ is found to be proportional to $|\, B^{(2)} |^2$:

$$|\, B^{(2)} |^2 \approx \left(E_0 / 2\hbar\right)^4 \left| \sum_n \left[< 2s \,|\, p_z \,|\, n > \; < n \,|\, p_z \,|\, 1s > \right] / \left(\omega_n - \omega_{1s} - \omega_L\right)\right|^2$$
$$\times \left\{ 1/2 \left[\left(\omega_{2s} - \omega_{1s}\right) / 2 - \omega_L\right]^2 \right\} \sin^2\left[\left(\omega_{2s} - \omega_{1s}\right)/2 - \omega_L\right] t$$

Of course, this expression is only valid if the population of $|\, 2s >$ is not too large, i.e. at times t such that $t << 2\pi / \left[\left(\omega_{2s} - \omega_{1s}\right) / 2 - \omega_L\right]$.

We can make two comments:

— The damping of the population of state $|\, 2s >$ should be taken into account by inserting term $(- \gamma B/2)$ into the evolution equation of B. For interaction times which are very large compared to $2/\gamma$, we would then obtain $|\, B^{(2)} |^2$ = constant, meaning that the spontaneous emission no longer changes with time.

— Just like saturated absorption spectroscopy, this spectroscopy is afflicted by Doppler effect. But, as said, the use of two laser waves propagating in opposite directions is a good method to do away with this undesired broadening.

VII.4.4 Problem VII.4: Vectorial representation of a two-level system acted on by a linearly polarized optical wave ; introduction to photon echoes

1. Let us consider a large set of atoms of the same kind. These are placed in a sinusoidal electromagnetic field of constant amplitude E_0, polarized linearly along 0z and having angular frequency ω_L. The value of ω_L is close to that of frequency ω_0 corresponding to the transition between two eigenstates $|\, a >$ and $|\, b >$ of the Hamiltonian of the atoms H_0 ($\omega_0 = \omega_b - \omega_a$). This is an allowed transition with a transition dipole moment p, assumed to be real $\left(p = < a \,|\, p_z \,|\, b >\right)$. The wave "lifts" the atoms, which are initially in state $|\, a >$, into a nonstationary state $|\, \psi(t) > = a(t) \,|\, a > + \, b(t) \,|\, b >$.

(a) Find the equations governing the evolution of a(t) and b(t) with time. Simplify these using the rotating wave approximation (the RWA), and use this approximation during the remainder of this problem.

(b) Without solving the equations found in (a), give a physical interpretation of the terms a (t) a*(t) and b (t) b*(t). Calculate the mean atomic dipole moment $< p_z >$ of state $|\, \psi(t) >$. Infer from this a physical meaning for term a(t) b*(t).

(c) Find the equations governing the evolution of the following real quantities:

$$R_x = ab^* \exp\left(-i\omega_L t\right) + cc$$

$$R_y = -i\, ab^* \exp\left(-i\omega_L t\right) + cc$$

$$R_z = bb^* - aa^*$$

Let $\mathbf{R} = R_x\,\mathbf{e}_x + R_y\,\mathbf{e}_y + R_z\,\mathbf{e}_z$, where $\mathbf{e}_x$, $\mathbf{e}_y$, $\mathbf{e}_z$ are an orthonormal base of three-dimensional space. Show that the above equations can be summarized by the following single vectorial equation:

$$\dot{\mathbf{R}} = \mathbf{R} \times \boldsymbol{\mathfrak{R}}', \text{ where } \boldsymbol{\mathfrak{R}}' \text{ is the vector defined by } \boldsymbol{\mathfrak{R}}' = \Omega\,\mathbf{e}_x - \left(\omega_0 - \omega_L\right)\mathbf{e}_z$$

($\Omega = p\,E_0 / \hbar$ is the Rabi frequency).

(d) Show that, starting from its initial state (atom in state $| a >$), the evolution of $\mathbf{R}$ with time can be described as a rotation of $\mathbf{R}$ around $-\boldsymbol{\mathfrak{R}}'$ at an angular frequency which we ask you to determine. Conclude from this that the atom regularly returns to the same state, and find at what angular frequency it does this. Find this angular frequency in the special case of resonance, i.e. $\omega_L = \omega_0$. How can you describe the evolution of $\mathbf{R}$ in this case ? Give a graphical representation of this evolution.

2. The system is submitted to a series of two quasi-resonant pulses. During the pulses, vector $\boldsymbol{\mathfrak{R}}'$ can be assimilated to vector $\Omega\mathbf{e}_x$. However, when the field vanishes (between pulses), the frequency difference $(\omega_0 - \omega_L)$ must be taken into account.

(a) First, from $t = 0$ to $t = t_p = \pi / 2\Omega$, the atomic system is submitted to a field of amplitude E_0. Make a graphic determination of the components of $\mathbf{R}$ corresponding to the state of the system at time $t = t_p$, remembering that the system was in state $| a >$ at $t = 0$.

(b) Next, the system evolves freely ($E_0 = 0$) until time $t = t' + t_p$ with $t' >> t_p$. What are the components of vector $\boldsymbol{\mathfrak{R}}'$, the free rotation of $\mathbf{R}$? Infer from these the components of $\mathbf{R}$ corresponding to the state of the system at time $t = t' + t_p$.

(c) Between $t = t' + t_p$ and $t = t' + 3t_p$, the system is again submitted to a field with amplitude E_0. Give a graphic representation of the state of the system at $t = t' + 3t_p$. For $t > t' + 3t_p$, the system evolves freely. Give the state of the system at any time t. What does one notice when $t = 2t' + 3t_p$?

(d) Now we assume that the set of atoms consists of equal parts of atoms with resonance frequency ω_0, and of atoms with resonance frequency ω'_0. The difference between ω_0 and ω'_0 is small so that in presence of the field, vector $\boldsymbol{\mathfrak{R}}'$ is the same for both classes of atoms, which we shall name 1 and 2. Time t' is such that $|\omega_0 - \omega'_0|\, t' >> 1$. Compare the evolution of these two classes of atoms as they are submitted to the series of two pulses described above. What happens at time $t = 2t' + 3t_p$? Show that the amplitude of $< p_z >$ goes through a maximum at that instant.

(e) Next, the set of atoms consists of a great number of classes whose resonance frequencies form a continuous band of frequencies. This band is centered around ω_0 and has a width $\Delta\omega_0$ with $\Delta\omega_0 << \omega_0$ and $\Delta\omega_0 t' >> 1$. Generalize the questions of (d) to this new set of atoms and answer them qualitatively. What happens at time $t = 2t' + 3t_p$? Can you give an example of such an ensemble of classes of atoms?

1. (a) The reader is asked to refer to Problem VII.2. Changing the variables as follows: a (t) = A (t) exp (– i ω_a t); b (t) = B (t) exp (– i ω_b t), along with using the RWA (with of course $\Omega = p E_0 / \hbar$), we find

$$\dot{A} \approx (i\,\Omega/2) \exp\left[i(\omega_L - \omega_0) t\right] B$$
$$\dot{B} \approx (i\,\Omega/2) \exp\left[-i(\omega_L - \omega_0) t\right] A$$

(b) a(t) a*(t) represents the probability of finding an atom in state | a >: It is the population of state | a >. In the same way, b(t) b*(t) represents the population of state | b >. The mean value of the z-component of the dipole moment is given by

$$< p_z > = < \psi \mid p_z \mid \psi > = \left(a^* < a \mid + b^* < b \mid\right) p_z \left(a \mid a > + b \mid b >\right)$$

(see Eq. I.1.9).

Operator p_z has only nondiagonal matrix elements, and therefore we can write $< p_z > = (ab^* + ba^*)\, p = 2\, \mathcal{R}(ab^*)\, p$. The term ab* represents the coherence of the two-level system. Its real part is proportional to the mean value of the induced dipole moment.

(c) Let us derive R_z with respect to time. This yields

$$\dot{R}_z = d\,(bb^* - aa^*) / dt = d\,(BB^* - AA^*) / dt = -\Omega R_y$$

In the same way, we calculate $\dot{R}_x = -(\omega_0 - \omega_L) R_y$ and $\dot{R}_y = \Omega R_z + (\omega_0 - \omega_L) R_x$. So R_x, R_y, and R_z appear as the 0x, 0y, and 0z components of a vector **R** governed by the equation $\dot{\mathbf{R}} = \boldsymbol{\mathfrak{R}}' \times \mathbf{R}$, where $\boldsymbol{\mathfrak{R}}'$ is a vector whose components on the 0x, 0y, 0z base are, respectively, Ω, 0, $-(\omega_0 - \omega_L)$.

(d) The vectorial equation $\dot{\mathbf{R}} = \boldsymbol{\mathfrak{R}}' \times \mathbf{R}$ describes a rotation characterized by the instantaneous rotation vector $\boldsymbol{\mathfrak{R}}'$. The length of vector $\boldsymbol{\mathfrak{R}}'$ is equal to the angular rotation frequency $\mathfrak{R}'$, given by

$$\mathfrak{R}' = \sqrt{\Omega^2 + (\omega_0 - \omega_L)^2}$$

We also have the relationship $d(\mathbf{R}\mathbf{R})/dt = 2\,\mathbf{R}\dot{\mathbf{R}} = 0$, implying $|\mathbf{R}|$ = constant. In other words, **R** is vector of constant length. Also, $d(\mathbf{R}\boldsymbol{\mathfrak{R}}')/dt = 0$, and therefore $|\mathbf{R}|\,|\boldsymbol{\mathfrak{R}}'| \cos(\mathbf{R}, \boldsymbol{\mathfrak{R}}')$ = constant. So we can see that the angle between **R** and $\boldsymbol{\mathfrak{R}}'$ is also constant.

In the case of resonance, $\boldsymbol{\mathfrak{R}}' = \Omega \mathbf{e}_x$. In this case, **R** turns around $\mathbf{e}_x$ at angular frequency $\mathfrak{R}' = \Omega$. Indeed, we obtain the set of equations $\ddot{R}_i + \Omega^2 R_i = 0$ whose solutions rotate at angular frequency Ω (i stands for any of the spatial coordinates x, y, or z).

At time t = 0 we obtain

$$R_z(0) = |b(0)|^2 - |a(0)|^2 = -D_0 = -1$$

where D_0 represents the population difference between states | a > and | b >.

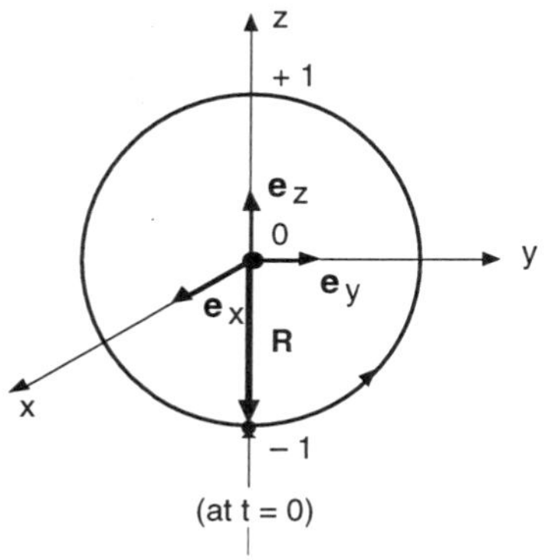

2. (a) At time $t = t_p = \pi / 2\Omega$, the system is in the state represented by vector $\mathbf{R}_{t_p} = \mathbf{e}_y$ (the rotation angle is equal to $\Omega t_p = \pi/2$). $R_{z,\,t_p} = 0$ and therefore the population difference vanishes. Coherence has reached a maximum.

(b) For t such that $t_p < t < t' + t_p$, the system follows its course freely. Only the z component of $\boldsymbol{\mathcal{R}}'$ is nonvanishing, and it is equal to $-(\omega_0 - \omega_L)$. So we have $\boldsymbol{\mathcal{R}}' = (\omega_L - \omega_0)\,\mathbf{e}_z$. **R** stays in the 0xy plane and turns at angular frequency $(\omega_0 - \omega_L)$. At time $(t' + t_p)$, vector **R** has turned by an angle of $(\omega_L - \omega_0)\,t'$. The components of $\mathbf{R}\,(t' + t_p)$ are

$$R_x\left(t' + t_p\right) = \cos\left[\left(\pi/2\right) + \left(\omega_L - \omega_0\right)t'\right] = -\sin\left[\left(\omega_L - \omega_0\right)t'\right]$$
$$R_y\left(t' + t_p\right) = \sin\left[\left(\pi/2\right) + \left(\omega_L - \omega_0\right)t'\right] = \cos\left[\left(\omega_L - \omega_0\right)t'\right]$$
$$R_z\left(t' + t_p\right) = 0$$

(c) Again we have $\boldsymbol{\mathcal{R}}' = \Omega\,\mathbf{e}_x$. $\Omega\,2t_p$ corresponds to an angle of π. $\mathbf{R}\,(t' + 3t_p)$ and $\mathbf{R}(t' + t_p)$ are symmetrical with respect to 0x, and we have

$$R_x\left(t' + 3t_p\right) = \cos\left[\left(-\pi/2\right) - \left(\omega_L - \omega_0\right)t'\right] = -\sin\left(\omega_L - \omega_0\right)t'$$
$$R_y\left(t' + 3t_p\right) = \sin\left[-\left(\pi/2\right) - \left(\omega_L - \omega_0\right)t'\right] = -\cos\left(\omega_L - \omega_0\right)t'$$
$$R_z\left(t' + 3t_p\right) = 0$$

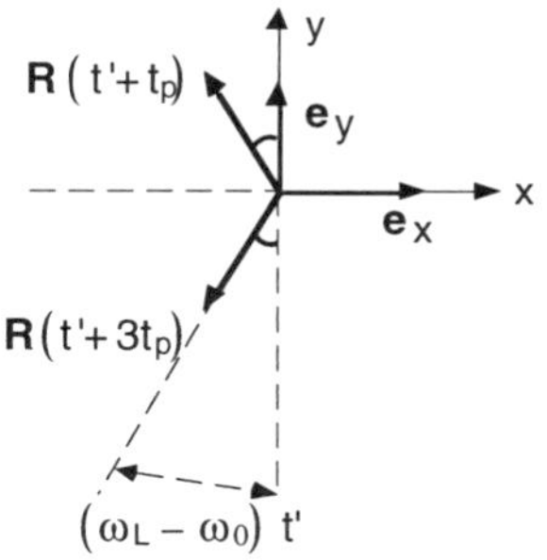

When field E_0 is turned off, the systems evolves freely in the x0y plane at angular frequency $-(\omega_0 - \omega_L)$, and is governed by the set of equations given by

$$R_x(t) = \cos\left[(\omega_L - \omega_0)(t - t' - 3\,t_p) - (\pi/2) - (\omega_L - \omega_0)\,t'\right]$$
$$= -\sin\left[(\omega_L - \omega_0)\,t' - (\omega_L - \omega_0)(t - t' - 3\,t_p)\right]$$
$$R_y(t) = \sin\left[(\omega_L - \omega_0)(t - t' - 3\,t_p) - (\pi/2) - (\omega_L - \omega_0)\,t'\right]$$
$$= -\cos\left[(\omega_L - \omega_0)(t - t' - 3\,t_p) - (\omega_L - \omega_0)\,t'\right]$$
$$R_z(t) = 0$$

And at time $t = (2t' + 3t_p)$ we obtain

$$R_x(2t' + 3t_p) = 0$$
$$R_y(2t' + 3t_p) = -1$$
$$R_z(2t' + 3t_p) = 0$$

In other words, at time $t = 2t' + 3t_p$ we have $\mathbf{R} = -\mathbf{e}_y$.

(d) During the pulses, the two classes of atoms follow a similar course, only $(\omega_L - \omega_0)\,t' \neq (\omega_L - \omega'_0)\,t'$. At times t such that $t > t' + t_p$, vectors $\mathbf{R}$ and $\mathbf{R}'$ point in different directions in the 0xy plane. However, at time $t = 2t' + 3t_p$, these vectors are superposed (they point in the same direction) on $-\mathbf{e}_y$. On the other hand, we have $2\,ab^* = \exp(i\omega_L t)(R_x + iR_y)$, so $2\mathcal{R}(ab^*) = R_x \cos\omega_L t - R_y \sin\omega_L t$. We find $< p_z^{(1)} > = p\left(R_x \cos\omega_L t - R_y \sin\omega_L t\right)$. Replacing R_x and R_y by their values yields

$$< p_z^{(1)} > = p \sin\left[\omega_L t + (\omega_L - \omega_0)(t - 2t' - 3t_p)\right]$$

In the same way, we find $< p_z^{(2)} > = p \sin\left[\omega_L t + (\omega_L - \omega'_0)(t - 2t' - 3t_p)\right]$. At time $t = 2t' + 3t_p$: $< p_z^{(1)} > = < p_z^{(2)} >$. The mean value of the z component of the induced dipole moment $\mathbf{p}$ can be written as

$$< p_z > = < p_z^{(1)} > + < p_z^{(2)} > = 2p \cos\left[(\omega_0 - \omega'_0)(t - 2t' - 3t_p)/2\right]$$
$$\times \sin\left[\omega_L t + (2\,\omega_L - \omega_0 - \omega'_0)(t - 2t' - 3t_p)/2\right]$$

The amplitude of $< p_z >$ goes through a maximum at $t = 2t' + 3t_p$.

(e) The mean values of $< p_z >$ for the different classes of atoms have random phases, so that their total sum will have a value of zero, except at time $t = 2t' + 3t_p$ since at this instant, all the mean values have the same phase. If there are N classes, $< p_z > = Np \sin(\omega_L t)$. This phenomenon is what is called the *photon echo*, which will be studied in greater detail in Chapter VIII. An example of such a set of atoms would be a gas in which Doppler effect induces a continuous distribution of resonance frequencies.

VII.4.5 Problem VII.5: Synchronized quantum beats

We wish to measure the frequency difference between two levels $|\,a>$ and $|\,b>$ of the hyperfine structure of an excited state of a molecule. To do this, the levels are filled by a laser pulse, or by a series of laser pulses, starting from the molecular ground state $|\,f>$, and we observe the spontaneous emission of levels $|\,a>$ and $|\,b>$. The energy of ground state $|\,f>$ is chosen to be zero. $E_a = \hbar\omega_a$ and $E_b = \hbar\omega_b$ are the energies of levels $|a\rangle$ and $|b\rangle$ ($\omega_a > \omega_b$). $p = \langle f|p_z|a\rangle = \langle f|p_z|b\rangle$ are the real matrix elements of operator p_z (**p** is the electric dipole moment). The laser field is linearly polarized in the 0z direction. Its angular frequency ω_L is such that $\omega_L = (\omega_a + \omega_b)/2$. Its amplitude E_0 is assumed to be constant during a laser pulse. The molecule is also assumed not to move (it is frozen) with respect to the laboratory reference frame.

1. Explain why states $|a\rangle$ and $|b\rangle$ must have the same parity, which must be opposite to the parity of ground state $|f\rangle$.

2. The molecule, which is initially in its ground state, is submitted to a laser pulse between times $t = 0$ and $t = t'$. Let $|\psi(t)\rangle$ be the wave vector describing the nonstationary state of the molecule at time t:

$$|\psi(t)\rangle = A(t)\exp(-i\omega_a t)|a\rangle + B(t)\exp(-i\omega_b t)|b\rangle + F(t)|f\rangle$$

Find the equations governing the evolution of A, B and F between 0 and t'. Calculate A and B at $t = t'$ using a first order perturbation method with respect to E_0 and using the rotating wave approximation. In what domain is this perturbation calculation justified? Give a physical reason explaining why the pulse duration must be chosen such that $t' \leq 1/(\omega_a - \omega_b)$.

3. Find the equations governing the evolution of A and B after the end of the pulse, for $t > t'$, then find $|\psi(t)\rangle$ for $t > t'$. Spontaneous emission between states $|a\rangle$ and $|b\rangle$ and ground state $|f\rangle$ depends on the matrix elements of p_z between $|\psi(t)\rangle$ and $|f\rangle$. It has an intensity $I_{sp}(t)$ equal to $I_{sp}(t) = \alpha|\langle f|p_z|\psi(t)\rangle|^2$, where α is a proportionality constant. Calculate $I_{sp}(t)$ for $t > t'$. Show that the value of $(\omega_a - \omega_b)$ can be found by analyzing the signal $I_{sp}(t)$. To make the calculations simpler, just factorize $|A(t')|^2$ in the expression for $I_{sp}(t)$ and calculate $B(t')/A(t')$.

4. A series of identical pulses is applied to the molecule at times t such that $0 < t < t'$, $t'' < t < t'' + t'$, ... $nt'' < t < nt'' + t'$ with $t'' \gg t'$. Still using a first-order perturbation method, calculate A and B after the second pulse, for t such that $t'' + t' < t < 2t''$, and after the nth pulse, for t such that $t > nt'' + t'$.

What is $I_{sp}(t)$ after n pulses? What do you notice if t'' is a multiple of $4\pi/(\omega_a - \omega_b)$?

Give a physical explanation for these results. Show that a series of n "synchronized" pulses [$t'' = 4\pi q/(\omega_a - \omega_b)$, where q is a whole number] is especially interesting when the excited state of the molecule is made up of a great number of hyperfine levels.

5. In the first question, levels $|a\rangle$ and $|b\rangle$ were assumed to have infinite lifetimes, implying that in the absence of laser field, $|A|^2$ and $|B|^2$ stay constant. Actually $|A|^2$ and $|B|^2$ relax exponentially with time. This is due, among others, to the coupling between the field and vacuum (spontaneous emission). Let us assume that the lifetimes of levels $|a\rangle$ and $|b\rangle$ are equal, and let $\gamma = 1/2\tau$ be the corresponding relaxation constant of A and B. Insert the correct relaxation term into the equations governing A and B so as to take into account the finite lifetimes of the levels. Show, without calculating explicitly, the consequences this term has on the results of questions 3 and 4.

1. Please report to Problem VII.3.

2. Postulate P6 (see Eq. I.1.10) can be written $i\hbar\,|\dot{\psi}> = (H_0 + V)\,|\psi>$. Using the expression given for $|\psi>$ and respectively left-multiplying by $<a|$, by $<b|$ and by $<f|$ (setting $\Omega = p\,E_0/\hbar$ for the Rabi frequency) yields

$$\dot{A} = i\,\Omega\cos(\omega_L t)\exp(i\,\omega_a t)\,F$$

$$\dot{B} = i\,\Omega\cos(\omega_L t)\exp(i\,\omega_b t)\,F$$

$$\dot{F} = i\,\Omega\cos(\omega_L t)\left[A\exp(-i\,\omega_a t) + B\exp(-i\,\omega_b t)\right]$$

At order zero of the perturbation, only the fundamental state has nonvanishing population at $t = 0$. Thus $F^{(0)}(0) = 1$; $A^{(0)}(0) = B^{(0)}(0) = 0$. At the first perturbation order, and using the RWA, we can write $\dot{A}^{(1)} \approx (i\,\Omega/2)\exp[i(\omega_a - \omega_L)t]$ which, upon integration and with $A^{(1)}(0) = 0$ gives us

$$A^{(1)} \approx [\Omega/2(\omega_a - \omega_L)]\ \{\exp[i(\omega_a - \omega_L)t] - 1\}.$$

In the same way, we find $B^{(1)} \approx [\Omega/2(\omega_b - \omega_L)]\ \{\exp[i(\omega_b - \omega_L)t] - 1\}$. And at time t' we have

$$A(t') \approx [\Omega/2(\omega_a - \omega_L)]\ \{\exp[i(\omega_a - \omega_L)t'] - 1\}$$

$$B(t') \approx [\Omega/2(\omega_b - \omega_L)]\ \{\exp[i(\omega_b - \omega_L)t'] - 1\}$$

Remembering that $\omega_L = (\omega_a + \omega_b)/2$, these equations can be rewritten as

$$A(t') \approx [\Omega/(\omega_a - \omega_b)]\ \{\exp[i(\omega_a - \omega_b)t'/2] - 1\}$$

$$B(t') \approx [-\Omega/(\omega_a - \omega_b)]\ \{\exp[-i(\omega_a - \omega_b)t'/2] - 1\}$$

The perturbation calculation is valid in the domain for which

$$|A(t')| = |B(t')| << 1$$

$$|A(t')| = |B(t')| = (\Omega\ t'/2)\sin[(\omega_a - \omega_b)t'/4]/[(\omega_a - \omega_b)t'/4].$$

For $t' << (\omega_a - \omega_b)$, the argument of the sine function becomes vanishingly small, so that the sine function tends to unity. Therefore very short pulses, whose spectrum is much broader than $(\omega_a - \omega_b)$, lead to high excitation probabilities of states $|a>$ and $|b>$.

3. When $t > t'$, $\dot{A} = \dot{B} = 0$. We therefore have $A(t) = A(t')$ and $B(t) = B(t')$. Thus $|\psi(t)> = A(t')\exp(-i\,\omega_a t)|a> + B(t')\exp(-i\,\omega_b t)|b> + |f>$, yielding

$$I_{sp}(t) \approx \alpha\,|A(t')\exp(-i\omega_a t)<f|p_z|a> + B(t')\exp(-i\omega_b t)<f|p_z|b>|^2$$

$$\approx \alpha\,p^2\,|A(t')\exp(-i\omega_a t) + B(t')\exp(-i\omega_b t)|^2$$

$$\approx \alpha\,p^2\ |A(t')|^2\,|1 + [B(t')/A(t')]\exp[i(\omega_a - \omega_b)t]|^2$$

But, $B(t')/A(t') = \exp[-i(\omega_a - \omega_b)t'/2]$ so that

$$I_{sp}(t) \approx \alpha\ p^2\ |A(t')|^2\,|1 + \exp[i(\omega_a - \omega_b)(t - t'/2)]|^2$$

$$\approx \alpha\ p^2\ |A(t')|^2[1 + \cos(\omega_a - \omega_b)(t - t'/2)]$$

with $|A(t')|^2 = (\Omega\ t'/2)^2\ \mathrm{sinc}^2[(\omega_a - \omega_b)t'/4]$.

I_{sp} as a function of time behaves like a sine function with period $2\pi/(\omega_a - \omega_b)$. Measuring this period thus leads directly to the frequency difference $(\omega_a - \omega_b)$.

4. The second pulse is felt by the atoms for all t such that $t'' < t < t'' + t'$. The equations governing A and B are exactly the same as in question 2, but the initial conditions are different. At time $t = (t'' + t')$ we obtain

$$A(t'' + t') \approx A(t')\{1 + \exp[i(\omega_a - \omega_b)t''/2]\}$$
$$B(t'' + t') \approx B(t')\{1 + \exp[-i(\omega_a - \omega_b)t''/2]\}$$

And, of course, after the second pulse and before the third one, i.e. for all times such that $t'' + t' < t < 2t''$, we have $A(t) = A(t'' + t')$ and $B(t) = B(t' + t'')$.

In the same way, for the n_{th} pulse ending at time $nt'' + t'$, we have for all $t > nt'' + t'$:

$$A(t) = A(nt'' + t') \approx A(t')\, 1 + \exp[i(\omega_a - \omega_b)t''/2] + \cdots + \exp[i(\omega_a - \omega_b)nt''/2]$$

or again $A(t) \approx A(t')\{1 - \exp[i(\omega_a - \omega_b)(n+1)t''/2]\} / \{1 - \exp[i(\omega_a - \omega_b)t''/2]\}$

(The reader will have recognized the sum of the terms of a geometric progression.)

Similarly, we would find

$$B(t) = B(nt'' + t') \approx B(t')\{1 - \exp[-i(\omega_a - \omega_b)(n+1)t''/2]\}$$
$$/\{1 - \exp[-i(\omega_a - \omega_b)t''/2]\}$$

Thus, after n pulses, spontaneous emission is equal to

$$I_{sp}(t) \approx \alpha\, p^2 |A(nt'' + t')|^2 |\, 1 + [B(nt'' + t')/A(nt'' + t')]\exp[i(\omega_a - \omega_b)t]\,|^2$$

with

$$|A(nt'' + t')|^2 = \sin^2[(n+1)(\omega_a - \omega_b)t''/4] / \sin^2[(\omega_a - \omega_b)t''/4]$$

and

$$B(nt'' + t')/A(nt'' + t') = \exp[-i(\omega_a - \omega_b)t'/2]\exp[-i(\omega_a - \omega_b)(n-1)t''/2]$$

At last we find that

$$I_{sp}(t) \approx 2\alpha\, p^2 |A(nt'' + t')|^2 \{1 + \cos(\omega_a - \omega_b)[t - t'/2 - (n-1)t''/2]\}$$

Again, we find a sinusoidal behavior for $I_{sp}(t)$, but now its amplitude depends on time t'' by means of function $|A(nt'' + t')|^2$, that is, it depends on the recurrence frequency of the laser pulses. This function has a maximum at $t'' = 4q\pi / (\omega_a - \omega_b)$ (with q a whole number), when it is equal to unity. Thus $I_{sp,\,max} \approx 4(n+1)^2 \alpha p^2$. So $I_{sp,\,max}$ increases as n^2 (for $n >> 1$). On the other hand, when the recurrence time t'' is different from $4q\pi / (\omega_a - \omega_b)$, the intensity is extremely weak. A maximum of the intensity of the signal is observed when there is synchronism between the recurrence frequency of the pulses and one of the resonance frequencies of the microsystem. This method is a way to set apart a pair of states ($|a\rangle$ and $|b\rangle$). It could be useful in case of a complex atomic hyperfine structure.

5. Now we have the following equations governing the evolution of A and B:

$$\dot{A} \approx (i\Omega/2)\exp[i(\omega_a - \omega_L)t] - \gamma A$$
$$\dot{B} \approx (i\Omega/2)\exp[i(\omega_b - \omega_L)t] - \gamma B$$

In the absence of light wave we have

$$|A|^2 = |A(0)|^2 \exp[-2\gamma t] = |A(0)|^2 \exp(-t/\tau)$$
$$|B|^2 = |B(0)|^2 \exp(-t/\tau)$$

The populations relax exponentially. The spontaneous emission $I_{sp}(t)$ will fade exponentially with time constant τ. The synchronization method is of use only if $nt'' << \tau$.

VII.4.6 Problem VII.6: Atom cooling by laser

Let us take a monokinetic atomic jet (i.e. a jet in which all atoms have the same velocity) with velocity $\mathbf{v}_0$ along the 0z axis. Using a laser wave propagating in the opposite, i.e. the – 0z, direction, we wish to slow down and stop these atoms.

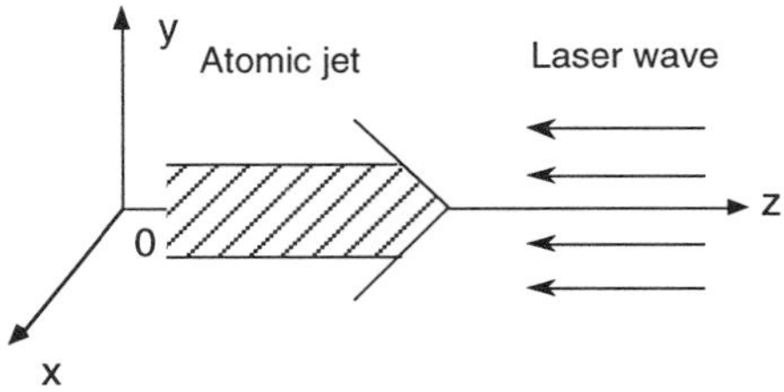

1. Let us study the wave-atom interaction using the semiclassical method. Let us also neglect any spontaneous emission of the atoms.

The jet is assumed to be perfect: in the absence of laser wave, all atoms travel at the same speed $\mathbf{v}_0$, parallel to 0z.

Let $|a\rangle$ and $|b\rangle$ be two eigenstates of the Hamiltonian H_0 of the atoms at rest. $|a\rangle$ represents the ground state, $\omega_0 = \omega_b - \omega_a$ is the frequency for the $|a\rangle \rightarrow |b\rangle$ transition, and $p = \langle a|p_x|b\rangle$ is the matrix element, assumed to be real, of the matrix representation of p_x on basis $|a\rangle, |b\rangle$. The laser field can be considered as a plane wave, linearly polarized in the 0x direction, monochromatic, and of angular frequency ω_L in the laboratory reference frame. Let E_0 represent the constant amplitude of its electric field.

(a) What is the angular frequency ω'_L of the wave as seen by the atoms in their own reference frame? (Assume $v_0/c << 1$ and take into account only first-order term in v_0/c of the Doppler effect.) Infer from this the laser frequency ω_L satisfying the resonance condition between the wave and transition $|a\rangle \rightarrow |b\rangle$. Throughout the remainder of this problem, assume this condition is satisfied.

(b) Starting at time t = 0, the field is applied to the atoms which are initially in state $|a\rangle$. Let $|\psi(t)\rangle$ be the wave function of an atom at time t in its own reference frame :

$$|\psi(t)\rangle = a(t)|a\rangle + b(t)|b\rangle.$$

Find a(t) and b(t) using the electric dipole- and the rotating wave approximations.

Conclude that, for an infinite interaction time between wave and atoms, the atoms have the same mean probability (averaged over time) of being in state $|a\rangle$ or in state $|b\rangle$.

(c) The interaction of each atom with the radiation can be conceived like a succession of absorptions and stimulated emissions of photons. Write the conservation of the total momentum of the system (atom + photon) for each of these two processes in the proper reference frame of the atom before interaction. The velocity of the atoms v_0 is nonrelativistic, so that all second-order terms in v_0/c can be neglected. Let m stand for the mass of one atom. Calculate the change in atomic speed induced by each of the two basic interaction processes.

Assume that after each single interaction, the change in the atom's velocity does not modify perceptibly the resonance condition with the laser wave. Find the total momentum balance for a complete absorption-stimulated emission cycle. What can one conclude concerning the effect of a laser wave on the atomic velocities of the jet, in case of infinite interaction time and neglecting spontaneous emission?

Numerical calculation: Laser wavelength λ_0 equals 600 nm; mass of one atom is 23 g mole^{-1}. Compare the change in speed induced by each process (absorption or stimulated emission) at speed $v_0 = 1000$ m s^{-1} of the jet atoms. Avogadro's number is 6.02×10^{23} mole^{-1}; Planck's constant h equals 6.6×10^{-34} Js.

2. Let us now take into account the phenomenon of spontaneous emission of state $|\, b >$. Let γ_b represent the probability per unit time that an atom in state $|\, b >$ spontaneously emit a photon.

(a) Without doing any calculations, say what condition E_0 must satisfy so that the number of photons emitted by stimulated emission per second and per atom is overwhelmingly greater than the number of spontaneously emitted photons per second and per atom. Let us now assume this condition is satisfied; thus we know that spontaneous emission does not significantly alter the evolution of the atom as determined in question 1.(b). Calculate N, the mean number of spontaneously emitted photons by one atom during one second.

(b) Spontaneously emitted photons are emitted in all directions, with the same probability for each direction. Infer from this the mean change in velocity of an atom undergoing an absorption-spontaneous emission cycle. Calculate the deceleration g' of an atom undergoing a great number of cycles per second (assume that resonance is maintained during the deceleration of the atom). *Numerical calculation:* $\gamma_b^{-1} = 10^{-8}$ s. Calculate g', the deceleration of the atoms; the stopping time of the atoms; and their stopping distance.

(c) If ω_L and ω_0 stay constant, the atom eventually gets out of resonance as it slows down. To maintain resonance, ω_0 is adjusted all along the 0z axis by using the Zeeman effect on the $|\, a > \rightarrow |\, b >$ transition. Assume that these states both have vanishing total spin and that the orbital kinetic moment for $|\, a >$ is J = 0 and for $|\, b >$ is J = 1. Let **B** be the magnetic field applied to the atoms of the jet; **B** points along the 0z axis and is constant in each plane z = constant. Find, as a function of the amplitude B_0 of field **B**, the change in frequency for the different possible transition frequencies ω_0. What transitions should be used, and what shape must one try to obtain for the field amplitude B_0 (z) along the z-axis in order to maintain resonance all the time during the slowing down of the atoms?

1. (a) Please refer to Section VII.3.2.4. Frequency ω'_L is given by $\omega'_L \approx \omega_L(1 + v_0/c)$. Resonance will occur for $\omega'_L = \omega_0$, that is, for $\omega_L \approx \omega_0(1 - v_0/c)$.

(b) Referring to Problem VII.2, let us assume resonance, change the variables to $A = a \exp(i\omega_a t)$ and $B = b \exp(i\omega_b t)$, and let us use the rotating wave-and electric dipole approximations. Then we can write $\dot{A} \approx (i\,\Omega/2)\,B$; $\dot{B} \approx (i\,\Omega/2)\,A$ ($\Omega = pE_0/\hbar$ is the Rabi angular frequency). Solving these differential equations leads to the probabilities of an atom being in state $\mathcal{P}_a(t) = |\,a(t)\,|^2 = \cos^2 \Omega\, t$ or in state

$| b>$, $\mathcal{P}_b(t) = |b(t)|^2 = \sin^2 \Omega t$. Averaged over time, the mean values of these probabilities are equal : $\overline{\mathcal{P}_a(t)} = \overline{\mathcal{P}_b(t)} = 1/2$.

(c) The conservation of the total momentum from the point of view of the reference frame of the atom can be written as follows:

absorption:
- atom + photon $\rightarrow$ excited atom
- $0 \quad -\hbar\omega'_L/c \quad mv^*$
- $\omega'_L = \omega_0 \quad ; \quad v^* = -\hbar\,\omega_0\,/mc$
- the atom slows down

stimulated emission:
- excited atom $\rightarrow$ photon + atom
- $0 \quad -\hbar\omega'_L/c \quad mv$
- $\omega'_L = \omega_0 \quad v = \hbar\,\omega_0\,/mc$
- the atom speeds up

The superscript * refers to the excited state. The wave vector of the light waves points in the negative z direction, but the velocity of the atoms points in the positive z direction. Altogether, after a complete cycle, the total "slowing" (v + v*) amounts to zero. The numerical calculation yields $v \approx 3 \times 10^{-2}$ ms^{-1} and $v/v_0 \approx 3 \times 10^{-5}$.

2. (a) The number of photons emitted via stimulated emission per unit time will be much larger than the number of photons emitted spontaneously provided $\tau_{st} << \tau_{sp}$, i.e. when $\Omega/2 >> \gamma_b$. For this to be true, we must have $E_0 >> 2\hbar\gamma_b/p$. During time dt, the atom spends in average dt/2 in state $| b>$. Consequently, $dN = \gamma_b\, dt/2$ and therefore $N = \gamma_b/2$.

(b) Let us write the balance for the absorption-spontaneous emission cycle (in the reference frame of the atom):

absorption: the same as above, with $\mathbf{v}^* = (-\hbar\,\omega_0\,/mc)\,\mathbf{e}_z$

spontaneous emission:
- excited atom $\rightarrow$ photon + atom
- $0 \quad (\hbar\omega_0/c)\,\mathbf{e}_i \quad m\mathbf{v}$
- $\mathbf{v} = (-\hbar\,\omega_0\,/mc)\,\mathbf{e}_i$

$\mathbf{e}_i$ is the unit vector pointing in the direction of the spontaneously emitted photon.

The resulting velocity after the cycle is $\mathbf{v}^* + \mathbf{v} = (-\hbar\omega_0/mc)\mathbf{e}_z - (\hbar\omega_0/mc)\mathbf{e}_i$. Spontaneous emission is spatially isotropic, so that the spatial average value of $\mathbf{e}_i$ vanishes. Hence, the atom is slowed down by a quantity $(-\hbar\omega_0/mc)$ per cycle. As in question 2 (a), the mean number of cycles per second is $\gamma_b/2$, and the mean deceleration is $g' = -\gamma_b\,\hbar\,\omega_0/2mc$. *Numerical calculation:* We find a deceleration of 1.5×10^6 m s^{-2}, a stopping time of 600 μs, and a stopping distance of 30 cm.

(c) If needed, the reader can refer to the first reference in Section I.6 for a thorough study of the Zeeman effect. State $|a>$ is nondegenerate (J = 0). On the other hand, the magnetic field lifts the threefold degeneracy of state $|b>$ for which J = 1. Therefore, three sublevels appear, corresponding respectively to $J_z = +1$; $J_z = 0$; $J_z = -1$, whose eigenfrequencies are $\omega_0^{(+1)} = \omega_0 - (q/2m_e)B_0$; $\omega_0^{(0)} = \omega_0$; $\omega_0^{(-1)} = \omega_0 + (q/2m_e)B_0$. q and m_e are respectively the rest charge and the rest mass of the electron. The laser frequency ω_L [with $\omega_L = \omega_0(1 - v_0/c)$] is constant. On the other hand, with the atoms being continuously decelerated, their velocity changes along the z-axis. To maintain equilibrium during the whole time of the slowing (i.e. cooling) procedure, the following equations must hold for all z:

$$\omega_L = \omega_0(1 - v_0/c) = \omega_0^{(\pm 1)}(z)[1 - v(z)/c]$$

with

$$\omega_0^{(\pm 1)} = \omega_0 \mp q B_0(z)/2m_e$$

But since we are dealing with a uniformly decelerated movement, $v^2(z) = v_0^2 - 2g'z$, yielding

$$\begin{aligned}\omega_0(1 - v_0/c) &= \omega_0^{(\pm 1)}\left[1 - \sqrt{v_0^2 - 2g'z}/c\right] \\ &= \omega_0[1 \mp q B_0(z)/2m_e\omega_0]\left(1 - \sqrt{v_0^2 - 2g'z}/c\right)\end{aligned}$$

Since $v_0 << c$, resonance will be maintained, provided that

$$-v_0/c = \mp q B_0(z)/(2m_e\omega_0) - \sqrt{v_0^2 - 2g'z}/c$$

which will be true if the field $B_0(z)$ has the following longitudinal shape:

$$B_0(z) = \pm(2m_e\omega_0 v_0/cq)\left(1 - \sqrt{1 - 2g'z/v_0^2}\right)$$

With such a setup, the Doppler-induced frequency shift is compensated by the field for the two transitions $|a> \rightarrow |b^{(+1)}>$ and $|a> \rightarrow |b^{(-1)}>$.

VII.4.7 Problem VII.7: Ramsey fringes

Let us take a continuous jet of atoms, of velocity $\mathbf{v}_0$ directed along the z-axis. Let $|a>$ and $|b>$ be two eigenstates of H_0, the Hamiltonian of the atoms at rest. $|a>$ is the ground state, and $\omega_0 = -\omega_a + \omega_b$ is the transition frequency for transition $|a> \rightarrow |b>$. $p = <a|p_x|b>$ is the matrix element, assumed to be real, of the operator corresponding to the x-component of dipole moment **p**.

A continuous laser wave interacts with the atoms. This wave is assumed to be a plane wave whose electric field **E** is polarized linearly along 0x. It is monochromatic with a tunable wavelength ω_L and its amplitude E_0 is constant. It propagates along 0y so that the Doppler shift can be neglected throughout this problem.

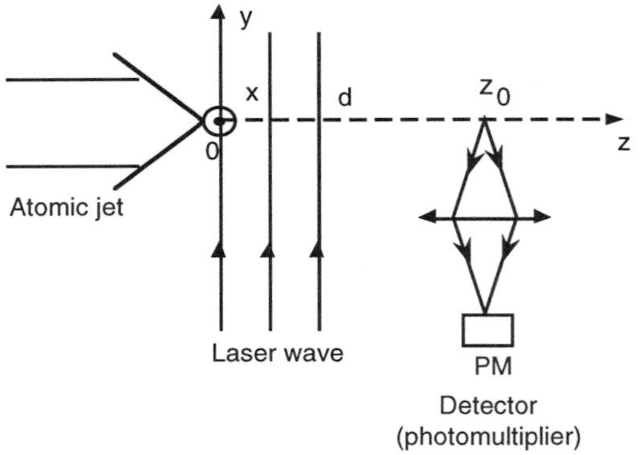

1. The laser beam, which crosses the atomic jet perpendicularly, has a cross section of d in the 0z direction. For z such that $0 < z < d$, the field has a value of E_0 and it vanishes for all other z. During this problem, we shall neglect the deviations from atomic velocity $\mathbf{v}_0$ which could arise after an absorption-spontaneous emission cycle.

(a) Let t_t be the time needed for one atom to cross the laser beam. Atoms arriving at $z = 0$ and $t = t_0$ are in their ground state $|a\rangle$. Their state at $z > 0$ is described by the wave vector $|\psi(t - t_0)\rangle = a(t - t_0)|a\rangle + b(t - t_0)|b\rangle$. Within the electric dipole- and the rotating wave approximations, find $a(t - t_0)$ and $b(t - t_0)$ using a first order perturbation calculation, first for t such that $0 < t - t_0 < t_t$, then for $t - t_0 > t_t$. In what domain is this perturbation calculation justified?

(b) At each instant t, the spontaneous emission S of the atoms is detected by a detector (PM) as they travel through plane $z = z_0$. This emission is observed in a direction parallel to 0y. Assume that S is proportional to the probability of finding an atom in state $|b\rangle$ at $z = z_0$. Explain why, at given ω_L and z_0, a constant signal (with respect to time) is detected on the PM? At resonance, what is the behavior of S if $0 < z_0 < d$ and for $z_0 > d$? How does S behave if z_0 is set at d_0 (with $d_0 > d$) and if frequency ω_L varies around ω_0? What can you conclude about the spectral resolution of this spectroscopic method for analyzing an atomic jet? Could this result have been predicted?

(c) Assume resonance, i.e. $\omega_L = \omega_0$, at $z_0 > d$, and assume that d varies slowly over such a domain that the perturbation calculation is no longer usable. Find S as a function of d. For what values of d is S vanishingly small? Give a physical interpretation of this result.

2. Now we go back to the domain where the perturbation calculation is justified. In order to improve the spectral resolution of this method, the atomic jet interacts a second time with the laser beam at coordinates z such that $L < z < L + d$. This second progressive wave, also traveling parallel to 0y, has the same amplitude E_0, the same frequency ω_L, and the same phase at $y = 0$ as the first wave; the two waves are coherent.

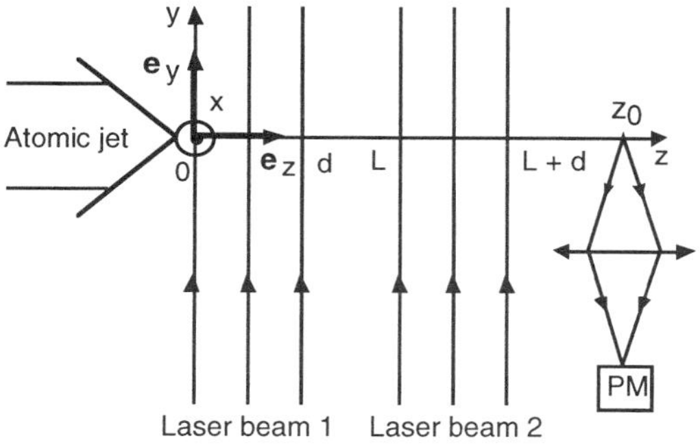

(a) First, assume that all the photons detected at z_0 are emitted only by atoms traveling at speed v_0 along the positive z-axis. Using a first-order perturbation method, find the variation of signal S, measured at $z_0 > L + d$, as ω_L varies around ω_0. Show that S as a function of $(\omega_L - \omega_0)$ displays sinusoidal oscillations whose amplitudes are modulated by the signal found in question 1. What is the period of these oscillations? Draw $S = f(\omega_L)$ graphically. Give a physical interpretation of the result. Show that the spectral resolution of the setup has improved.

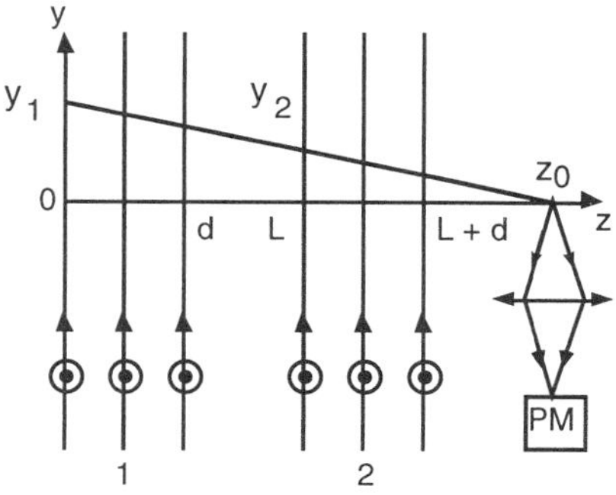

(b) Actually, the atomic jet is slightly divergent, and the photons detected at z_0 are emitted by atoms whose velocity $\mathbf{v} = v_0\,\mathbf{e}_z + v_y\,\mathbf{e}_y$ has a y-component which can be nonvanishing. The distribution of v_y is centered around 0 and has a width of $2\Delta v$. Let us look at a particular atom having a specific v_y. It will enter beam 1 and beam 2 at different y-coordinates: y_1 and y_2. Give the phase of the progressive waves 1 and 2 at y-coordinates y_1 and y_2, and express their difference as a function of v_y, v_0, ω_L, and L. Assume that $d \ll L$ and that, when the atom travels a distance d on the z-axis, the shift in y-coordinate is negligible. Using a first-order perturbation calculation, find the probability for an atom having z-coordinate $z = z_0$ to be in state $|\, b >$. How does this probability change when $|\, v_y \,|$ varies between 0 and Δv? From this, infer a condition on Δv such that the spectral resolution calculated in 2 (a) is not affected very much by the divergence of the jet. What becomes of S when Δv is very large?

(c) *Numerical calculation:* $d = 1$ mm, $L = 1$ m, $v_0 = 1000$ m s^{-1}, wavelength of the laser, λ_1, is 600 nm. Find the resolution R of signal S (ω_L), first for the one-beam interaction, then for the two-beam interaction, assuming a nondivergent atomic jet. To actually observe this resolution experimentally, what condition must the divergence of the jet, $\Delta v / v_0$, satisfy? Discuss this result. In the optical domain, how could one overcome the problem caused by the divergence of the atomic jet?

1. (a) Referring to Problems VII.2 and VII.3 and setting $t' = t - t_0$, $a(t') = A(t') \exp(-i\,\omega_a t')$ and $b(t') = B(t') \exp(-i\,\omega_b t')$, we write

$$\dot{A} \approx (i\,\Omega\,/2)\exp\left[\,i\left(\omega_L - \omega_0\right)t'\right]\exp\left(\,i\,\omega_L\,t_0\right)B$$

$$\dot{B} \approx (i\,\Omega\,/2)\exp\left[-\,i\left(\omega_L - \omega_0\right)t'\right]\exp\left(-\,i\,\omega_L\,t_0\right)A,\ \text{with}\ \Omega = pE_0/\hbar$$

With $A^{(0)} = 1$; $B^{(0)} = 0$, for the first-order terms of the perturbation calculation we obtain

$$\dot{B}^{(1)} \approx (i\,\Omega\,/2)\exp\left[-\,i\left(\omega_L - \omega_0\right)t'\right]\exp\left(-\,i\,\omega_L\,t_0\right)$$

So for $0 < t' < t_t$, after integrating $\dot{B}^{(1)}(t')$ we find

$$B^{(1)}(t') \approx \left[\,i\Omega\,\exp\left(-\,i\,\omega_L\,t_0\right)t'/2\right]\exp\left[-\,i\left(\omega_L - \omega_0\right)t'/2\right]$$
$$\times \sin\left[\left(\omega_L - \omega_0\right)t'/2\right]/\left[\left(\omega_L - \omega_0\right)t'/2\right]$$

For $t' > t_t$, we have $B^{(1)}(t') = B^{(1)}(t_t) =$ constant. The perturbation calculation holds if $\left|B^{(1)}(t')\right| << 1$ for all t', which is true provided that $t' << 2/\Omega$.

(b) The atoms are at z-coordinate z_0 at time $t = t_0 + z_0/v_0$, i.e. at $t' = z_0/v_0$. At that instant, all atoms are in the same state. Signal S stays constant with time. For $0 < z_0 < d$, $S \approx \left|B^{(1)}(z_0/v_0)\right|^2$. In case of resonance ($\omega_L = \omega_0$), we find $S_r \approx \left(\Omega\,z_0/2v_0\right)^2$.

For $z_0 > d$, $S_r \approx \left|B^{(1)}(d/v_0)\right|^2 = \left(\Omega\,d/2v_0\right)^2$. Away from resonance and for a given $z_0 > d$ we find $S \approx S_r\left\{\sin\text{card}\left[\left(\omega_L - \omega_0\right)d_0/2v_0\right]\right\}^2$. The spectrum width $\Delta\omega$ of this function limits the spectral resolution ($\Delta\omega = 2\,\pi\,v_0/d_0$) of the signal. The resolution decreases when v_0 increases and/or when d_0 decreases. This result was predictable since for the atoms, it is as if they encountered a laser pulse of duration (d/v_0), i.e., with a frequency width of (v_0/d). The resolution decreases when this frequency width increases.

(c) At resonance, the differential equations of 1 can be solved without using the perturbation method (see Problem VII.2). In this case, we find $S \approx \sin^2(\Omega d/2v_0)$. S vanishes when $\Omega d/2v_0 = n\pi$ (where n is a whole number), i.e. for values of the beamwidth $d_n = 2n\,\pi v_0/\Omega$. If one varies d, the population of state $|\,b>$ oscillates at angular frequency Ω/v_0. When $d = d_n$, all atoms are in ground state $|\,a>$ so that spontaneous emission from states $|\,b>$ vanishes.

2. (a) Let $L/v_0 = t''$. For t' such that $t_t < t' < t''$, we have $B^{(1)}(t') \approx \left[\Omega\,\exp\left[-\,i\,\omega_L\,t_0\right]/2\left(\omega_0 - \omega_L\right)\right]\left[\exp\left(\omega_0 - \omega_L\right)t' - 1\,\right]$ [this was found in answering question 1. (a)]. For $t'' < t' < t'' + t_t$, the atoms interact with the second laser beam so that we can write the same differential equation for $B^{(1)}(t')$ as in question 1. (a). Let us integrate this equation with the new limit conditions: at $t' = t''$, $B^{(1)}(t'') = B^{(1)}(t_t)$. This yields $B^{(1)}(t') \approx B^{(1)}(t_t)\left[1 + \exp\left[-\,i\left(\omega_L - \omega_0\right)t''\right]\right]$

$$S \approx 4\left|B^{(1)}(t_t)\right|^2\cos^2\left[\left(\omega_0 - \omega_L\right)t''/2\right]$$

S oscillates with a period oft" $= L/v_0$, inside an envelope curve of width $2\pi/t_t = 2\pi\,v_0/d$. (see accompanying diagram).

The resolution is now given by the width of function $\cos^2[\,(\omega_0 - \omega_L)\,t''/2]$, which is $\delta\,\omega = 2\pi/t'' = 2\pi\,v_0/L$. SinceL > d, the resolution has been improved. The fringes which are observed in this set up are called Ramsey fringes. They are due to the interaction between the second laser

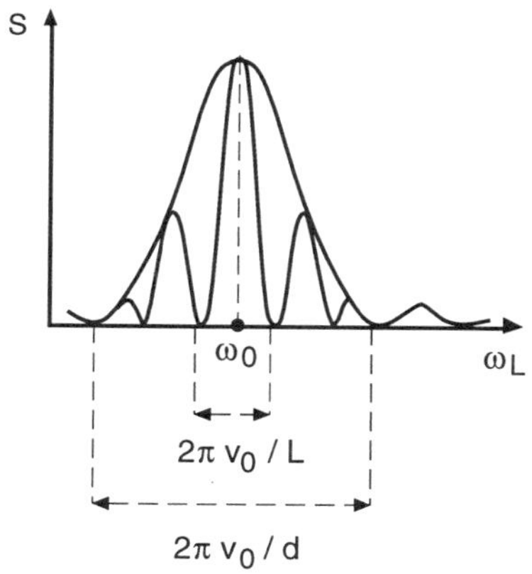

wave with the dipole moment induced by the first laser wave. The phase of this induced dipole moment turns freely between d and L, while its amplitude remains quasi-constant (except for some damping). When the electric field of the second wave is in phase with the dipole moment, constructive interference occurs, producing the observed phenomenon.

b) The phase difference between points y_1 and y_2 is $\Delta\varphi = k\,(y_2 - y_1) = kv_y\, L / v_0$; **k** is the wave vector (parallel to 0y) of the laser wave. This phase difference must be taken into account when calculating $B^{(1)}(t')$. For $t' > t'' + t_t$, we have $B^{(1)}(t') = B^{(1)}(t_t)\left\{1 + \exp\left[-\,i\left(\omega_L - \omega_0\right)t'' + k\left(y_2 - y_1\right)\right]\right\}$. Therefore

$$S \approx 4\left|B^{(1)}(t_t)\right|^2 \cos^2\left[\left(\omega_0 - \omega_L\right)t''/2 + k\, v_y\, L / 2v_0\right]$$

When $|v_y|$ varies between 0 and Δv, S varies as $\cos^2 [2\pi\, v_y / T]$, with a period of $T = v_0\, \lambda_L/L$. The resolution will not change provided that $\Delta v << T$—in other words, provided that $\Delta v / v_0 << \lambda_L/L$. On the other hand, if $\Delta v/v_0 >> \lambda_L/L$, the Ramsey fringes disappear and one observes spontaneous emission with an intensity of $S \approx 2\left|B^{(1)}(t_t)\right|^2$.

(c) The numerical calculations lead to the following spectral resolutions:

— For a single beam: $R \approx \nu/\Delta\nu = \nu d_0 / v_0 \approx 5 \times 10^8$

— For two beams: $R \approx \nu/\Delta\nu = \nu L / v_0 \approx 5 \times 10^{11}$

In order to actually observe this last value of the resolution, the divergence of the beam must be such that $\Delta v/v_0 << 6 \times 10^{-7}$. Technically, it seems impossible to produce a jet whose divergence would be smaller than 0.6 μrd.

To observe Ramsey's fringes and to reach the corresponding resolution in the optical domain, several methods have been suggested. They all use a third interaction zone between jet and laser wave (see reference 5.5).

VII.5 Bibliography and further reading

VII.5.1 The vector theory of the laser is explained in:

LE FLOCH, A., AND LE NAOUR, R. Polarization Effects in Zeeman Lasers with x-y-types Loss Anisotropies, *Phys. Rev. A*, 4A, 1971, p. 290.

VII.5.2 Saturated absorption can be studied in the following works:

LEVENSON, M. D. *Introduction to Nonlinear Laser Spectroscopy*, Academic Press, New York, 1982.

LETOKHOV, V. S., AND CHEBOTAYEV, V. P. *Nonlinear Laser Spectroscopy*, Springer Verlag, Berlin, 1977.

VII.5.3 Atom cooling by laser is described in:

COHEN-TANNOUDJI , C., AND PHILLIPS, W. P. New Mechanisms for Laser Cooling, *Physics Today*, 43 (10), October 1990, p. 33.

LEA , S., CLAIRON, A., SALOMON, C., LAURENT, P., LOUNIS, B., REICHEL, J., NADIR, A., AND SANTARELLI, G. Laser Cooling and Trapping of Atoms: New Tools for Ultra-stable Cesium Clocks, *Physica Scripta* 51, 1994, p. 78.

VII.5.4 Manipulation of objects by means of a laser can be found in:

CHU, S. Laser Trapping of Neutral Particles, *Scientific American*, 266 (2), February 1992, p. 70.

VII.5.5 Techniques for observing Ramsey Fringes in the optical domain can be looked for in:

COUILLAUD, B., AND DUCASSE, A. New Methods in High-Resolution Laser Spectroscopy, in BEYER, H. J., AND KLEINPOPPEN, H. *Progress in Spectroscopy, Part C,* Plenum Press, New York, 1984.

CHAPTER VIII

Application of the Vector of State Formalism to Laser-Matter Interaction: nth-Order Optical Susceptibility

In this chapter, we reach the ultimate part of our work, namely the most general description of laser-molecule interaction. Here, we wish to generalize the arguments of Chapter VII to the two following cases: (i) *nonstationary interactions* and (ii) *nonresonant interactions*.

(i) Nonstationary interactions are relevant to all of *pulsed-laser optics*. The reader has been given a preliminary feeling of these in our presentation of *induced transparency* (see Problem VII.2) and of *photon echoes* (see Problem VII.4) which we kept deliberately short. We shall study them more in detail at the end of this chapter.

(ii) Nonresonant interactions arise when the laser frequency ω_L is different from any of the eigenfrequencies associated to the various electronic transitions of the microsystems. So it is impossible to focus on two particular states, as we did in Chapter VII, and the two-level model can no longer be used. This, of course, is the most general case. We have already shown (see Chapter II) that in this case, and provided that the laser intensities are not too strong, the polarization induced by the laser can be decomposed as a sum of terms oscillating at frequencies ω_L, $2\omega_L$, ..., $n\omega_L$, ..., that is, at whole multiples of the applied frequency ω_L. These *harmonic frequencies* can, in turn, create *harmonic optical waves* with wavelengths which are fractions of the original laser wavelength. Thus, an Nd^{3+}/YAG laser (see Chapter IV) which emits at 1060 nm—*invisible* light therefore!—gives rise to a *green laser wave* at 530 nm, for instance by interacting with a *transparent* crystal.

The spectacular visual effect of this kind of experiment—it was done as early as 1961 by Franken—accounts a lot for the extraordinary fascination of the scientific community for what would later become *nonlinear optics*.

A comprehensive description of this field, which in any case would be exceedingly difficult because of continuous new developments of the fundamental as well as of the applied aspects, is not the purpose of this book. However, the interested reader will find a few excellent works on the subject listed in the bibliography at the end of the chapter. We chose rather to present nth-order susceptibility by way of the vector-of-state formalism. We shall then show how the simple fact of expressing this nth-order susceptibility leads to a complete description of quite a large number of observed phenomena. Last, we shall illustrate a few of the most exemplary of these effects.

VIII.1 Stationary interaction

VIII.1.1 Microscopic nth-order susceptibility

Let us look at an electric laser field **E**(**r**,t) which is applied at time $t_0 = -\infty$ to a microscopic system. In the absence of perturbations, the Hamiltonian of the microsystem is H_0 and it does not depend on time. We assume that initially, at time t_0, only state $| a >$, the state of lowest energy, is filled. We also assume for the time being that $\gamma_{ab} = \Delta\omega_0 = 0$, meaning that we assume that the energy levels of the microsytem are discrete and infinitely thin.

Let **p**(**r**,t) be the total polarization of the system, permanent as well as induced. Let us then calculate the quantum mean value of the αth component of **p**. Omitting the spatial dependence of the vector of state, we find at time t (see Eq. I.1.9):

$$< \mathrm{p}_\alpha (t) > = < \psi (t) | p_\alpha | \psi (t) > \qquad \text{(VIII.1.1)}$$

For reasons described in Section I.2.2, let us change to the interaction representation, via Eq. I.2.12, of operator p_α:

$$p^{\mathrm{I}}(t) = U_0^{-1}(t, t_0)\, p_\alpha\, U_0(t, t_0) \qquad \text{(VIII.1.2)}$$

Setting $| \psi (t_0) > = | a >$ yields

$$< \mathrm{p}_\alpha (t) > = < a \left| (U^{\mathrm{I}})^{-1}(t, t_0) \right| p^{\mathrm{I}}_\alpha (t) \left| U^{\mathrm{I}}(t, t_0) \right| a > \qquad \text{(VIII.1.3)}$$

Now we can use Dyson's formalism and, more precisely, Eqs. I.2.15 and I.2.16 to express the time-evolution operator in the interaction representation. We obtain

$$< \mathrm{p}_\alpha (t) > = < a | p^{\mathrm{I}}_\alpha (t) | a > + \sum_q \sum_r < a \left| (U^{\mathrm{I}}_q)^{-1}(t, t_0) \right| p^{\mathrm{I}}_\alpha (t) \left| U^{\mathrm{I}}_r (t, t_0) \right| a > \qquad \text{(VIII.1.4)}$$

The first term, $< a | p^{\mathrm{I}}_{\alpha}(t) | a >$, represents the permanent polarization (it is the zero-order term of the perturbation induced by a laser field). We shall from now on ignore this term to deal exclusively with induced polarization. From the double summation making up the second term, we can extract the induced polarization of order $n = q + r$; thus the nth-order term of the mean value of the αth component of the mean polarization can be written as

$$< p^{(n)}_{\alpha}(t) > = \sum_{\substack{q,r \\ q+r=n}} < a \left| \left(U^{\mathrm{I}}_{q} \right)^{-1}(t, t_0) \right| p^{\mathrm{I}}_{\alpha}(t) \left| U^{\mathrm{I}}_{r}(t, t_0) \right| a > \qquad \text{(VIII.1.5)}$$

Now Eq. I.2.16 shows that

$$U^{\mathrm{I}}_{q}(t, t_0) = (i\hbar)^{-q} \int_{t_0}^{t} dt_1 \cdots \int_{t_0}^{t_{q-1}} dt_q\, H'^{\mathrm{I}}(t_q) \cdots H'^{\mathrm{I}}(t_1)$$

with

$$H'^{\mathrm{I}}(t_q) = -\,\mathbf{E}(t_q)\, \boldsymbol{p}^{\mathrm{I}}(t_q) = -\,E_{\beta_j}(t_q)\, p_{\beta_j}(t_q) \qquad \text{(VIII.1.6)}$$

(β_j represent the spatial coordinates of the space in which the functions are defined.)

E(t) can be written as a Fourier integral:

$$\mathbf{E}(\mathbf{r}, t) = \int_{\omega} \mathbf{E}_0(\mathbf{r}, \omega) \exp(-i\omega t)\, d\omega \qquad \text{(VIII.1.7)}$$

Then, omitting the spatial dependency of $\mathbf{E}_0$ (i.e. calculating $< p_{\alpha} >$ at vanishing wave vector), we get

$$U^{\mathrm{I}}_{q}(t, t_0) = (i\hbar)^{-q}(-1)^{q} \sum_{\beta_j, \omega_j} \int_{t_0}^{t} dt_1 \cdots \int_{t_0}^{t_{q-1}} dt_q$$

$$\times \int_{\omega_q} d\omega_q \cdots \int_{\omega_1} d\omega_1 \exp\left[-i(\omega_1 t_1 + \cdots + \omega_q t_q)\right]$$

$$\times E_{0\beta_q}(\omega_q) \cdots E_{0\beta_1}(\omega_1)\, p^{\mathrm{I}}_{\beta_q}(t_q) \cdots p^{\mathrm{I}}_{\beta_1}(t_1)$$

[The factor $(-1)^q$ gives the sign of the product of the q Hamiltonians given by Eq. VIII.1.6; the double sum $\sum_{\beta_j, \omega_j}$ is to remind of the double summation which must be done over the β_j and the ω_j].

The reader may have noticed that this double sum is a way to express the sum over all the possible permutations over the β; let us call these $\mathcal{P}_{\beta}$, combined with a simultaneous permutation over all ω. With this notation, we can write

$$U_q^I(t, t_0) = \sum_{\mathcal{P}_\beta} \hbar^{-q} i^q \int_{\omega_q} d\omega_q \cdots \int_{\omega_1} d\omega_1\, E_{0\beta_1}(\omega_1) \cdots E_{\beta_q}(\omega_q)$$

$$\times \int_{t_0}^{t} dt_1 \cdots \int_{t_0}^{t_{q-1}} dt_q \exp\left[-i(\omega_1 t_1 + \cdots + \omega_q t_q)\, p_{\beta_q}^I(t_q) \cdots p_{\beta_1}^I(t_1)\right]$$

Let us calculate $U_r^I(t, t_0)$ in the same way. The final result is

$$< p_\alpha^{(n)}(t) > = \int_{\omega_n} d\omega_n \cdots \int_{\omega_1} d\omega_1 \exp[-i(\omega_1 + \cdots + \omega_n)t]\, E_{\beta_1}(\omega_1) \cdots E_{\beta_n}(\omega_n)$$

$$\times \sum_{\substack{q,r \\ q+r=n}} \sum_{\mathcal{P}_\beta} (\hbar)^{-n} i^{r-q} \int_{t_0}^{t} dt_1 \cdots \int_{t_0}^{t_{n-1}} dt_n \qquad \text{(VIII.1.8)}$$

$$\times \exp[-i\omega_1(t_1 - t)] \cdots \exp[-i\omega_n(t_n - t)]$$

$$\times < a \left| p_{\beta_1}^I(t_1) \cdots p_{\beta_q}^I(t_q)\, p_\alpha^I(t)\, p_{\beta_n}^I(t_n) \cdots p_{\beta_{q+1}}^I(t_{q+1}) \right| a >$$

As it is, this formula already gives us valuable information on six points:

(a) *The polarization will oscillate at frequency* $(\omega_1 + \cdots + \omega_n)$, the sum of the n laser frequencies appearing in the expression.

(b) It will be *proportional to the product of the n components* of the electric field appearing in the expression.

(c) The two last lines of the right-hand side of Eq. VIII.1.8 form the *nth-order microscopic susceptibility*, $\chi_{mic}^{(n)}$. Formally, this susceptibility appears as

→ *a tensor of rank* (n + 1) (there are n indices β_j plus index α). It can be expressed as

$$\chi_{mic}^{(n)} = \chi_{\alpha\beta_1 \cdots \beta_n}^{(n)} \qquad \text{(VIII.1.9)}$$

→ *a sum of* (n + 1)! *terms* . Summing over $\mathcal{P}_\beta$ involves n! terms, while the sum over q and r such that (q + r) = n has (n + 1) terms. $\chi_{mic}^{(n)}$ is indeed a sum of (n + 1)! terms.

→ *explicitly dependent on the n frequencies* $\omega_1, \omega_2 \ldots \omega_n$. This can be expressed by

$$\chi_{mic}^{(n)} = \chi_{\alpha\beta_1 \cdots \beta_n}^{(n)}(\omega_1, \omega_2 \cdots \omega_n) \qquad \text{(VIII.1.10)}$$

As a matter of fact, as was explained earlier (see Section II.3.1.1) we want to use the extra frequency $\omega_\Sigma = -(\omega_1 + \cdots + \omega_n)$, which is the "sum frequency" with a minus sign. This little trick allows us to write the susceptibility as a tensor depending on (n+1) frequencies whose sum adds to zero, so that it can finally be written

$$\chi^{(n)}_{\text{mic}} = \chi^{(n)}_{\alpha\beta_1\ldots\beta_n}(\omega_\Sigma, \omega_1, \omega_2 \ldots \omega_n) \qquad \text{(VIII.1.11)}$$

However, $\chi^{(n)}_{\text{mic}}$ is not yet determined. The integration over the n ordered times appearing in the last two lines of Eq. VIII.1.8 must be actually calculated. We shall leave this to the reader, merely pointing out the successive phases of the calculation, and the physical consequences of the results where necessary.

First, one must calculate the quantum mean value <...> appearing in Eq. VIII.1.8 by going back to the Schrödinger representation. One must take Eq. VIII.1.2 and left-multiply it by $U_0(t,t_0)$ then right-multiply it by $U_0^{-1}(t,t_0)$. Thus one obtains operators of the form $\exp[\pm i H_0 (t_j - t_0)/\hbar]$ as is shown by Eqs. I.2.7 and I.2.8. Applying these operators to vectors $\langle a|$ and $|a\rangle$ (found at the beginning and at the end of the expression for the quantum mean) leads to the exponential function $\exp[i\omega_a(t_1 - t_{q+1})]$. In this way, the first eigenfrequency of the system is revealed. Next, one must break up the product of exponential operators by using the completeness relation (see Section I.3.2), i.e., by inserting the unit operator $\mathbb{1} = \sum_j |b_j\rangle\langle b_j|$, in which the vectors $|b_j\rangle$ designate the whole basis of eigenvectors of H_0, including $|a\rangle$. This kind of procedure is very often used in quantum mechanics. It is a way of projecting the changing state of the system—which initially is in state $|a\rangle$—onto the whole set of eigenstates of H_0. We can draw two physical conclusions from this way of calculating:

(d) $\chi^{(n)}_{\text{mic}}$ mixes *all the eigenstates* of H_0.

(e) $\chi^{(n)}_{\text{mic}}$ will depend on the differences between the eigenvalues of H_0, i.e. on *the differences between the eigenfrequencies* of the microsystem.

With this information, it is possible to integrate over all the successive times, and ultimately one obtains

$$\chi^{(n)}_{\text{mic}} = \left(\varepsilon_0^{-1}\hbar^{-n}\right) \sum_{\substack{q,r \\ (q+r=n)}} \sum_{\mathcal{P}_\beta} \sum_{b_j}$$

$$\times \frac{\langle a|p_{\beta_1}|b_1\rangle\langle b_1|p_{\beta_2}|b_2\rangle\cdots\langle b_q|p_\alpha|b_{q+1}\rangle\cdots\langle b_n|p_{\beta_n}|a\rangle}{(\omega_{a b_1} + \omega_1)\cdots(\omega_{a b_q} + \omega_1 + \cdots + \omega_q)(\omega_{a b_{q+1}} - \omega_{q+1} - \cdots - \omega_n)\cdots(\omega_{a b_n} - \omega_n)}$$

(VIII.1.12)

To be of use, this formula must first be simplified. Indeed, we can see that by writing the successive terms corresponding to (q = 0, r = n), (q = 1, r = n – 1),

TABLE VIII.1.1 — **Spatiotemporal permutations** $\mathcal{P}(\alpha\beta, \omega)$.

First term	Last term	Permutation
(0, n)	(1, n – 1)	$(\alpha, \omega_\Sigma) \to (\beta_1, \omega_1)$
(0, n)	(2, n – 2)	$(\alpha, \omega_\Sigma) \to (\beta_2, \omega_2)$
.	.	.
.	.	.
.	.	.
(0, n)	(n, 0)	$(\alpha, \omega_\Sigma) \to (\beta_n, \omega_n)$

and so on, $\chi_{mic}^{(n)}$ can be built up by starting from the term $\chi_{mic}^{(0,\, n)}$ (corresponding to q = 0, r = n), and doing spatiotemporal permutations. Let us call these permutations $\mathcal{P}_{(\alpha\beta,\, \omega)}$. (see Table VIII.1.1.)

To make things clearer, let us write the first two terms of $\chi_{mic}^{(n)}$:

$$\chi_{mic}^{(0,\, n)} = \left(\varepsilon_0^{-1}\, \hbar^{-n}\right) \sum_{\mathcal{P}_\beta} \sum_{b_j} \frac{<a|p_\alpha|b_1><b_1|p_{\beta_1}|b_2>\cdots<b_n|p_{\beta_n}|a>}{(\omega_{a\,b_1}+\omega_\Sigma)\cdots(\omega_{a\,b_n}-\omega_n)}$$

$$\chi_{mic}^{(1,\, n-1)} = \left(\varepsilon_0^{-1}\, \hbar^{-n}\right) \sum_{\mathcal{P}_\beta} \sum_{b_j} \frac{<a|p_{\beta_1}|b_1><b_1|p_\alpha|b_2>\cdots<b_n|p_{\beta_n}|a>}{(\omega_{a\,b_1}+\omega_1)\cdots(\omega_{a\,b_n}-\omega_n)}$$

So we can simplify the writing of Eq. VIII.1.12, yielding

$$\chi_{mic}^{(n)} = \left(\varepsilon_0^{-1}\, \hbar^{-n}\right) \sum_{\mathcal{P}_{\alpha\beta,\,\omega}} \sum_{b_j} \frac{<a|p_\alpha|b_1><b_1|p_{\beta_1}|b_2>\cdots<b_n|p_{\beta_n}|a>}{(\omega_{a\,b_1}+\omega_\Sigma)\cdots(\omega_{a\,b_n}-\omega_n)} \qquad \text{(VIII.1.13)}$$

$\mathcal{P}_{\alpha\beta,\,\omega}$ is a generalized space-time symmetry relation due to Butcher (see references in Section VIII.4.1).

A last difficulty remains in Eq. VIII.1.13, in the summation over all the eigenstates of H_0: If the exciting frequencies $\omega_1, \omega_2, \ldots, \omega_n$ (and their combinations) are very different from frequencies $\omega_{ab_1} \ldots \omega_{ab_n}$, the denominator of Eq.VIII.1.13 never vanishes, so that the summing is possible. In this case, $\chi_{mic}^{(n)}$ is a real quantity. If not, the sum is impossible. In order to avoid this difficulty, the axis of real numbers is slightly pushed over, away from the poles of function $\sum_n \omega_n$ by writing $\chi_{mic}^{(n)}$ as

$$\chi^{(n)}_{mic} = \left(\varepsilon_0^{-1}\ \hbar^{-n}\right)\left(\lim \gamma \to 0^+\right) \sum_{\mathcal{P}_{\alpha\beta,\,\omega}} \sum_{b_j}$$

$$\times \frac{<a|p_\alpha|b_1><b_1|p_{\beta_1}|a_2>\ldots<b_n|p_{\beta_n}|a>}{\left(\omega_{ab_1} - \omega_1 \ldots - \omega_n - i\gamma\right)\left(\omega_{ab_n} - \omega_n - i\gamma\right)} \qquad \text{(VIII.1.14)}$$

The limit 0^+ is imposed by the causality principle. But in fact, adding a small complex quantity to the first equation does nothing more than taking into account the finite width $\Delta\omega$ of the transitions of the microsystem. These were initially assumed to be infinitely sharp (zero width) so as to simplify the mathematics. Let us, for instance, assume a distribution of energy levels with a Lorentzian shape as follows:

$$f(\omega'_{ab_j}) = \left(\Delta\omega_{ab_j} / \pi\right) / \left[\ (\omega'_{ab_j} - \omega_{ab_j})^2 + \Delta\omega^2_{ab_j}\right]$$

centered on ω_{ab_j}. Then we can calculate the term $\chi^{(0,1)}_{mic}$ of the first-order susceptibility:

$$\chi^{(0,1)}_{mic} = \quad (\varepsilon_0\, \hbar)^{-1} \sum_{b_j} < a\,|\,p_{\alpha}|\ b_j > < b_j\,|\,p_{\beta_1}|\ a >$$

$$\times \left(\lim \varepsilon \to 0^+\right) \int_{-\infty}^{+\infty} \{\ [f(\omega'_{ab_j}) / (\omega'_{ab_j} - \omega_1 - i\gamma)\ ]\ \}\ d\omega'_{ab_j}$$

But the integral can also be written as

$$\left(\lim \varepsilon \to 0^+\right) \left[\ \left(\Delta\omega_{ab_j} / \pi\right) \int_{-\infty}^{+\infty} dz / (z - x)(z - y)(z - u)\right]$$

with $x = \omega_1 + i\varepsilon$, $y = \omega_{ab_j} - i\,\Delta\omega_{ab_j}$, and $u = \omega_{ab_j} + i\,\Delta\omega_{ab_j}$

Integration using the residue method is straightforward and yields $1 / \left(\omega_{ab_j} - \omega_1 - i\,\Delta\omega_{ab_j}\right)$. Finally, we obtain

$$\chi^{(0,1)}_{mic} = (\varepsilon_0\, \hbar)^{-1} \sum_{b_j} \frac{< a|\,p_{\alpha}|\ b_j > < b_j\,|\,p_{\beta_1}|\ a >}{\omega_{ab_j} - \omega_1 - i\,\Delta\omega_{ab_j}}$$

And the most general term $\chi^{(1)}_{mic}$ is found to be

$$\chi^{(1)}_{mic} = \chi^{(0,1)}_{mic} + \chi^{(1,0)}_{mic} =$$

$$(\varepsilon_0\, \hbar)^{-1} \sum_{b_j} \left\{ \frac{< a\,|\,p_{\alpha}|\ b_j > < b_j\,|\,p_{\beta_1}|\ a >}{\omega_{ab_j} - \omega_1 - i\,\Delta\omega_{ab_j}} + \frac{< a\,|\,p_{\beta_1}|\ b_j > < b_j\,|\,p_{\alpha}|\ a >}{\omega_{ab_j} + \omega_1 + i\,\Delta\omega_{ab_j}} \right\}$$

The reader may check that within the framework of the model used in Chapter VII, i.e. when $\omega_1 \approx \omega_{ab_1}$, the above expression is reduced to its first term only and can thus be written

$$\chi^{(1)}_{mic} = (\varepsilon_0 \hbar)^{-1} \frac{< a|p_{\alpha}|b_1 >< b_1|p_{\beta_1}|a >}{\omega_{ab_1} + \omega_1 - i\,\Delta\omega_{ab_1}}$$

Here we recognize Eq. VII.2.6 in which we would have $N = D_0 = 1$, $\omega_0 = \omega_{ab_1}$, and $\Delta\omega_0 = \gamma_{ab} = \Delta\omega_{ab_j}$. From all this, we can conclude that

(f) $\chi^{(n)}_{mic}$ is a *complex quantity.*

We have therefore determined a general expression for the microscopic nth-order susceptibility. The general space-time symmetry relation makes it possible to condense the expression to a form that is easy to use in practical applications. Problems VIII.1, 2, 3 will make the reader more familiar with its use.

We would like to draw attention to one point: We treated the problem in the state-vector formalism. In the first and in the third references, the reader will find an equivalent perturbation treatment using the density-matrix formalism, but limited to first-and second-order susceptibilities. A summary of the main characteristics of the nth-order microscopic susceptibility $\chi^{(n)}_{mic}$ is shown in Inset VIII.1.

And now let us examine some of the consequences of the existence of $\chi^{(n)}_{mic}$.

VIII.1.2 Macroscopic nth-order polarization

Earlier in this book, in Inset III.2, we defined the conditions governing the passage from microscopic to macroscopic descriptions. Therefore, let us associate $\chi^{(n)}_{mic}$ to $\chi^{(n)}_{mac}$ by means of Eq. III.3.1. Now we can determine the αth component of the nth-order macroscopic polarization $\mathbf{P}^{(n)}$, using Eq. VIII.1.8 :

$$P^{(n)}_{\alpha}(\mathbf{r}, t) = \int_{\omega_1} \cdots \int_{\omega_n} d\omega_1 \cdots d\omega_n \exp[-i(\omega_1 + \cdots + \omega_n)t]$$
$$\times E_{0\beta_1}(\mathbf{r}, \omega_1) \cdots E_{0\beta_n}(\mathbf{r}, \omega_n)\ \chi^{(n)}_{mac}(\omega_\Sigma, \omega_1 \cdots \omega_n) \quad \text{(VIII.1.15)}$$

This equation shows that

$$P^{(n)}_{\alpha}(\mathbf{r}, -\omega_\Sigma) = \chi^{(n)}_{mac}(\omega_\Sigma, \omega_1 \cdots \omega_n) E_{0\beta_1}(\mathbf{r}, \omega_1) \cdots E_{0\beta_n}(\mathbf{r}, \omega_n) \quad \text{(VIII.1.16)}$$

And if the Fourier components of the incident fields are represented by plane waves, we obtain

$$E_{0\beta_j}(\mathbf{r}, \omega_j) = E_{0\beta_j}(\omega_j) \exp i\,\mathbf{k}_j\,\mathbf{r} \quad \text{(VIII.1.17)}$$

yielding

$$P^{(n)}_{\alpha}(r, -\omega_\Sigma) = \chi^{(n)}_{mac}(\omega_\Sigma, \omega_1 \cdots \omega_n) E_{0\beta_1}(\omega_1) \cdots E_{0\beta_j}(\omega_j) \exp\left(i \sum \mathbf{k}_j \mathbf{r}\right) \quad \text{(VIII.1.18)}$$

Inset VIII.1 ***nth-order microscopic susceptibility*** $\chi_{mic}^{(n)}$ ***and its properties***

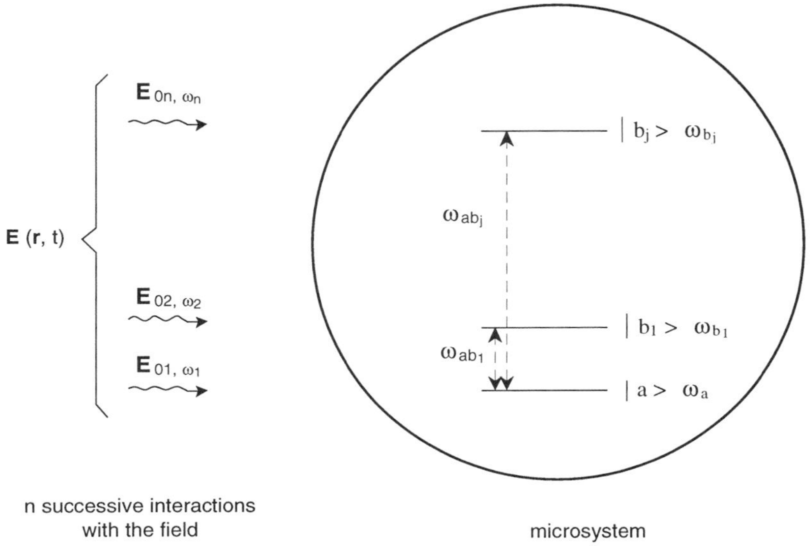

1. $\chi_{mic}^{(n)}$ contains (n + 1)! terms.

2. $\chi_{mic}^{(n)}$ is a tensor of rank (n + 1).

3. $\chi_{mic}^{(n)}$ depends on the frequencies ω_p (and their combinations) of the n successive interactions with the field.

4. $\chi_{mic}^{(n)}$ mixes all the eigenstates $| b_j >$ of the unperturbed microsystem.

5. $\chi_{mic}^{(n)}$ depends on the frequency differences $\omega_{ab_j} = \omega_{b_j} - \omega_a$ between the eigenfrequencies of the unperturbed microsystem.

6. $\chi_{mic}^{(n)}$ is a complex quantity.

So the nth-order polarization is seen to oscillate at a frequency which is the "sum frequency" *of all the frequencies* involved in the n interactions. This is in fact a consequence of the *conservation of energy* during the interaction of the wave(s) with the medium. As was pointed out during the description of Lamb's theory (see Section VII.3.1), nth-order polarization can play the part of a source term in Maxwell's equations and create a resultant wave with wave vector $\mathbf{k}$ and with frequency ω.

If the exciting field consists in fact of a set of monochromatic waves of frequencies ω_i, the n frequencies $\omega_1, \omega_2, \ldots, \omega_n$ of the n interactions will be taken among these ω_i frequencies (some of them may be taken several times, others not at all). The resulting wave created by this medium will then have frequency $\omega = -\omega_\Sigma = \omega_1 + \omega_2 + \cdots \omega_n$. One must remember, however, that this equality, which results from the conservation of energy, is strictly true only because the exciting waves are assumed to be monochromatic waves which, by definition, occupy an infinite time-space (from $t = -\infty$ to $t = +\infty$). A similar restriction concerns the wave vector $\mathbf{k}$ and the conservation of momentum. If the nonlinear medium is assumed to have infinite spatial dimensions and if the exciting monochromatic waves are assumed to be plane waves with wave vectors $\mathbf{k}_i$, then $\mathbf{k}$, the wave vector of the resulting wave, must necessarily be equal to the sum of the first: $\mathbf{k} = \sum_j \mathbf{k}_j$. This relation, called *phase synchronization*, means that all elementary waves emitted by different points of the nonlinear medium via $\mathbf{p}^{(n)}(\mathbf{r},t)$ interfere constructively. Now if this is not true for one of the point sources of the medium of infinite size, then one can always find another point source emitting an elementary wave whose phase is opposite to that of the first point, so that only a wave with wave vector $\mathbf{k} = \sum_j \mathbf{k}_j$ can be created by this infinite medium.

In practice, if the relation $\omega = -\omega_\Sigma$ can often be considered as correct (when the spectrum of the exciting waves is narrow enough so that monochromatic waves can be considered excellent approximations), it is usually more difficult to neglect the finite dimensions of the nonlinear medium. When the nonlinear medium is small, then each point behaves like a dipole oscillating at frequency $-\omega_\Sigma$ and, of course, waves with wave vectors other than $\mathbf{k} \neq \sum_j \mathbf{k}_j$ will be emitted. But if a sizable amount of the exciting energy is transferred to the induced wave, it will be possible to observe a next-to-perfect phase synchronization with $\mathbf{k} = \sum_j \mathbf{k}_j$. It's when this constructive interference occurs that the intensity of the induced wave will be greatest.

Presently, we shall study more in detail four consequences of nonlinear optics, chosen for their illustrative value. The first example is that of stimulated scattering (Rayleigh, Brillouin, and Raman). In this case, phase synchronization is not a fundamental problem, which we shall demonstrate in the study of *Rayleigh scattering of the second harmonic* (RSSH). The second example is the *optical Kerr effect* (OKE) in which phase-synchronization is automatically true.

Nevertheless, if one wants to optimize the amplitude of a nonlinearly induced wave, phase-synchronization is absolutely essential. This is the case, among others, for the generation of harmonic waves with "sum" or "difference" frequencies, for coherent anti-Stokes Raman scattering (CARS), and more generally for generating wave mixing. We shall restrict ourselves to the study of *phase conjugation by degenerate four-wave mixing* (DFWM), and to that of *second-order harmonic wave generation.* But we would like to point out that the structural definition of $\chi^{(n)}_{mic}$ (Eq. VIII.1.14) makes it very easy to handle a great number of other effects of nonlinear optics (Problems VIII.2 and VIII.3 will illustrate this point nicely). Last, Problems VIII.8, VIII.9, and VIII.10 deal with an extremely spectacular effect of nonlinear optics, namely *self-focusing* of powerful laser waves (and its applications to *optical bi-stability*).

VIII.1.3 Example of a low-transfer nonlinear interaction: scattering of the second harmonic of light (Terhune, 1968)

We have already shown (in Inset III.1) that Rayleigh scattering is characterized by an intensity which is proportional to the square of the optical field E and to a term which can take the abstract form of $\alpha^{(e)}_{\alpha\beta}\,\alpha^{(e)}_{\alpha\beta}$. Basically, we are dealing with a *binary* phenomenon, in which two particles interact, each of which has an *effective* polarizability $\alpha^{(e)}_{\alpha\beta}$, which can be written as

$$\alpha^{(e)}_{\alpha\beta} = \alpha_{\alpha\beta}(-\omega_L, \omega_L) + \beta_{\alpha\beta\gamma}(-2\omega_L, \omega_L, \omega_L)\, E_\gamma(\omega_L) + \gamma_{\alpha\beta\gamma\delta}(-2\omega_L, \omega_L, \omega_L, 0)\, E_\gamma(\omega_L)\, F_\delta(0) \qquad \text{(VIII.1.19)}$$

We can see that the last term involves F(0), the zero-frequency internal field prevailing inside the medium, and which was described in Section II.5.

Now if we calculate the statistical mean at the zero order of the perturbation, we obtain, on top of the dominant term $< \alpha_{\alpha\beta}(-\omega_L, \omega_L)\, \alpha_{\alpha\beta}(-\omega_L, \omega_L) >_0$ oscillating at frequency ω_L, two extra terms. These can be written, in a purposely simplified way, as follows:

$$< \beta^{(p)}_{\alpha\beta\gamma}(-2\omega_L, \omega_L, \omega_L)\, \beta^{(q)}_{\alpha\beta\eta}(-2\omega_L, \omega_L, \omega_L) >_0 E_\gamma(\omega_L)\, E_\eta(\omega_L) \qquad \text{(VIII.1.20)}$$

and

$$< \gamma^{(p)}_{\alpha\beta\gamma\delta}(-2\omega_L, \omega_L, \omega_L, 0)\, \gamma^{(q)}_{\alpha\beta\eta\mu}(-2\omega_L, \omega_L, \omega_L, 0) >_0 < F^{(p)}_\delta(0)\; F^{(q)}_\mu(0) >_0 \times E_\gamma(\omega_L)\, E_\eta(\omega_L) \qquad \text{(VIII.1.21)}$$

The two terms have several common characteristics, namely:

— They involve statistical mean values taken over *even-ranked* tensors (rank 6 in the first term, and rank 8 in the second). We proved that in such a case the mean values are nonvanishing [see Section III.1.2.1(c)]: This means that the corresponding contributions, obviously depending on *bimolecular* interactions (involving molecules p and q), actually exist.

— They generate scattering at frequency $2\omega_L$, in other words, of the *second harmonic wave* of the applied laser wave.

— The scattered intensity is proportional to the fourth power of the applied optic field, i.e. to the square of the applied optical intensity.

This harmonic scattering, discovered only with the advent of the laser, is also known as *hyper-Rayleigh scattering*. Clearly, it is only the first terms of a power expansion of a multiharmonic scattering of the laser wave.

— Both terms describe processes involving three photons; two incident photons of frequency ω_L give rise to a scattered photon with the double frequency $2\omega_L$.

We can distinguish two very different cases:

(a) When the microsystems, in liquid or gaseous phase, are not centrosymmetrical, the first contribution (Eq. VIII.1.20) dominates by far since the second contribution (Eq. VIII.1.21) involves a microscopic susceptibility whose order is larger by one than that of the first term. In these cases, hyper-Rayleigh scattering is used to measure second-order microscopic susceptibilities.

(b) When the microsystems are centrosymmetrical, the microscopic susceptibility $\beta_{\alpha\beta\gamma}$ vanishes, and so only the second term, contribution VIII.1.21, remains. In this case, second-harmonic scattering is proportional to $<F_\delta^p(0)\, F_\delta^q(0)>_0$, i.e. to the square of the internal field of a dense fluid at thermal equilibrium. This contribution does not exist for diluted gases. As far as we know, this is one of a very few instances where the internal field has a *direct* physical effect. Its effects usually go no further than to add corrections to applied optical fields.

For example, expression II.5.8 shows that for a set of quadrupolar molecules belonging to class D_{6h} (as is the case of benzene, see Appendix II.8.3.2) we obtain

$$< F_\delta^{(p)}(0)\, F_\delta^{(q)}(0) > \;\equiv\; Q_\perp^2 < \sum_{(q\neq p)=1}^{N} r_{pq}^{-8} >_0 \qquad \text{(VIII.1.22)}$$

r_{pq} is the distance between molecules p and q; N is the number of molecules per unit volume, and $Q_\perp$ is that component of the molecular quadrupole moment which is perpendicular to the sixth-order axis. Thus, if we know the value of the third-order susceptibility (and if the mean value $< >_0$ can be calculated by other means), measuring harmonic scattering gives access to the value of Q. Or again, it can give access to *radial intermolecular correlations* if Q is known from other experiments.

Besides hyper-Rayleigh scattering, other tri-photonic processes exist, involving the vibrational phonons of the microsystems, giving rise, for instance, to hyper-Raman scattering. Moreover, frequency-resolved harmonic scattering experiments have led to interesting results about molecular movements within dense fluids which complement those obtained by conventional optical spectroscopy. Nevertheless, and notwithstanding its fascinating nature, harmonic scattering,—as all other processes involving only low-transfer nonlinear interactions—is very difficult to put to use experimentally since the signals to be observed are extremely weak.

We are now going to look at three processes involving stronger energy transfers.

VIII.1.4 Examples of high-transfer nonlinear interactions

VIII.1.4.1 Interaction with natural phase synchronism: the optical Kerr effect (OKE) (Buckingham, 1956; Mayer, 1964)

The geometry of the experiment is described by Figure VIII.1.1.

The frequencies of the pump wave ($\mathbf{E}_{pu}$) and of the probe wave ($\mathbf{E}_{pr}$) are, respectively, ω_{pu} and ω_{pr}. The optical Kerr effect is a third-order process, revealed at frequency ω_{pr}. *The experiment measures the quadratic effect of the pump field on the polarization of the probe field.* The experiment involves a four-wave mixing, with the fourth wave being that component of the created wave whose polarization is perpendicular to the incident polarization and whose wave vector is a linear combination of the pump and the probe wave vectors. The first part of this chapter has shown the existence of a macroscopic susceptibility $\chi^{(3)}_{ijkl}$. This susceptibility is used to describe the Kerr effect, and it involves successively the frequencies $-\omega_{pr}$, ω_{pr}, ω_{pu}, and $-\omega_{pu}$ (whose total sum adds up to zero). The corresponding polarization takes the form

$$P^{(3)}_i(\omega_{pr}) = \chi^{(3)}_{ijkl}(-\omega_{pr}, \omega_{pr}, \omega_{pu}, -\omega_{pu})\, E_j(\omega_{pr})\, E_k(\omega_{pu})\, E^*_l(-\omega_{pu}) \qquad \text{(VIII.1.23)}$$

If we restrict ourselves to the study of OKE in isotropic macroscopic systems, the tensor $\chi^{(3)}_{ijkl}$ is completely symmetrical [as defined in Section III.1.2.1(b)] and can be written as the following sum:

$$\chi^{(3)}_{ijkl} = \chi^{(3)}\,\delta_{ij}\,\delta_{kl} + \chi^{(3)}\,\delta_{ik}\,\delta_{jl} + \chi^{(3)}\,\delta_{il}\,\delta_{jk} \qquad \text{(VIII.1.24)}$$

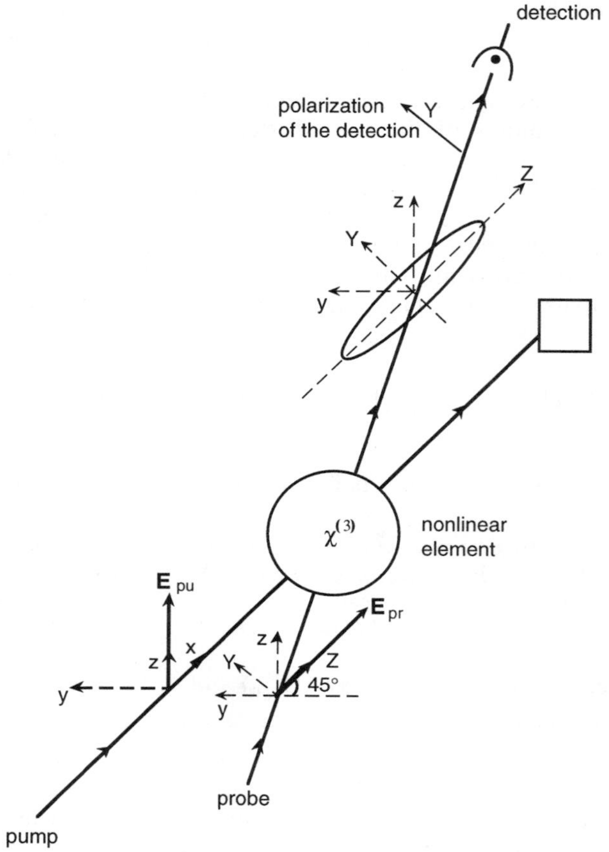

Figure VIII.1.1: Optical Kerr effect: Diagram of the principle.

where $\chi^{(3)}$ is the nondiagonal element $\chi^{(3)} = \chi_{xxyy} = \chi_{xxzz} = \chi_{yyzz}$ of the completely symmetrical nonlinear susceptibility tensor of rank 3 (see Appendix II.8.1.3). This yields

$$P_i^{(3)}(\omega_{pr}) = \chi^{(3)}(-\omega_{pr}, \omega_{pr}, \omega_{pu}, -\omega_{pu})\Big[E_i(\omega_{pr})\, E_j(\omega_{pu})\, E_j^*(-\omega_{pu}) + E_j(\omega_{pr})\, E_i(\omega_{pu})\, E_j^*(-\omega_{pu}) + E_j(\omega_{pr})\, E_j(\omega_{pu})\, E_i^*(-\omega_{pu})\Big] \quad \text{(VIII.1.25)}$$

But (see Eq. II.2.14) we proved that

$$\varepsilon_{ij}^{(r)} - \delta_{ij} = P_i / \varepsilon_0 E_j$$

Now let $\Delta\varepsilon_{ij}^{(r)}$ stand for the increment of the relative dielectric constant relative to the third-order induced polarization $\left(\Delta\varepsilon_{ij}^{(r)} = P_i^{(3)}/\varepsilon_0 E_j\right)$. Then we can calculate the increments $\Delta\varepsilon_{zz}^{(r)}$ parallel to the applied pump field (the z-axis), as well as $\Delta\,\varepsilon_{yy}^{(r)}$ parallel to the y-axis:

$$\Delta\,\varepsilon_{ij}^{(r)}(\omega_{pr}) = \left[\chi^{(3)}(-\,\omega_{pr}\,,\,\omega_{pr}\,,\,\omega_{pu}\,,\,-\,\omega_{pu})\,/\,\varepsilon_0\right]$$

$$\times\left[\delta_{ij}\,E_k(\omega_{pu})\,E_k^*(-\,\omega_{pu}) + E_i(\omega_{pu})\,E_j^*(-\,\omega_{pu}) + E_j(\omega_{pu})\,E_i^*(-\,\omega_{pu})\right] \qquad \text{(VIII.1.26)}$$

and :

$$\Delta\,\varepsilon_{zz}^{(r)}(\omega_{pr}) - \Delta\,\varepsilon_{yy}^{(r)}(\omega_{pr})$$

$$= 2\,\chi^{(3)}(-\,\omega_{pr},\,\omega_{pr},\,\omega_{pu},\,-\,\omega_{pu})\,E_z(\omega_{pu})\,E_z^*(-\,\omega_{pu})\,/\,\varepsilon_0 \qquad \text{(VIII.1.27)}$$

The left-hand side of this last equation is approximately equal to $2n(\omega_{pr})\,\Delta n(\omega_{pr})$ where $\Delta n(\omega_{pr}) = n_{zz}(\omega_{pr}) - n_{yy}(\omega_{pr})$. So this shows that the pump laser induced a variation of the refractive index at the frequency of the probe wave. This variation is such that

$$\Delta n\,(\omega_{pr}) \approx \chi^{(3)}(-\,\omega_{pr},\,\omega_{pr},\,\omega_{pu},\,-\,\omega_{pu})\,|\,E_z(\omega_{pu})\,|^2\,/\,n\,\varepsilon_0 \qquad \text{(VIII.1.28)}$$

The variation of the refractive index is *proportional to the intensity* of the applied pump wave. This is indeed a Kerr effect (see Problem III.2), but one that is induced by an optical pump wave. The reader may refer to Section VIII.4.2 to find a complete bibliography of this effect, as well as for a description of a large number of experiments developed since 1964.

Comment 1: We assumed tensor $\chi_{ijkl}^{(3)}$ to be completely symmetrical. This approximation is appropriate in a great majority of experimental situations, i.e. when frequencies ω_{pu} and ω_{pr} (as well as their combinations $\omega_{pu} \pm \omega_{pr}$) are very different from the eigenfrequencies of the system studied in the experiment. However, there are some rare exceptions, reported in literature, of *resonant optical Kerr effect* (resonant OKE induced by the Raman effect for instance). In this case, and for an isotropic system, $\chi_{ijkl}^{(3)}$ consists of three independent elements and Eq. VIII.1.24 must be rewritten as

$$\chi_i^{(3)} = \chi_{iijj}^{(3)}\,\delta_{ij}\,\delta_{kl} + \chi_{ijij}^{(3)}\,\delta_{ik}\,\delta_{jl} + \chi_{ijji}^{(3)}\,\delta_{il}\,\delta_{jk} \qquad \text{(VIII.1.29)}$$

(where indices i and j, such that $i \neq j$, represent any of the three coordinates x, y, and z). The reader may check that in this case the birefringence is expressed in a very general way by

$$\Delta n\,(\omega_{pr}) = \left(\chi_{ijij}^{(3)} + \chi_{ijji}^{(3)}\right)|\,E_z(\omega_{pu})\,|^2\,/\,2n\,\varepsilon_0 \qquad \text{(VIII.1.30)}$$

It is clear that for a completely symmetrical $\chi^{(3)}$ we have

TABLE VIII.1.2 — **Various microscopic contributions to static Kerr effect (SKE) and to optical Kerr effect (OKE), their names, and their causes.**

n'	$\chi_{mic}^{(n')}$	n"	Perturbation energy	Order of the statistical mean value	OKE	SKE	Name of the contribution	Cause of the contribution
3	$\gamma_{\alpha\beta\gamma\delta}$	0	0	0	Yes	Yes	Voigt	Molecular deformations
2	$\beta_{\alpha\beta\gamma}$	1	$-\mathbf{pE}$	1	No	Yes	Born 1	Orientation of the permanent dipole moments
1	$\alpha_{\alpha\beta}$	2	$-\mathbf{pE}$	2	No	Yes	Born 2	Orientation of the permanent dipole moments
			$-\alpha E^2/2$	1	Yes	Yes	Langevin	Orientation of the induced dipole moments

$$\chi_{ijij}^{(3)} + \chi_{ijji}^{(3)} = 2\,\chi^{(3)}$$

so that Eq. VIII.1.30 takes the form of Eq. VIII.1.28.

Comment 2: When $\omega_{pr} = \omega_{pu}$, Eq. VIII.1.28 (or VIII.1.30) describe *induced self-birefringence*. So, an elliptically polarized laser beam induces *circular birefringence* in an isotropic medium, and the axis of its elliptical polarization turns by an angle proportional to $\chi^{(3)}$. An elliptically polarized laser wave undergoes a change of polarization upon interacting with the medium. This change can be considerable in media characterized by large values of $\chi^{(3)}$.

Let us turn to the physical nature of the microscopic phenomena involved in the OKE. To have a clearer idea of what is going on, let's apply Eq. III.3.3 to the case n = 3 and list the results in a table (see Table VIII.1.2.).

First, we see that neither of the two Born contributions (see Problem III.2) contributes to optical Kerr effect. This is because the Born contributions are due to an orientation of the microsystems by field E via the permanent dipole moment **p**. Now the orienting moment $\mathbf{p} \times \mathbf{E}$ is a vector which changes its orientation at the frequency of the laser wave, and therefore its mean value over time vanishes.

On the other hand, we can see two new contributions which appear as well in optical Kerr effect as in static Kerr effect.

(a) The Voigt contribution (1908)

The Voigt contribution assumes that one takes a statistical mean value at the zero perturbation order. It can therefore reveal only microscopic processes which *do not depend on temperature*, such as *molecular deformability* (Born, 1933) and electric-field-induced *hyperpolarizability* (Buckingham and Pople, 1955).

For an isotropic microsystem, first-order susceptibility, limited to the second-order in the field, takes the form

$$\alpha_{\alpha\beta} = \overline{\alpha}\,\delta_{\alpha\beta} + \gamma_{ND}\left[\delta_{\alpha\beta}\,\delta_{\gamma\delta} + \delta_{\alpha\gamma}\,\delta_{\beta\delta} + \delta_{\alpha\delta}\,\delta_{\beta\gamma}\right] E_\gamma\, E_\delta \qquad \text{(VIII.1.31)}$$

where γ_{ND} is the nondiagonal element of the third-order microscopic susceptibility.

Equation VIII.1.31 can be rewritten as

$$\alpha_{\alpha\beta} = \overline{\alpha}\,\delta_{\alpha\beta} + \gamma_{ND}\left[\delta_{\alpha\beta}\,E_\gamma\,E_\gamma + 2\,E_\alpha\,E_\beta\right]$$

Now if we would apply a field polarized linearly along the z-axis, we would obtain

$$\begin{aligned} \alpha_{xx} &= \alpha_{yy} = \alpha_\perp = \overline{\alpha} + \gamma_{ND}\,E_z^2 \\ \alpha_{zz} &= \alpha_{//} = \overline{\alpha} + 3\,\gamma_{ND}\,E_z^2 \end{aligned} \qquad \text{(VIII.1.32)}$$

Thus, the applied field induces increments to the mean first-order polarizability which are

$$\Delta\alpha_{//} = \alpha_{//} - \overline{\alpha} = 3\,\Delta\alpha_\perp = 3\left(\alpha_\perp - \overline{\alpha}\right) = 3\,\gamma_{ND}\,E_z^2$$

or again

$$\Delta\alpha_{//}\,/\,\Delta\alpha_\perp = +\,3 \qquad \text{(VIII.1.33)}$$

This equation, which is illustrated in Figure VIII.1.2, appears as a useful *signature*, identifying a specific deformation induced by the laser wave. Moreover, this signature should not depend on temperature.

Now that we defined the experimental identification criteria, we still need to calculate the contribution itself. To do this, let us proceed in the same way as in Problem III.2 and write

$$\Delta n = \rho\,\gamma_{ND}\left(-\,\omega_{pr}, \omega_{pr}, \omega_{pu}, -\,\omega_{pu}\right) E_2^2\left[\left(n^2 + 2\right)/3\right]^4/2n\varepsilon_0 \qquad \text{(VIII.1.34)}$$

where ρ is the density of the particles and n is the mean refractive index of the medium.

(b) Langevin contribution (1910)

This contribution is due to reorientation of the microsystems.

It assumes that one takes a statistical mean value at the first perturbation order of the field, therefore it reveals microscopic phenomena varying with temperature as 1/T, such as molecular orientation (Langevin, 1910) caused by the coupling of the induced moment $\alpha\mathbf{E}$ with the field. In the latter case, the coupling moment has a

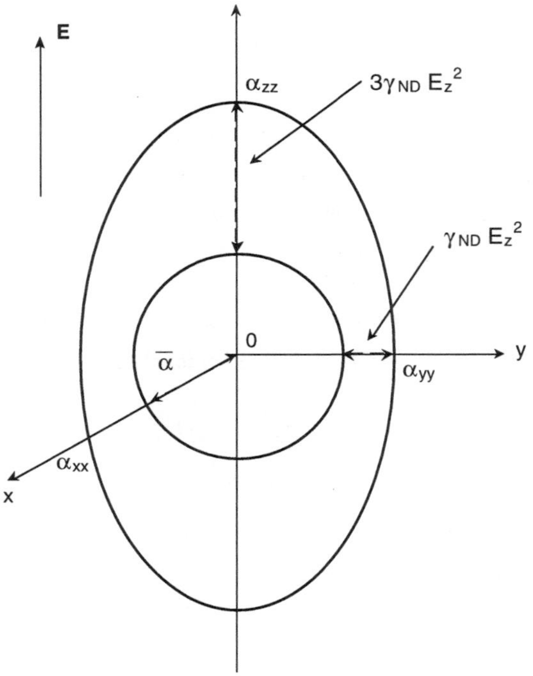

Figure VIII.1.2 : Molecular deformation induced by an electric field.

nonvanishing mean value over time, and optical-field-induced orientation indeed occurs. As before, for a microscopic system having a symmetry of revolution around direction 3, we shall write (see Problem II.3)

$$\alpha_{\alpha\beta} = \overline{\alpha}\, \delta_{\alpha\beta} + \gamma\left(c_{\alpha 3}\, c_{\beta 3} - \delta_{\alpha\beta} / 3\right)$$

where $\gamma = \alpha_{//} - \alpha_{\perp}$ is the anisotropy of the first-order microscopic susceptibility (subscript // refers to direction 3).

The equation is a definition of the natural anisotropy of the microsystem. Now if direction 3 coincides with the z-axis, we have $c_{zz}^2 = 1$ and $c_{x3}^2 = c_{y3}^2 = 0$, yielding

$$\alpha_{zz} = \alpha_{//} = \overline{\alpha} + 2\,\gamma / 3$$

$$\alpha_{xx} = \alpha_{yy} = \alpha_{\perp} = \overline{\alpha} - \gamma / 3$$

Figure VIII.1.3 shows the natural anisotropy of molecules presenting axial symmetry. We observe that

$$\Delta\alpha_{//} = \alpha_{//} - \overline{\alpha} = -\,2\,\Delta\alpha_{\perp} = -\,2\left(\alpha_{\perp} - \overline{\alpha}\right)$$

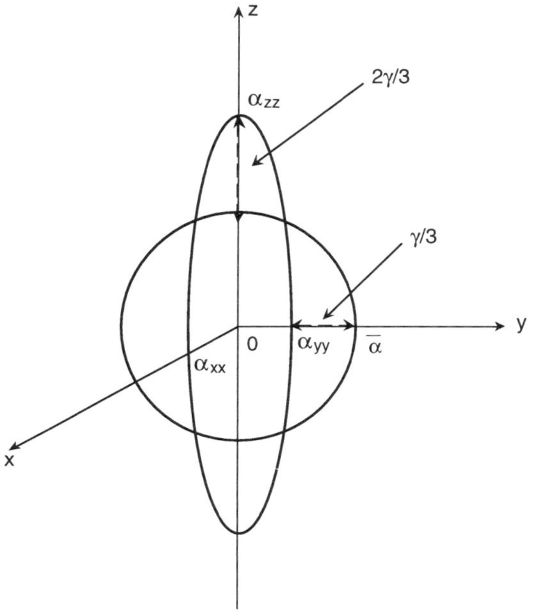

Figure VIII.1.3 : Natural anisotropy of so-called Langevin molecules (molecules with axial symmetry).

In other words

$$\Delta\alpha_{//} / \Delta\alpha_{\perp} = -2 \tag{VIII.1.35}$$

Equation VIII.1.35 gives a characterization of the *natural anisotropy* of axially symmetrical molecules, also known as *Langevin molecules*. Let us calculate the contribution of this anisotropy to optical Kerr effect.

Setting $\chi^{(1)} = \overline{\alpha}$, with of course $\gamma = \alpha_{//} - \alpha_{\perp}$, we obtain

$$n_{zz}^2 - 1 = (\rho/\varepsilon_0)\left[\overline{\alpha} + \gamma\left(< c_{z3}^2 >_1 - 1/3\right)\right]$$

$$n_{yy}^2 - 1 = (\rho/\varepsilon_0)\left[\overline{\alpha} + \gamma\left(< c_{y3}^2 >_1 - 1/3\right)\right]$$

This yields

$$\Delta n = n_{zz} - n_{yy} \approx \left[\rho\gamma(-\omega_s, \omega_s) / 2\varepsilon_0\right]\left(< c_{z3}^2 >_1 - < c_{y3}^2 >_1\right)$$

We must now determine mean values at the first order of the perturbation. Using Table III.1.1 and Problem III.1, we find

$$< c_{z3}^2 >_1 = 2\,\gamma(-\,\omega_p,\,\omega_p)\,E_z^2\,/\,45\,kT$$

$$< c_{y3}^2 >_1 = -\,\gamma(-\,\omega_p,\,\omega_p)\,E_z^2\,/\,45\,kT$$

The reader will notice that the ratio -2 (Havelook) between the zz and the yy contributions, together with the 1/T temperature dependency, form the signature of an orientational origin of optical Kerr effect. With this in mind, we find the following relation:

$$\Delta n = \rho\,\gamma(-\,\omega_{pr},\,\omega_{pr})\,\gamma(-\omega_{pu},\,\omega_{pu})\,E_2^2\,[(n^2+2)\,/\,3]^4\,/\,30\,n\varepsilon_0\,kT \qquad \text{(VIII.1.36)}$$

Induced birefringence is proportional to the square of the anisotropy γ of the first-order microscopic susceptibility. It therefore vanishes for so-called "isotropic" molecules. But actually, even in these kinds of "isotropic" fluids a nonvanishing contribution does exist which is due to *induced redistribution*. Induced redistribution has to do with radial correlations between molecules from which the system acquires a kind of formal, many-particle anisotropy. The two contribution we just studied (Voigt and Langevin) are usually said to be "fast": molecular, electronic, and nuclear deformations occur in the femto- and in the picosecond domain, while molecular reorientation takes a few tens of picoseconds for most gaseous, liquid, and solid media. Yet the susceptibilities we are dealing with are so weak that they usually lead to relatively weak signals. Nevertheless, we would like to draw attention to the case of liquid crystals in which very intense effects have been observed which critically depend on temperature and which are rather slow. These have been studied in detail for a few decades now (see references in Section VIII.4.2), and they seem to be due to a *cooperative* reorientation of the molecules in some of their phases.

It is clear that other physical mechanisms exist which are responsible for changes in the quadratic indices in the applied field. They can be due, for instance:

— to a laser-induced temperature rise,

— to indirect changes of the concentration of mixtures, due to an induced temperature rise (Soret effect),

— to changes in the density (electrostriction).

As said earlier, the effect common to all these processes is to perturb the propagation of laser waves in the medium. And since a laser beam has an inhomogeneous radial structure (see Chapter V), index gradients arise, which in turn result in the spectacular effects of *self-focusing* and *self-defocusing*, and sometimes even in the actual *self-guided propagation* of the optical wave. These phenomena, in which considerable energy transfers are involved, will be dealt with in Problems VIII.3 and VIII.4, and it will be shown how they can be used to obtain *optical bi-stability.*

VIII.1.4.2 Interaction with induced phase synchronism

(a) Phase conjugation by degenerate four-wave mixing (DFWM) (Hellwarth, 1977)

Let us look at the four-wave interaction described in Figure VIII.1.4. The name "degenerate" is due to the fact that the two pump waves E_{p1} and E_{p2}, and the probe wave E_{pr} share the same angular frequency ω_L. As a matter of fact, the mixing is done starting from a single laser source, with frequency ω_L, which is separated in three components: two intense pump waves and a probe wave of weaker energy.

The induced polarization can be written as

$$P^{(3)}(\omega_L) = \chi^{(3)}(-\omega_L, \omega_L, \omega_L, -\omega_L)\, E_{p1}(\omega_L)\, E_{p2}(\omega_L)\, E^*_{pr}(-\omega_L) \qquad \text{(VIII.1.37)}$$

Writing the polarization this way shows an essential and original feature of this interaction. If, as shown in the figure, the wave vectors $\mathbf{k}_{p1}$ and $\mathbf{k}_{p2}$ are such that $\mathbf{k}_{p1} + \mathbf{k}_{p2} = 0$, then an interaction which results in an output wave E_{si} with wave vector $\mathbf{k}_{si} = -\mathbf{k}_{pr}$, i.e. propagating in the direction *opposite* to the probe wave, conserves not only the total energy, but also the total momentum of the exchanged photons, and this for any angle $(\mathbf{k}_{pr}, \mathbf{k}_{p2})$. This means that we are in the case $\mathbf{k} = \sum_j \mathbf{k}_j$, which we discussed in the beginning of this section. Therefore the energy transfer from the pump waves to the output wave (the signal wave) can be expected to reach high values, even in a medium of limited dimensions. *It will be possible to amplify the probe wave.* We shall see (see Problem VIII.6) that in this case, and for a rather special kind of laser resonator, the medium can (simultaneously) serve as closing mirror or (and) as amplifying medium. Equation VIII.1.37 also indicates that a change in the probe frequency will result in a loss of phase synchronization, and therefore in a weakening of the intensity of the created signal. Seen from this viewpoint, DFWM acts like a *reflection optical filter.*

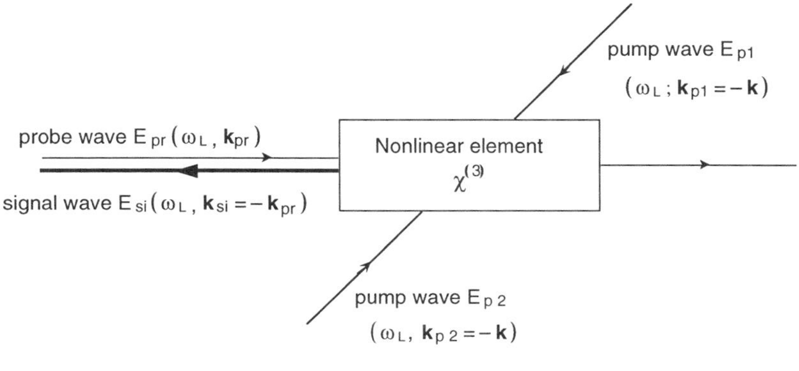

Figure VIII.1.4: Degenerate four-wave mixing: principle.

Let us analyze the properties of the signal wave. First, let us write the incident fields in the form of Eq. IV.2.3:

$$\mathbf{E} = (1/2)\, \varepsilon \exp(i\phi) + cc$$

with

$$\phi = \omega t - \mathbf{k}\, \mathbf{r} + \varphi$$

Equation VIII.1.37 then becomes

$$P^{(3)}(\omega_L) = (1/8)\, \chi^{(3)}(-\omega_L, \omega_L, \omega_L, -\omega_L)\, \varepsilon_{p1} \exp(i\phi_{p1}) \times \varepsilon_{p2} \exp(i\phi_{p2})\, \varepsilon_{pr}^{*} \exp(-i\phi_{pr}) + c\,c \qquad \text{(VIII.1.38)}$$

or again

$$P^{(3)}(\omega_L) = (1/8)\, \chi^{(3)}\, \varepsilon_{p1}\, \varepsilon_{p2}\, \varepsilon_{pr}^{*} \exp[i(\omega_L t + \mathbf{k}_{pr}\, \mathbf{r} + \varphi_{p1} + \varphi_{p2} - \varphi_{pr})] + c\,c$$

This equation shows that the signal wave will oscillate at angular frequency ω_L with a wave vector $\mathbf{k_{si}} = -\,\mathbf{k_{pr}}$, and a phase φ_{si} such that

$$\varphi_{si} = \varphi_{p1} + \varphi_{p2} - \varphi_{pr} \qquad \text{(VIII.1.39)}$$

That is, it varies in fact as $(-\varphi_{pr})$.

Thus, no matter what wave vector the probe wave has, the emitted signal will have a *conjugate phase* with respect to the probe phase. This explains the name *conjugated phase mirror* (CPM) used for such a system consisting of a nonlinear medium illuminated by two pump waves.

A second way to interpret this kind of degenerate four-wave mixing is shown in Figure VIII.1.5. It is a holographical interpretation. The probe wave and pump wave 1 (case a) or 2 (case b) give rise to two gratings, whether amplitude or phase gratings. Their spacing Λ_1 or Λ_2, respectively, is given by

$$\Lambda_1 = \lambda_L / 2\cos(\theta/2)\ , \ \ \Lambda_2 = \lambda_L / 2\sin(\theta/2)$$

λ_L is the wavelength of the laser used in this experiment. The illustration shows holographic registration, where the probe wave plays the part of the object wave while the pump wave plays the part of reference wave. These gratings diffract the other pump wave: pump wave 2 in case a, and pump wave 1 in case b. The second pump wave then plays the part of the restoring wave, so that the restored signal wave corresponds to the conjugated signal of DFWM.

Just by looking at the figure, we can see that diffraction is at its maximum in the direction of the signal since $\mathbf{k}_{si}$ is symmetrical (except for its sign) to $\mathbf{k}_{p2}$ or to $\mathbf{k}_{p1}$ with respect to the normal to the observed fringes. This is indeed holography, in which registering and restoring occur globally. It is an experiment in *real-time holography*.

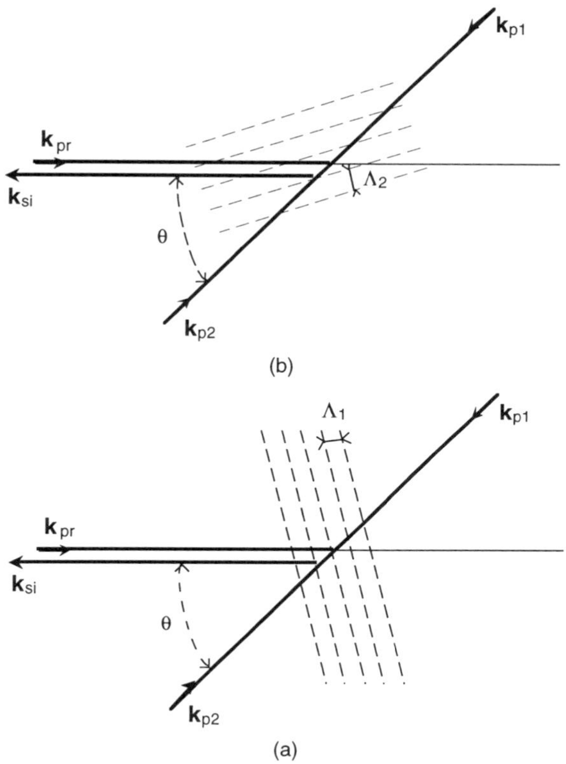

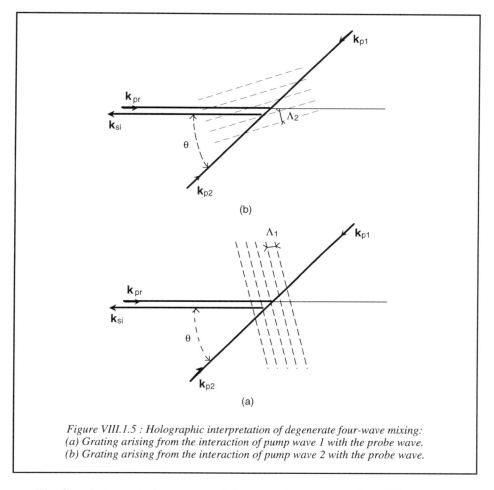

Figure VIII.1.5 : Holographic interpretation of degenerate four-wave mixing: (a) Grating arising from the interaction of pump wave 1 with the probe wave. (b) Grating arising from the interaction of pump wave 2 with the probe wave.

The first degenerate four-wave mixing experiments were done 25 years ago on dye solutions, on solids (silicon), and on liquids (carbon disulfide). Since that time, a great number of original results have been obtained, in applied physics as well as in fundamental physics.

— In applied physics, in addition to the setup we already mentioned, it is interesting to point out the possibility of correcting optical waves travelling through "perturbing" objects. Indeed, if a wave passes a second time through a perturbing medium after reflection on a standard mirror, the perturbations are multiplied by two. On the other hand, passing a second time through this medium after reflection on a conjugate phase mirror corrects the initial perturbations, so that the wave is restored to its initial purity. This is called *wave front rectification* and is obtained thanks to the property of phase conjugation.

— In fundamental physics, a great number of various informations has been obtained and published. A detailed bibliography as well as some general descriptions will be found in the references of Section VIII.4.3.

The last three processes we examined are third-order processes. We would like to end this illustration by describing a second-order process, namely the generation of second-harmonic light waves.

(b) Generation of second-harmonic light (Franken, 1962)

Second-harmonic light generation is a special case of the generation of waves with sum and difference frequencies (see Problem VIII.1). It is described by the second-order polarization $P^{(2)}(2\omega_L)$ given by

$$P_i^{(2)}(2\omega_L) = \chi_{ijk}^{(2)}(-2\omega_L, \omega_L, \omega_L)\, E_j(\omega_L)\, E_k(\omega_L) \qquad \text{(VIII.1.40)}$$

Let us now insert this nonlinear polarization in Maxwell's equations, and especially into the wave equation (see Eq. VII.3.1) relative to a medium (assumed to be nonconductor, and to have a vanishing free electron density). This wave equation is

$$\Delta \mathbf{E} - (1/c^2)\, \partial^2 \mathbf{E}/\partial t^2 = \mu_0\, \partial^2 \mathbf{P}/\partial t^2 \qquad \text{(VIII.1.41)}$$

In this equation, we can write the Fourier expansions of field E and polarization P. In our problem, we shall deal with two fields which can be approximated by plane waves:

— the exciting field, called the "fundamental wave" $\mathbf{E}_1(\omega_L)$

$$\mathbf{E}_1(\omega_L) = \mathbf{e}_1 A_1 \exp[\, i\, (\mathbf{k}_1 \mathbf{r} - \omega_L t)] \qquad \text{(VIII.1.42)}$$

$\mathbf{k}_1$ is the wave vector of the fundamental wave, $\mathbf{e}_1$ is the unit vector pointing in the direction of its linear polarization.

— the generated field of the "second-harmonic wave" $\mathbf{E}_2(2\omega_L)$

$$\mathbf{E}_2(2\omega_L) = \mathbf{e}_2 A_2 \exp[\, i\, (\mathbf{k}_2 \mathbf{r} - 2\omega_L t)] \qquad \text{(VIII.1.43)}$$

We are mostly concerned with the propagation equation of the second-harmonic wave, which we can find starting with Eq. VIII.1.41 and by equating in its two members those terms with frequency $2\omega_L$. At this frequency, the polarization can be decomposed in a linear and a nonlinear part:

$$\mathbf{P}(2\omega_L) = \mathbf{P}_L(2\omega_L) + \mathbf{P}_{NL}(2\omega_L) \qquad \text{(VIII.1.44)}$$

where $\mathbf{P}_{NL}(2\omega_L)$ is given by expression VIII.1.40 in which the incident field components are defined by Eq. VIII.1.42. We finally obtain:

$$\Delta \mathbf{E}_2(2\omega_L, \mathbf{r}) + \left(4\, \varepsilon r\, \omega_L^2 / c^2\right) \mathbf{E}_2(2\omega_L, \mathbf{r}) = -4\mu_0\, \omega_L^2\, \mathbf{P}_{NL}(2\omega_L, \mathbf{r}) \qquad \text{(VIII.1.45)}$$

It is relatively simple to solve this equation if we assume that the two waves propagate along the z-axis and that

$$\partial^2 A_2 / \partial z^2 << k_2 \, \partial A_2 / \partial z \qquad \text{(VIII.1.46)}$$

In fact, this approximation assumes that $\lambda_2 \nabla A_2 << A_2$, in other words, it assumes that the nonlinear interaction is weak enough for the amplitude and the phase of the wave to change only slightly over one wavelength (an approximation called the « slow variation envelope »).

The left-hand side of Eq. VIII.1.45 can be calculated with the help of VIII.1.43. Subsequently equating the result with the expression obtained by substituting Eqs. VIII.1.40 and VIII.1.42 into the right-hand side of Eq. VIII.1.45 yields

$$\partial A_2 / \partial z \approx \left[i\mu_0 \, \omega_L \, c / n_2 \cos^2 \alpha_{2\omega_L} \right] \left(\mathbf{e}_2 \, \chi^{(2)} \, \mathbf{e}_1 \, \mathbf{e}_1 \right) A_1^2 \exp\left(i\Delta kz \right) \qquad \text{(VIII.1.47)}$$

In this equation we have used the following notation:

— n_2 is the refractive index of the medium at frequency $2\omega_L$.

— $\alpha_{2\omega_L}$ is the angle between the Poynting vector and the wave vector of the second-harmonic wave.

— $\mathbf{e}_2\chi^{(2)}\mathbf{e}_1\mathbf{e}_1$ defines an effective second-order susceptibility χ_e which takes into account the polarizations of the fundamental and of the second-harmonic waves.

— $\Delta k = 2\,k_1 - k_2 = (4\pi / \lambda_L)\,[n_1\,(\omega_L) - n_2\,(2\omega_L)]$ (VIII.1.48)

Δk is a measure of the lack of phase-synchronization when $n_1(\omega_L) \neq n_2(2\omega_L)$, of the waves traveling along the z-axis.

When $\Delta k \neq 0$, there is, on average, no gain ($\partial A_2/\partial z \approx 0$) in a medium of infinite length. Then, the energy transfer from the fundamental wave to the second-harmonic wave is very weak.

As a result, it appears that the generation of a relatively intense second-harmonic wave is possible only:

(a) if the illuminated medium has a *nonvanishing macroscopic second-order susceptibility* , i.e.:

— if the natural symmetry of the medium is such that $\chi^{(2)}$ is nonvanishing (so there must be no center of symmetry, of course), or

— if, notwithstanding the medium being centrosymmetric, this symmetry is broken because the experimenter has applied some external anisotropic *stress* (for instance, a zero-frequency electric or magnetic field).

Here, as opposed to the case of harmonic scattering, the internal field cannot give rise to a break in symmetry since $< F > = 0$ (the mean value of F vanishes, but not that of F^2). However, a relatively weak generated wave, probably of quadrupolar origin, has been observed in some liquids. In this case, the polarization involved is

$$P_i^{(2)}(2\omega_L) = \chi_{ijkl}^{(3)}(-2\omega_L, \omega_L, 0, \omega_L)\, E_j(\omega_L)\, \nabla_k\, E_l(\omega_L)$$

(b) if the experimenter chooses an experimental configuration effectively leading to phase synchronization. The two mainly used setups are described in Inset VIII.2.

Integration of Eq. VIII.1.47 gives the intensity of the second-harmonic wave. Assuming that $A_2(0) = 0$ and $A_1(z) \approx A_1(0)$, which amounts to neglecting the intensity changes of the fundamental wave as it travels through the medium (weak conversion), we obtain

$$A_2(z) \approx \left(i\, \mu_0\, \omega_L\, c \,/\, n_2 \cos^2 \alpha_{2\omega_L}\right) \chi_e\, A_1^2(0) \exp\left(i\Delta kz\right)\left[\sin^2\left(\Delta kz/2\right) / \left(\Delta k/2\right)\right] \quad \text{(VIII.1.49)}$$

And the intensity of the second harmonic wave is proportional to $| A_2(z) |^2$, given by

$$| A_2(z) |^2 \approx \left(\mu_0^2\, \omega_L^2\, c^2 \,/\, n_2^2 \cos^4 \alpha_{2\omega_L}\right) \chi_e^2 \left| A_1^2(0) \right|^2 \left[\sin\left(\Delta kz/2\right) / \left(\Delta k/2\right)\right]^2 \quad \text{(VIII.1.50)}$$

The generated wave varies periodically with z (Maker fringes with a period of $2\pi/\Delta k$). $\pi/\Delta k$ defines the coherence length l_c of the system: It is the *optimal* length of a medium for a given value of Δk. The generated wave has the greatest intensity for $\Delta k = 0$, i.e. for an infinite coherence length. Its intensity is proportional to the square of the laser intensity and to the square of the macroscopic second-order susceptibility.

We have to add that for samples of small length, $l << l_c$, the generated wave theoretically varies as l^2. However, our calculations assumed that the fundamental wave stays quasi-constant; this approximation is no longer true for high conversion rates. Therefore a decrease of the yield quickly appears.

The discovery of media with large second-order susceptibilities χ_e^2 makes for a rich field of research, in inorganic as well as in organic chemistry. Considerable progress has been made and conversion yields, which were on the order of a few percent in the early 1960, now reach several tens of percent. These yields have been greatly improved by inserting the medium inside the laser cavity and by the use of focalized Gaussian beams. Second-harmonic wave generation is a valuable means of obtaining laser waves in the blue, violet, and ultraviolet parts of the spectrum. Many frequency-doubled waves are used as pump waves in laser setups. This is the case, for instance, of the Nd^{3+}/YAG laser which delivers green light at 530 nm by means of frequency doubling. This laser is widely used in fundamental and in applied research.

Inset VIII.2 ***How to obtain phase synchronization (of type I) in negative uniaxial optical media***

The methods are based on the polarization and frequency dispersion properties of the medium.

First method (Giordmaine, 1962): Choosing a propagation direction at a given temperature.

The extraordinary refractive index $n_2(2\omega_L)$ (of the second-harmonic wave) is equal to the ordinary refractive index $n_1(\omega_L)$ (of the fundamental wave) in direction θ (with respect to the optical axis) chosen as shown in the accompanying diagram.

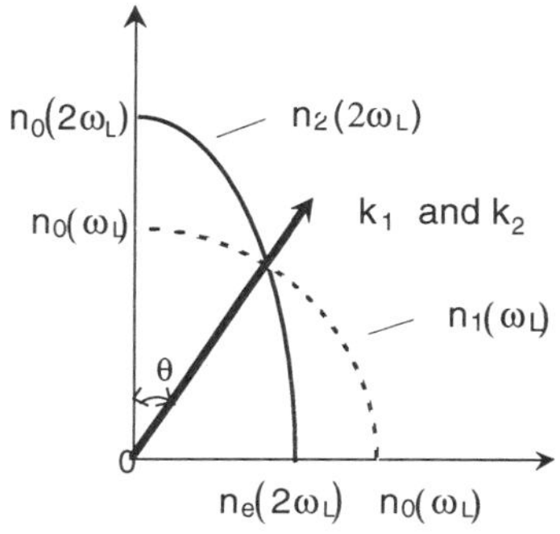

Drawbacks:

— Conversion is very sensitive to the value of θ.

— Conversion is limited because of the natural birefringence of the medium.

Second method (Boyd, 1968): Thermal adjustment of the refractive indices at 90° of the optical axis.

The temperature is chosen in such a way as to have $n_e(2\omega_L) = n_0(\omega_L)$ at 90° of the optical axis.

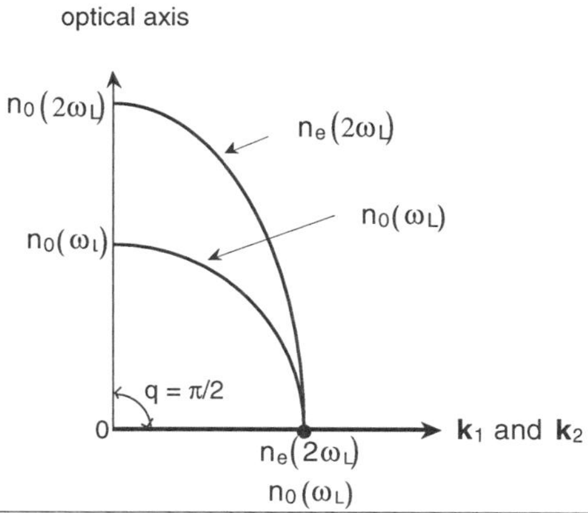

Advantages: Conversion is not very sensitive to the exact value of angle θ (focusing is possible).

In fact, both methods are used in practically all of present-day experiments.

It is also possible to obtain third-harmonic waves. To deal with this case, one needs only insert an extra polarization P_{NL} in the right-hand side of Eq. VIII.1.41, which is such that

$$P_{NL} = P_i^{(3)}(3\omega_L) = \chi_{ijkl}^{(3)}(-3\omega_L\,;\,\omega_L\,;\,\omega_L\,;\,\omega_L)\,E_j(\omega_L)\,E_k(\omega_L)\,E_l(\omega_L) \qquad \text{(VIII.1.51)}$$

The intensity of the third-harmonic wave is proportional to the third power of the intensity of the fundamental wave and to the square of $\chi^{(3)}$. It varies as $[\sin(\Delta kz/2)/(\Delta k/2)]^2$, where

$$\Delta k = 3k_1 - k_2 = (6\pi/\lambda_L)[n_1(\omega_L) - n_2(3\omega_L)] \qquad \text{(VIII.1.52)}$$

This intensity is much weaker than that of the second-harmonic wave (with comparable laser intensities).

Last, we would like to point out that harmonic waves can be generated not only by transmission as described above, but also by reflection on surfaces and on interfaces.

The advances observed lately in the field of harmonic generation of laser waves —research seems to be directed toward the synthesis of media which are both amplifying and frequency-doubling—are explained and described in the references in Section VIII.4.4, to which we refer the interested reader for further study.

We are about to start on the last section of this chapter, dedicated to a deliberately succinct description of nonstationary interactions. We shall illustrate it by the example of photon echoes.

VIII.2 Nonstationary resonant interaction: photon echo

We want to give a simple description of the photon echo. A first approach to the phenomenon was made in Problem VII.4.

Let us look at the transition between two levels $|a\rangle$ and $|b\rangle$ of a microsystem. This transition presents some inhomogeneous broadening due to the existence of different atomic classes.

The photon echo method makes it possible to study at the same time the inhomogeneous width of the transition and the homogeneous relaxation constant. To do so, a first pulse, the so-called $\pi/2$ pulse, is used to create coherences within the different classes of the medium. After the passage of this pulse, the coherences evolve freely during a length of time t'. Because the various classes all have different resonance frequencies, their coherences dephase very rapidly with respect to one another. No coherent light wave can be emitted by the medium. A second pulse, called π pulse, lasting twice as long as the first one, is then sent onto the medium in order to reverse the rotation direction of the phases of the coherences. After another length of time t', the coherences of all the classes are in phase again and the medium emits a coherent light wave. This constitutes the photon echo.

The time width of the echo corresponds to the inhomogeneous width of the transition. The way its amplitude changes with t' makes it possible to determine the relaxation of each coherence created by the first, the $\pi/2$, pulse during a time lasting 2t'. Analyzing this amplitude change therefore yields a determination of γ_{ab}, the homogeneous relaxation constant.

A simplistic image can be sketched to make the process appear more tangible: A group of runners are running on a circular arena in the same direction but each at a different, though constant speed. At time t = 0, and at the flagman's signal (corresponding to the $\pi/2$ pulse), they all start to run from the origin O of the track. After a time t', due to their differing speeds, they are distributed all over the track. A new signal from the flagman (corresponding to the π pulse) tells them to turn around and start running in the opposite direction. After another interval of time t', they end up all together at the origin O, i.e. at the starting blocks of the track (corresponding to a coherence of all phases), except for those who abandoned before reaching O (damping of the coherences).

We can give a simple theoretical treatment of the problem, in the state-vector formalism for example, without introducing the damping constant γ_{ab} so as not to overburden the calculations. First, let us consider a particular class of the medium, having resonance frequency $\omega'_0 = \omega_b - \omega_a$ ($\hbar\omega_a$ and $\hbar\omega_b$ are the respective energies of the two levels $|a>$ and $|b>$ we are dealing with). We assume that initially the system is in its ground state $|a>$ (so $|\psi(0)> = |a>$).

The first ($\pi/2$) pulse has a duration of $t_{\pi/2} = \pi\hbar/2pE_0$, where p is the matrix element of the transition dipole moment of the $|a> \rightarrow |b>$ transition, and E_0 is the amplitude of the applied field of frequency ω_L. The pulse projects the system in a state which is a mixture of states $|a>$ and $|b>$ (see Problem VII.2) such that the probability of being in $|a>$ or in $|b>$ is the same (see Problem VII.2). This state can be found easily by using the evolution equation (postulate P_6) with $H = H_0 - \boldsymbol{p}\mathbf{E}$. Calculations yield

$$|\psi(t_{\pi/2})> = (\sqrt{2}/2)\left[\exp(-i\omega_a t_{\pi/2})|a> + i\exp(-i\omega_b t_{\pi/2})|b>\right] \qquad \text{(VIII.2.1)}$$

For t between $t_{\pi/2}$ and t'($t_{\pi/2} < t < t'$), the systems evolves freely in such a way as to keep $|c_a(t)| = |c_b(t)| = \sqrt{2}/2$ and so as to satisfy postulate P_6 (with $H = H_0$). At time t', the system is in the following state:

$$|\psi(t')> = (\sqrt{2}/2)\left[\exp(-i\omega_a t')|a> + i\exp(-i\omega_b t')|b>\right] \qquad \text{(VIII.2.2)}$$

in which we neglected $t_{\pi/2}$ with respect to t'. Applying the pulse causes a permutation of coefficient $c_a(t)$ with $c_b(t)$ except for a factor i:

$$|\psi(t' + t_\pi)> = (i\sqrt{2}/2)\left\{\exp\left[-i\omega_b(t' + t_\pi)\right]|a> + \exp\left[-i\omega_a(t' + t_\pi)\right]|b>\right\} \qquad \text{(VIII.2.3)}$$

Neglecting t_π with respect to t', the free evolution of the system after the π pulse is described by

$$| \psi (t) > = (i\sqrt{2} / 2) \{\exp [- i (\omega_b - \omega_a) t'] \exp (- i\omega_a t) | a > \\ + \exp [- i (\omega_a - \omega_b) t'] \exp (- i\omega_b t) | b >\}$$

The mean polarization of this class of the medium is simple to calculate:

$$< p (\omega'_0) > \; = < \psi (t) | \boldsymbol{p} | \psi (t) > = (p/2) \{\exp [i (\omega_b - \omega_a) t] \exp [i (\omega_a - \omega_b) 2t'] + cc\} \\ = p \cos [\omega'_0 (t - 2t')]$$

(VIII.2.4)

The global polarization $< P >$ is found by integrating $< p(\omega'_0) >$ over all the classes composing the inhomogeneous broadening. Because of the sine shape of $< p(\omega'_0) >$, except for values of t around 2t', it is easy to understand that integrating over ω'_0 yields zero. To take an example, let us consider a gaseous medium at thermal equilibrium. Here, the classes are classes of atomic velocities, and the inhomogeneous broadening is the Doppler broadening (see Section VII.3.2.3). In case of a Maxwellian distribution of the velocities, with mean velocity v_0, the number of atoms whose resonance-frequency (in the laboratory reference frame) is comprised between ω'_0 and $\omega'_0 + d\omega'_0$ is given by

$$dN = (1/\sqrt{\pi}\, kv_0) \exp [- (\omega'_0 - \omega_0)^2 / (kv_0)^2] \, d\omega'_0 \qquad \text{(VIII.2.5)}$$

where k and ω_0 are respectively the modulus of the wave-vector and the atomic resonance frequency not shifted by Doppler effect. The total polarization $< P >$ is then given by the following weighed integral:

$$< P > = \int_0^{\infty} < p (\omega'_0) > dN \approx p \cos [\omega_0 (t - 2t')] \exp [- (k^2 v_0^2 / 4) (t - 2t')^2]$$

(VIII.2.6)

The shape of the time envelope of the photon echo as calculated by this simplified approach is Gaussian. Its width $T_2^* = (kv_0)^{-1}$ is equal to the inverse of the inhomogeneous width of the transition. Inserting the relaxation of the coherences would have added to this expression a time evolution of the amplitude of the echo proportional to exp $[-2\gamma_{ab}t']$, which in turn determines time constant $T_2 = \gamma_{ab}{}^{-1}$. Accordingly, photon echoes give access to both time constants T_2 and T_2^*. other techniques exist (optical-nutation, incoherent resonance, optical free induction decay, ...) using *transient coherent responses*, also induced by laser pulses sent onto the medium. These give access not only to T_2 and T_2^*, but also to T_1, the relaxation constant of the populations (see Chapter IV). All these techniques, which the reader will find described in the references in Section VIII.4.1, are now extensively used in molecular physics.

VIII.3 Problems and outlined solutions

VIII.3.1 Problem VIII.1: Second-order optical susceptibility

1. Write the *condensed* (three terms) general expression of the second-order optical susceptibility $\chi^{(2)}_{\alpha\beta_1\beta_2}\left[-(\omega_1+\omega_2)\,;\,\omega_1\,;\,\omega_2\right]$ [in fact, each term is double because of the permutation on (β_1, β_2)].

2. Show that

$$\chi^{(2)}_{\alpha\beta_1\beta_2}\left[-(\omega_1+\omega_2)\,;\,\omega_1\,;\,\omega_2\right] = \chi^{(2)}_{\beta_1\alpha\beta_2}\left[\omega_1\,;\,-(\omega_1+\omega_2)\,;\,\omega_2\right]$$
$$= \chi^{(2)}_{\beta_2\beta_1\alpha}\left[\omega_2\,;\,\omega_1\,;\,-(\omega_1+\omega_2)\right]$$

3. What physical effects correspond to the three terms of question 1?

1. The general expression for the susceptibility $\chi^{(2)}_{\alpha\beta_1\beta_2}$ is

$$\chi^{(2)}_{\alpha\beta_1\beta_2}\left[-(\omega_1+\omega_2)\,;\,\omega_1\,;\,\omega_2\right] = \varepsilon_0^{-1}\,\hbar^{-2}\,\mathcal{P}_\beta \sum_{bj} \left\{ \frac{<a|p_\alpha|b_1><b_1|p_{\beta_1}|b_2><b_2|p_{\beta_2}|a>}{(\omega_{ab_1}-\omega_1-\omega_2)(\omega_{ab_2}-\omega_2)} + \frac{<a|p_{\beta_1}|b_1><b_1|p_\alpha|b_2><b_2|p_{\beta_2}|a>}{(\omega_{ab_1}+\omega_1)(\omega_{ab_2}-\omega_2)} + \frac{<a|p_{\beta_2}|b_1><b_1|p_{\beta_1}|b_2><b_2|p_\alpha|a>}{(\omega_{ab_1}+\omega_2)(\omega_{ab_2}+\omega_1+\omega_2)} \right\}$$

2. Permutation $\mathcal{P}_{\alpha\beta,\omega}$ operates a circular permutation on the three components of $\chi^{(2)}_{\alpha\beta_1\beta_2}\left[-(\omega_1+\omega_2)\,;\,\omega_1\,;\,\omega_2\right]$ without changing their sum. The first permutation gives $\chi^{(2)}_{\beta_1\alpha\beta_2}\left[\omega_1\,;\,-(\omega_1+\omega_2)\,;\,\omega_2\right]$ and the second gives $\chi^{(2)}_{\beta_2\beta_1\alpha}\left[\omega_2\,;\,\omega_1\,;\,-(\omega_1+\omega_2)\right]$. The three expressions of $\chi^{(2)}$ are equal.

3. (a) The term $\chi^{(2)}_{\alpha\beta_1\beta_2}\left[-(\omega_1+\omega_2)\,;\,\omega_1\,;\,\omega_2\right]$
This term describes the mixing of waves ω_1 and ω_2 and the creation of the "sum" wave with frequency $(\omega_1 + \omega_2)$.

If we have $\omega_1 = \omega_2 = \omega_L$, we are describing the generation of a second-harmonic wave at frequency $2\omega_L$. If ω_1 is a frequency of the optical domain and ω_2 is a microwave frequency, we obtain a phase modulation of the optical wave. The amplitude of the optical wave can be modulated with the help of polarizers (Pockel's effect).

The inverse situation can also arise, in which a nonlinear medium illuminated by two laser waves of frequencies ω_1 and $\omega_3 = (\omega_1 + \omega_2)$ gives rise to a "difference" wave, whose frequency is ω_2, together with a simultaneous amplification of waves ω_1 and ω_3. This property is used in the construction of optical parametric amplifiers (OPA) and of optical parametric oscillators (OPO).

(b) Terms $\chi^{(2)}_{\beta_1\alpha\beta_2}[\omega_1 ; -(\omega_1+\omega_2) ; \omega_2]$ and $\chi^{(2)}_{\beta_1\beta_2\alpha}[\omega_2 ; \omega_1 ; -(\omega_1+\omega_2)]$ These terms describe the effect of optical wave rectification. If ω_1 is an optical frequency and ω_2 a radio frequency, the second term describes the interaction between the two waves ω_1 and $(\omega_1+\omega_2)$ yielding a low frequency modulation at ω_2 (optical heterodyning).

VIII.3.2 Problem VIII.2: Third-order optical susceptibility

1. Give the *condensed* general expression (four terms) of the susceptibility $\chi^{(3)}_{\alpha\beta_1\beta_2\beta_3}[-(\omega_1+\omega_2+\omega_3) ; \omega_1 ; \omega_2 ; \omega_3]$ [each term is the sum of six terms obtained by the permutations of $(\beta_1, \beta_2, \beta_3)$].

2. Express the generalized space-time law of symmetry.

3. What physical effects could correspond to the terms of question 1?

1. The expression of $\chi^{(n)}$ for n = 3 can be written as

$$\chi^{(3)}_{\alpha\beta_1\beta_2\beta_3}(\omega_\Sigma ; \omega_1 ; \omega_2 ; \omega_3) = \hbar^{-3}\sum_{b_i}\sum_{\mathcal{P}_\beta}$$

$$\times \frac{\langle a|p_\alpha|b_1\rangle\langle b_1|p_{\beta_1}|b_2\rangle\langle b_2|p_{\beta_2}|b_3\rangle\langle b_3|p_{\beta_3}|a\rangle}{(\omega_{ab_1}-\omega_1-\omega_2-\omega_3)(\omega_{ab_2}-\omega_2-\omega_3)(\omega_{ab_3}-\omega_3)}$$

$$+ \frac{\langle a|p_{\beta_1}|b_1\rangle\langle b_1|p_\alpha|b_2\rangle\langle b_2|p_{\beta_2}|b_3\rangle\langle b_3|p_{\beta_3}|a\rangle}{(\omega_{ab_1}+\omega_1)(\omega_{ab_2}-\omega_2-\omega_3)(\omega_{ab_3}-\omega_3)}$$

$$+ \frac{\langle a|p_{\beta_2}|b_1\rangle\langle b_1|p_{\beta_1}|b_2\rangle\langle b_2|p_\alpha|b_3\rangle\langle b_3|p_{\beta_3}|a\rangle}{(\omega_{ab_1}+\omega_2)(\omega_{ab_2}+\omega_1+\omega_2)(\omega_{ab_3}-\omega_3)}$$

$$+ \frac{\langle a|p_{\beta_3}|b_1\rangle\langle b_1|p_{\beta_2}|b_2\rangle\langle b_2|p_{\beta_1}|b_3\rangle\langle b_3|p_\alpha|a\rangle}{(\omega_{ab_1}+\omega_3)(\omega_{ab_2}+\omega_1+\omega_3)(\omega_{ab_3}+\omega_1+\omega_2+\omega_3)}$$

2. Let's express the law of space-time symmetry:

$$\begin{aligned}\chi^{(3)}_{\alpha\beta_1\beta_2\beta_3}(\omega_\Sigma ; \omega_1 ; \omega_2 ; \omega_3) &= \chi^{(3)}_{\beta_1\alpha\beta_2\beta_3}[\omega_1 ; -(\omega_1+\omega_2+\omega_3) ; \omega_2 ; \omega_3] \\ &= \chi^{(3)}_{\beta_1\beta_2\alpha\beta_3}[\omega_2 ; \omega_1 ; -(\omega_1+\omega_2+\omega_3) ; \omega_3] \\ &= \chi^{(3)}_{\beta_1\beta_2\beta_3\alpha}[\omega_3 ; \omega_1 ; \omega_2 ; -(\omega_1+\omega_2+\omega_3)]\end{aligned}$$

3. These terms correspond to a large number of effects:

(a) If we have $\omega_1 = \omega_2 = \omega_3 = \omega_L$, the induced polarization oscillates at frequency $-\omega_\Sigma = 3\omega_L$. This corresponds to third-harmonic scattering or to third-harmonic wave generation.

(b) If $\omega_1 \neq \omega_2 \neq \omega_3$: three different laser waves generate a sum wave.

(c) If $\omega_1 = \omega_2 = \omega_L$ and $\omega_3 = 0$, then $P_i^{(3)}(2\omega_L) = \chi_{ijkl}(-2\omega_L\ ;\ \omega_L\ ;\ \omega_L\ ;\ 0)$, and there is second-harmonic generation (or scattering) induced by a static electric field.

(d) If $\omega_2 = -\omega_3 = \omega_L$ and $\omega_1 = \omega'_L$, the polarization oscillates at frequency ω'_L. The susceptibility $\chi^{(3)}_{ijkl}(-\omega'_L\ ;\ \omega'_L\ ;\ \omega_L\ ;\ -\omega_L)$ describes the optical birefringence.

VIII.3.3 Problem VIII.3: Self-focusing of laser waves and optical bi-stability without cavity

In this problem, we take a nonlinear medium in which self-focusing of the laser beam occurs. This medium then gives rise to an optically bi-stable system without using energy stored in a cavity.

The laser beam

1. The laser cavity consists of a plane mirror M_1 having a reflection coefficient of $r_1 = 1$, and of an output mirror M_2, with radius of curvature R_2 and reflection coefficient $r_2 = 0.95$ for the wave intensity. Assume that M_2 does not have any lenslike effect on the beam. The manufacturer of the laser indicates a divergence of the output beam of $\theta = 0.25$ mrd. Find the radius ω_0 and the position of the beam waist. The optical path L between the mirrors measures 1.50 m. Infer from this the radius of curvature R_2 of mirror M_2. The output power of the laser can vary between 0 and 2 W. What is the maximum value of the intensity of the light present inside the cavity?

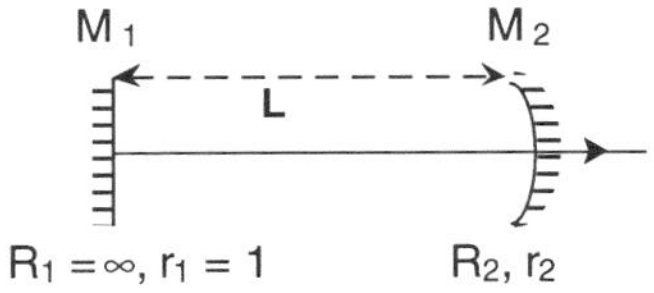

2. To study the nonlinear behavior of the medium, we want to be able to vary the size of the radius of the beam waist at the entrance of the medium, ω_1, between 1 and 10 μm. To this end, we use a microlens L_1 with a focal distance of $f_1 = 30$ mm.

(a) L_1 directly focus the output beam of the Ar^+ laser. Let L_0 be the distance between M_1 and L_1. Between what two values must L_0 vary to obtain the desired variation of beam waist ω_1? Comment on this result.

(b) It is better to use a second lens L_2, with a focal distance $f_2 = 500$ mm, placed at a distance L_0 of M_1, in order to create an intermediary beam waist between the laser beam waist and L_1. Calculate the position and the dimension ω_2 of this beam waist. Using the inside corner of a mirror-coated cube placed as shown on the diagram, one can shorten or lengthen distance L_2 by a translation of the corner. Between what values must L_2 vary to achieve the desired variation of 1 to 10 μm onω_1? What is the change in position of ω_1 as it undergoes this variation in size?

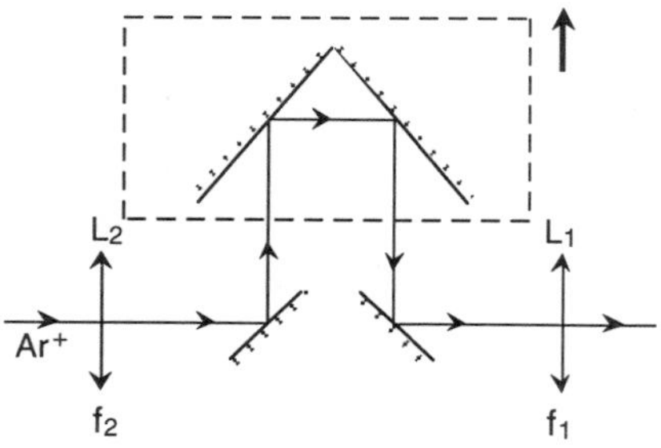

Bi-stability associated to the self-focusing of a laser beam by a nonlinear medium

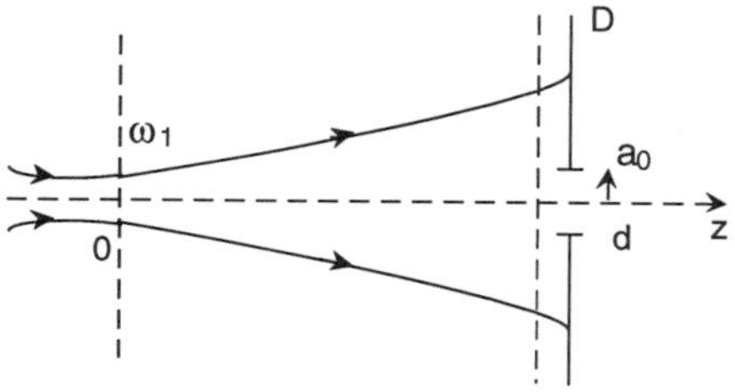

3. The beam waist ω_1 of the laser is positioned at z = 0, the front face of the nonlinear medium. The medium is contained in a cell of thickness d. We shall assume that the beam keeps a Gaussian shape during its propagation, but that the law governing the propagation of the radius of the beam is modified by the medium. Let I(r,z) be the intensity of the beam in a plane of ordinate z and at a distance r from the propagation axis: $I(r, z) = [\alpha P/\omega^2(z)] \exp[-2r^2/\omega^2(z)]$, where P is the power of the beam, $\omega(z)$ its radius at z, and α is a constant. The modification brought about by the medium is due to the law governing the propagation of light in the medium. Without demonstration, this law will be assumed to be given by

$$\omega^2(z) = \omega_1^2 \left\{1 + \left(1 - P/P_{cr}\right)\left[z^2\lambda^2/\left(\pi\,\omega_1^2\right)^2\right]\right\} \quad \text{for } P < P_{cr}$$

$$\omega(z) = \omega_1 = \text{constant} \qquad \text{for } P \geq P_{cr}$$

P_{cr} is a constant characteristic of the medium. The wave, in fact, induces a modification of the refractive index of the medium with $\Delta n = n_2 I$. The critical power P_{cr} is inversely proportional to the nonlinear refractive index n_2.

(a) $P_i \ll P_{cr}$. We want to obtain a beam waist at the back face of the cell with a radius $\omega(d) = 100$ µm. Calculate the corresponding thickness of the cell, d, for $\omega_1 = 1$ µm and for $\omega_1 = 10$ µm. Now let us assume the cell has the thickness corresponding to $\omega_1 = 10$ µm and let us place a pinhole diaphragm D at z = d. The center of the pinhole is on the axis of the beam and its radius is $a_0 = 10$ µm. We define a ratio T as the ratio between the transmitted power P_t and the incident power P_i: $T = P_t/P_i$. Calculate T.

(b) Let's assume P_i can take on any value. Calculate how T varies as a function of P_i. Do a numerical calculation and give a graphic representation of $T = f(P_i)$. Give a detailed representation of the region $P_i \leq P_{cr}$ for the two extreme values of ω_1, using the corresponding values of L_2 calculated in question 2b.

4. A mirror M, with an intensity reflection coefficient of $r^2 = 0.99$ is placed at $z = d$, close to diaphragm D. This mirror reflects the wave back onto itself. Assume that the total power inside the nonlinear medium is $P = P_i + r^2\, TP_i$.

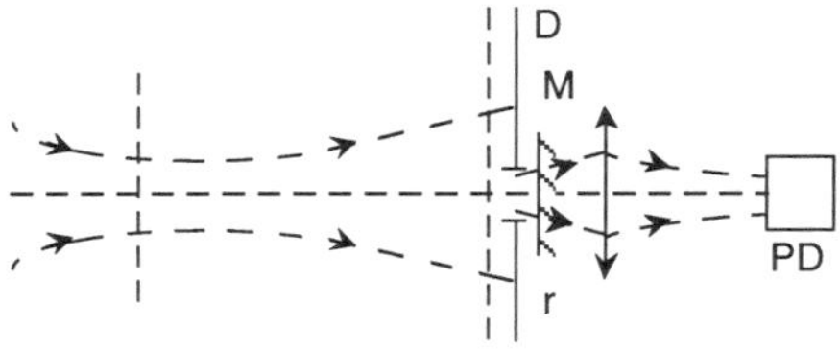

(a) Discuss this approximation.

(b) Let us choose $\omega_1 = 1\ \mu m$ along with the appropriate length for d. Assume that $r^2 \approx 1$ for this question. Show on the same graph the law $T = f(P)$ found in 3b, and the linear law given above which defines T in the new experimental configuration (for given values of P_i and of P_{cr}). With the help of this graph, show that as P_i increases, starting from 0, one comes upon values of P_i for which there exist several values of T, provided that P_i/P_{cr} stay within a range of values you are asked to determine. Calculate the values numerically and give a graphic representation of the variations of T as a function of P_i as P_i is made to increase and then to decrease again. Show the existence of a hysteresis phenomenon.

(c) Let us take again $r^2 = 0.99$. The power P' transmitted by mirror M is collected on a photodiode PD. Calculate P as a function of P_i. Draw a graph of the variations of P as a function of P_i which shows the observed hysteresis loop.

1. The radius of the beam waist is $\omega_0 = \lambda/\pi\theta$ (see Section IV.2.2.2). The numerical values yield $\omega_0 \approx 655\ \mu m$. The beam waist is located on M_1, the plane mirror. The equations derived in Chapter IV showed that $\omega_0^2 = (\lambda/\pi)\sqrt{L(R-L)}$. Calculating the radius of curvature of mirror M_2 gives $R_2 \approx 6.08$ m. The maximum power within the cavity is 40 watts (20 times greater than the maximum output power).

2.(a) Here, we use Eq. V.2.7: $\omega_1^2 = \omega_0^2 f_1^2 / \left[(L_0 - f_1)^2 + (\pi \omega_0^2 / \lambda)^2 \right]$. For $\omega_1 = 1\ \mu m$ we find $L_0 \approx 19.44$ m; for $\omega_1 = 10\ \mu m$ we find an impossibility. The greatest value of the beam waist which can be obtained occurs when $L_0 = f_1$ (but this condition is impossible to fulfill experimentally) and is equal to $\omega_1 = f\lambda / \pi\omega_0 \approx 7.5\ \mu m$.

(b) The value of the intermediate beam waist is given by $\omega_2^2 = \omega_0^2 f_2^2 / \left[(L_0 - f_2)^2 + (\pi \omega_0^2 / \lambda)^2 \right]$. We obtain $\omega_2 \approx 100\ \mu m$. Its position is given by the relation $L_2' - f_2 = (L_0 - f_2)\, f_2^2 / \left[(L_0 - f_2)^2 + (\pi \omega_0^2 / \lambda)^2 \right]$ yielding $L'_2 \approx 0.546$ m.

We take up the calculations again, starting from the intermediate beam waist of 100 μm, and we find that $L_{2\,max} = 3.03$ m and $L_{2\,min} = 0.324$ m. The corresponding positions of beam waist ω_1 are $L_{1,\,min} = 30.3$ mm and $L_{1,\,max} = 32.9$ mm. Therefore the beam waist moves by 2.6 mm as it increases in size from 1 μm to 10 μm.

3.(a) For $P_i << P_{cr}$, we obtain the classical magnification law:

$\omega^2(d) \approx \omega_1^2\left[1 + d^2\lambda^2/(\pi\,\omega_1^2)^2\right]$. With ω (d) = 100 μm, we can calculate the values of d corresponding to ω_1 = 1 μm and to ω_1 = 10 μm. We find, respectively, d_1 = 610 μm and d_1 = 6.1 μm.

The transmitted power is $P_t = \int_0^{a_0} \left[\alpha\, P/\omega^2(d)\right] \exp\left[-2r^2/\omega^2(d)\right] 2\pi r\, dr$. The incident power is written in the same way, except that the upper integration limit is ∞ instead of a_0.

We conclude that P_i = KP (where K is a constant). But since $P = P_i$, K = 1.
Using this value, we can finally write $P_t = P_i\left\{1 - \exp\left[-2a_0^2/\omega^2(d)\right]\right\}$ and $T = 1 - \exp\left[-2a_0^2/\omega^2(d)\right]$. Now $a_0/\omega(d) = 0.1$. This yields T $\approx$ 0.02. At weak intensities, the diaphragm cuts off about 98 % of the power.

(b) For any power smaller than the critical power we have

$$T' = 1 - \exp\left[-2a_0^2/\omega_1^2\left\{1 + (1 - P/P_{cr})\left[d^2\lambda^2/(\pi\,\omega_1^2)^2\right]\right\}\right].$$

For $P \geq P_{cr}$ we have $T'' = 1 - \exp\left[-2a_0^2/\omega_1^2\right]$.

Let us express T' and T'' numerically for $\omega_1^{(1)}$ = 1 μm and for $\omega_1^{(2)}$ = 10 μm.

$$\begin{cases} T_1' = 1 - \exp\left[-200/1 + 9840\,(1 - P/P_{cr})\right] \\ T_2' = 1 - \exp\left[-2/1 + 98.40\,(1 - P/P_{cr})\right] \end{cases}$$

$$\begin{cases} T_1'' = 1 - \exp[-200] \approx 1 \\ T_2'' = 1 - \exp[-2] \approx 0.87 \end{cases}$$

The table beneath lists some values of T_1' and T_2' for values of P very close to P_{cr},. The graph illustrated these results.

P/P_{cr}	T_1'	T_2'
0.9	0.18	0.17
0.99	0.87	0.64
0.999	1	0.86

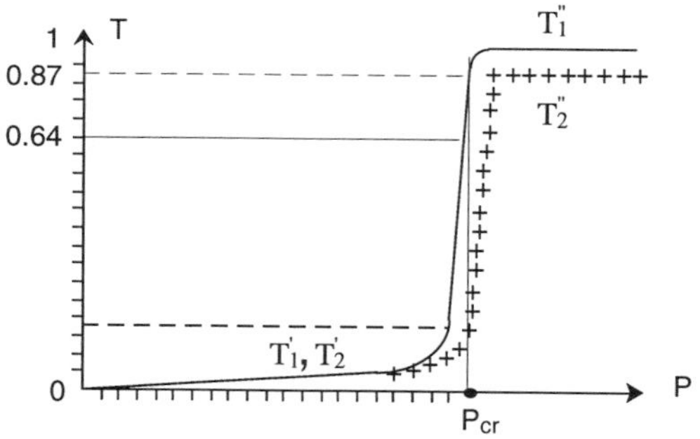

Self-focusing, which occurs in the immediate vicinity of P_{cr}, confines the laser beam radially so that it passes through the pinhole without any significant attenuation.

4.(a) Positioning mirror M at d creates a stationary wave in the medium, with a distribution of nodes and loops which are not accounted for in our description. The radius of curvature of the mirror must be adapted to the structure of the Gaussian wave.

(b) We have $T \approx (P - P_i) / P_i \approx -1 + P / P_i$.

The graphic representation of graph (T'_1, T''_1) and of $-1 + P/P_i$ shows the existence of intersections, i.e. of solutions, here represented by dots, of the inequality

$$P_{cr} / 2 \leq P_i \leq P_{cr}$$

in other words, of

$$1/2 \leq P_i/P_{cr} \leq 1$$

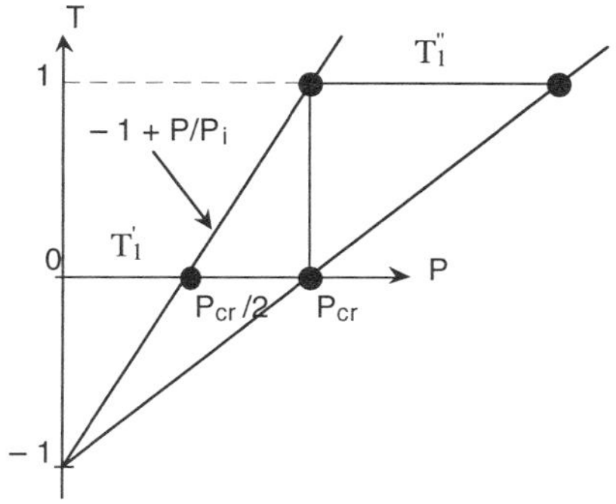

For $P_i < P_{cr} / 2$: $T = 0$
For $P_i = P_{cr} / 2$: $T = 0$ and $T = 1$
For $P_{cr} / 2 \leq P_i \leq P_{cr}$, there are two stable solutions: $T = 0$ and $T = 1$
For $P_i > P_{cr}$: $T = 1$
The hysteresis loop is shown on the graph.

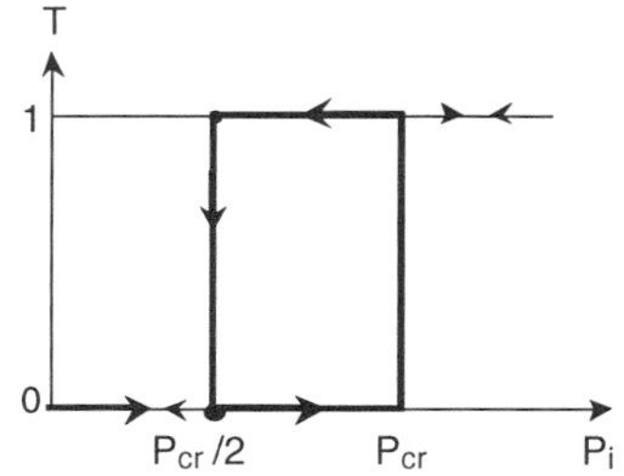

(c) We have
$P' = (1 - r^2)\, TP_i \approx 0.01\, TP_i$.

The accompanying graph shows P' as a function of P_i and illustrates the hysteresis phenomenon as registered by the photodiode PD.

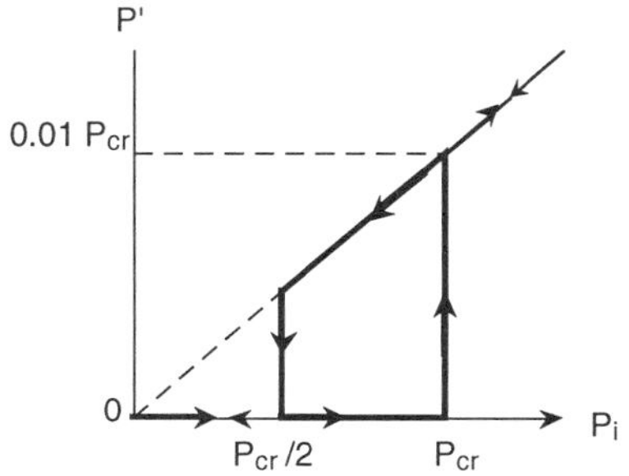

VIII.3.4 Problem VIII.4: Self-focusing and optical bi-stability inside a laser cavity

We place a thin ($d \ll L_1$) cell containing a fluid with refractive index n at point A of the ring cavity shown in the accompanying. The cell is set at Brewster angle so as to avoid losses by reflection of the stored optical wave (which is polarized in the incident plane).

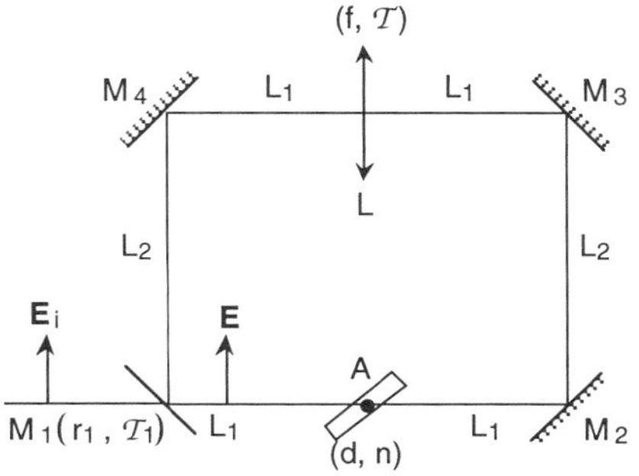

(1) The refractive index n is assumed to be constant inside the cell. The field reflection coefficient of mirror M_1 is r_1 such that $r_1 = T$. When the cell is empty, the incident frequency ν_i is resonant with an eigenmode of the cavity of frequency ν_q. Assume the waves inside the cavity to be plane waves and calculate the dephasing ϕ induced by inserting the medium in the cell for one transit of the wave through the ring.

Conclude from this that the intensity I stored in the cavity in presence of the medium can be expressed as $I = I_0 / [1 + F \sin^2 (\phi/2)]$, where F and I_0 are functions of I_i and of r_1 which you are asked to determine.

Refractive index n is assumed to be very close to 1. What is the largest value of (n – 1) one can tolerate if one does not want the intensity of the stored energy to differ by more than 5/100 from its maximum value?

Numerical calculation: d = 2 mm, λ = 514.5 nm , r_1 = 0.95.

(2) The refractive index of the medium is not homogeneous. It decreases as the distance from the axis increases: $n(r) = n_0 [1 - r^2 / a^2]$, where r is the distance between the point under consideration and the axis of the resonator, and "a" is a constant characteristic of the medium.

We wish to study how a paraxial (close to the axis) ray of light propagates in the medium. Remember that the laws of optical geometry in an inhomogeneous medium lead to the following equation for the propagation of a ray of light which makes at most a slight angle with the z-axis: $d^2r/dz^2 \approx (1/n_0)\,[dn(r)/dr]$.

Find the transfer matrix ABCD of a paraxial ray of light as it travels through the medium of thickness d.

Now we want to study how a Gaussian beam propagates through this medium. If q_1 denotes the complex beam parameter at the entrance (the front face) of the medium, calculate its value q_2 at the back side. Show that by choosing appropriate values for q_0 and q_1, the Gaussian wave propagates through the medium while keeping its characteristic parameters (same beam radius, same radius of curvature). Conclude from this that there is a specific value of parameter "a" (which we ask you to determine) for which the cavity can stay in the same TEM_{00} mode. We remind you that the complex beam parameter q of a Gaussian beam transforms following the ABCD law as $q_2(z) = \dfrac{A(z)q_i(0) + B(z)}{C(z)q_i(0) + D(z)}$, where A(z), B(z), C(z), and D(z) are elements of the transfer matrix ABCD governing the transformation of paraxial rays from z = 0 to z.

(3) Here we again assume the refractive index of the medium to be homogeneous and we also assume that the waves which propagate inside the cavity can be treated like plane waves. This time, however, we assume the refractive index to vary linearly with the intensity traveling through the medium: n = 1 + bI, where b is a characteristic value of the medium. The length of the cavity is taken so as to minimize the amount of stored intensity I when the cell is empty. Show that the dephasing induced by filling the cell with the medium varies linearly with the intensity I reigning inside the cavity. Conclude from this that, for a given incident intensity I_i, it is possible to determine the intensity inside the cavity by finding the intersection of two curves relating I to ϕ. Find these two relations. Show qualitatively, using a approximate graph of these curves, that there

is only one solution for small values of I_i, while several solutions may exist for large values of I_i. Discuss this result. We want to see what happens during a sequence in which we slowly change I_i, starting from $I_i = 0$, and increase it until it reaches values very much larger than those for which the graph showed several solutions; then we decrease I_i again continuously until $I_i = 0$. Give a qualitative description of the behavior of I, the intensity inside the cavity, during the course of this sequence.

1. The induced dephasing ϕ is such that $\phi = 2\pi\,(n-1)\,d/\lambda$. Inside the cavity, the field E can be expressed as $E = \mathcal{T}_1\, E_i + \mathcal{T}\, r_1\, E \exp(-i\phi)$, yielding $E = E_i \sqrt{1 - r_1^2}\,/\left[1 - r_1^2 \exp(-i\phi)\right]$ $(\mathcal{T} = r_1)$. Therefore we have the intensity: $I = |E|^2 = E_i^2\left(1 - r_1^2\right)/\left[1 - r_1^2 \exp(-i\phi)\right]$ $\times\left[1 - r_1^2 \exp(i\phi)\right] = \left[E_i^2/\left(1 - r_1^2\right)\right]\left\{1/\left[1 + 4\, r_1^2 \sin^2(\phi/2)/\left(1 - r_1^2\right)^2\right]\right\}$ $= I_0/\left[1 + F \sin^2(\phi/2)\right]$ with $I_0 = E_i^2/\left(1 - r_1^2\right)$ and $F = 4\, r_1^2/\left(1 - r_1^2\right)^2$. We see that a variation of $(n-1)$ of about 10^{-6} induces an intensity fluctuation of $\Delta I / I = 5 \times 10^{-2}$, therefore the stored intensity is extremely sensitive to fluctuations of the refractive index and also to fluctuations of the length of the cavity.

2. We find $dn/dr = -2\, n_0\, r / a^2$; $d^2r / dz^2 \approx -2r / a^2$. The general solution of this second-order differential equation with constant coefficients is of the form

$$r = A \cos\left(\sqrt{2}\, z / a\right) + B \sin\left(\sqrt{2}\, z / a\right)$$

At $z = 0$: $r = r_0$ so $A = r_0$ and $r' = r'_0$ so $B = a\, r'_0 / \sqrt{2}$. Thus

$$r_0 = r_0 \cos\left(\sqrt{2}\, z / a\right) + \left(r'_0\, a / \sqrt{2}\right) \sin\left(\sqrt{2}\, z / a\right)$$

$$r_d = r_0 \cos\left(\sqrt{2}\, d / a\right) + \left(r'_0\, a / \sqrt{2}\right) \sin\left(\sqrt{2}\, d / a\right)$$

$$r'_d = \left(-\, r_0 \sqrt{2} / a\right) \sin\left(\sqrt{2}\, d / a\right) + r'_0 \cos\left(\sqrt{2}\, d / a\right)$$

And the transfer matrix takes the form (see Inset IV.2)

$$\begin{pmatrix} \cos\left(\sqrt{2}\, d / a\right) & \left(a / \sqrt{2}\right) \cos\left(\sqrt{2}\, d / a\right) \\ -\left(\sqrt{2} / a\right) \sin\left(\sqrt{2}\, d / a\right) & \cos\left(\sqrt{2}\, d / a\right) \end{pmatrix}$$

At last, we find

$$q_2 = \frac{\cos\left(\sqrt{2}\, d / a\right) q_1 + \left(a / \sqrt{2}\right) \sin\left(\sqrt{2}\, d / a\right)}{-\left(\sqrt{2} / a\right) \sin\left(\sqrt{2}\, d / a\right) q_1 + \cos\left(\sqrt{2}\, d / a\right)}$$

We want to have $q_2 = q_1 = q_0$ for all d, so we must have $q_0 = i\, a / \sqrt{2}$ which is the only root having a physical meaning (see Eq. IV.2.59). Placing the beam waist at the entrance (the front face) of the cell and with $q_0 = i\, a / \sqrt{2} = i\pi\, \omega_0 / \lambda$ (see Eq. IV.2.60), we obtain a Gaussian beam which propagates in the medium with the same radius ω_0. If L_1 and L_2 are very large compared to d, then the beam waist of the cavity, now located inside the nonlinear medium, stays the same.

3. We have $\phi = 2\pi(n - 1)\, d/\lambda = 2\pi b d I/\lambda$. Moreover, if ϕ' is the total dephasing, $\phi' = \phi'' + \phi$, where $\phi'' = (2k + 1)\,\pi$ (corresponding to the smallest intensity in the absence of the medium), then the intensity inside the cavity is related to ϕ' by $I = I_0 / [1 + F \sin^2(\phi'/2)]$, in other words, $I = I_0 / [1 + F \cos^2(\phi/2)]$. The intensity stored inside the cavity is found by looking for values of ϕ which satisfy $\lambda\phi/2\pi b d = I_0 / [1 + F \cos^2(\phi/2)]$.

Let us do this graphically

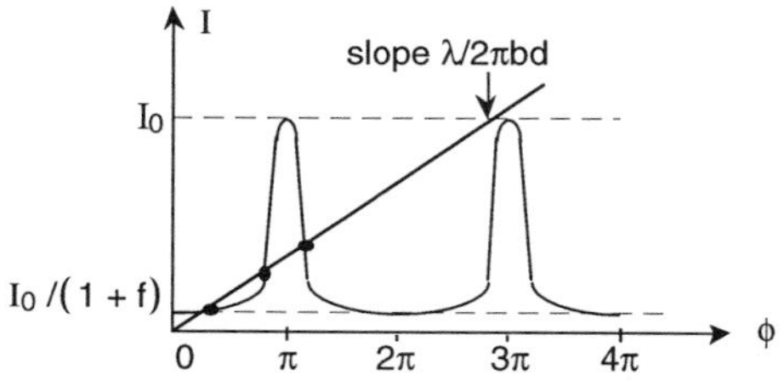

Large values of I_i:
several solutions

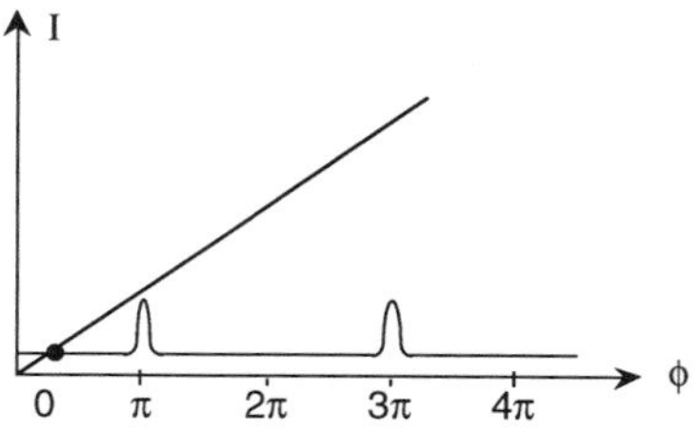

Small values of I_i:
one solution

Graphs for increasing I_i (I_i is proportional to I_0):

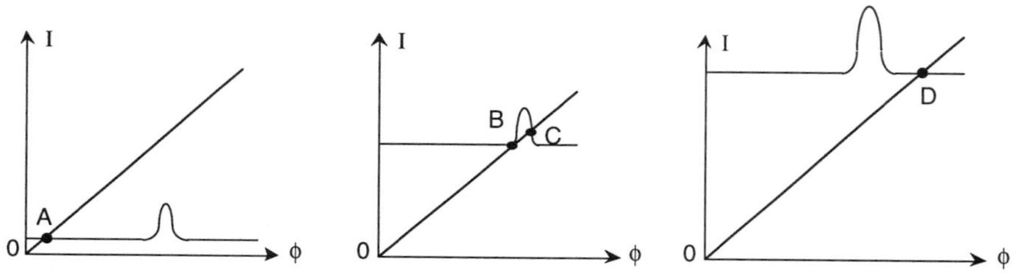

Graphs for decreasing I_i:

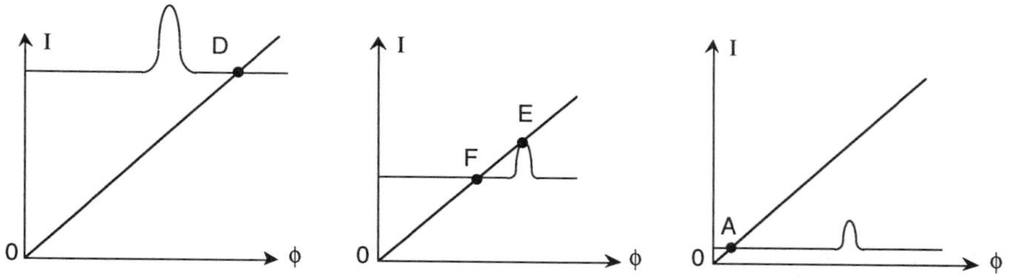

Graph of the variations of I plotted against I_0:

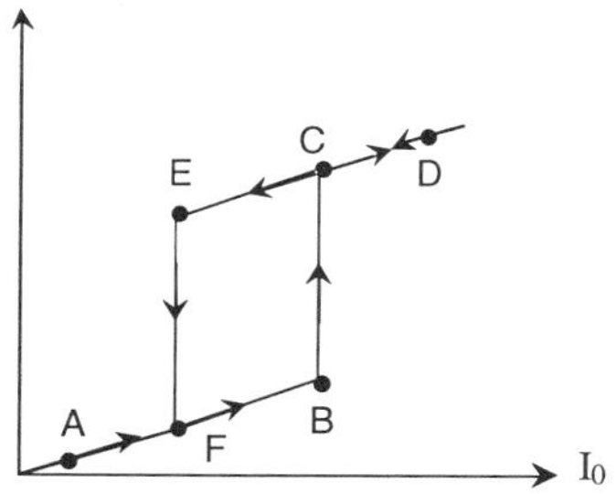

VIII.3.5 Problem VIII.5: The use of phase conjugation in a Ti/sapphire laser (linear cavity)

In the accompanying diagram, you recognize the structure of the Ti/sapphire laser with linear cavity seen earlier, in Problem IV.5. The CPM (conjugated phase mirror) is placed against the amplifying medium and has a field amplification coefficient of a. The mirror can be treated as a plane mirror which reflects a complex incident field ε_i in such a way that the reflected complex field ε'_r is equal to $\varepsilon'_r = r_c \exp(i\varphi_e)\, \varepsilon_i^*$.

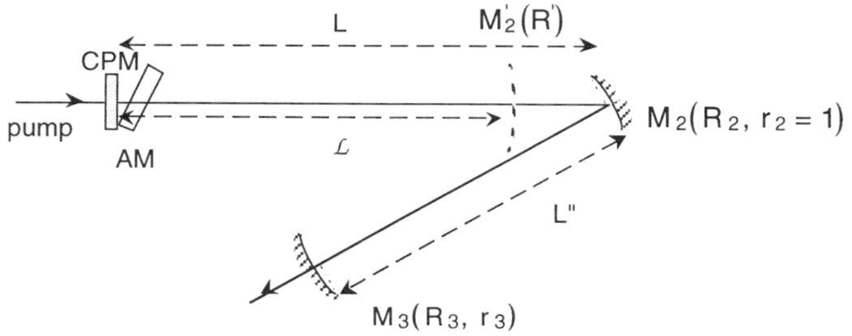

1. Using the plane wave approximation inside the cavity, give the conditions leading to self-oscillation of the field. To do this, express the fact that the field must be equal to itself after *two* back-and-forth trips through the cavity. Does the notion of longitudinal mode still make sense here? What determines the frequency of the self-oscillation? What determines the intensity of the self-oscillating wave?

2. We want to determine the transverse characteristics of the resonant mode of the cavity.

(a) Show that the conjugated phase mirror transforms a Gaussian wave in the following way: If q_1 is the complex beam parameter of the incident beam, the beam parameter of the reflected beam q_r is given by

$$q_r = \frac{Aq_1^* + B}{Cq_1^* + D} \quad \text{where matrix} \begin{pmatrix} A & B \\ C & D \end{pmatrix} \text{ is equal to } \begin{pmatrix} 1 & 0 \\ 0 & -1 \end{pmatrix}$$

(b) As defined in Problem IV.5, use an equivalent mirror M'_2 to replace the set of mirrors M_2, M_3. Assume that the amplifying medium does not perturb the resonant mode of the empty cavity.

Let q_0 be the beam parameter of the beam after reflection on the equivalent mirror M'_2 Follow the behavior of this parameter as the wave travels back and forth through the cavity two times, and express the self-consistency of q_0 in a steady-state regime. Is it possible to derive the characteristic parameters of a self-oscillating Gaussian mode from the characteristic parameters of the cavity? Discuss this result.

1. Let us follow the evolution of the field ε_i as it goes through two back-and-forth trips in the cavity (the various active elements are indicated over the arrows):

$$\varepsilon_i \xrightarrow{AM} a\,\varepsilon_i \xrightarrow{1^{st}\text{ trip through air}} a \exp\left[i2k\left(L+L''\right)\right]\ \varepsilon_i$$

$$\xrightarrow{M_3} r_3\, a \exp\left[i2k\left(L+L''\right)\right]\ \varepsilon_i \xrightarrow{AM} r_3\, a^2 \exp\left[i2k\left(L+L''\right)\right]\ \varepsilon_i$$

$$\xrightarrow{CPM} r_3\, r_c\, a^2 \exp\left(i\varphi_c\right) \exp\left[-i2k\left(L+L''\right)\right]\ \varepsilon_i^*$$

$$\xrightarrow{2^{nd}\text{ trip through air}} r_3\, r_c\, a^2 \exp\left(i\varphi_c\right)\varepsilon_i^* \xrightarrow{M_3} r_3^2\, r_c\, a^2 \exp\left(i\varphi_c\right)\varepsilon_i^*$$

$$\xrightarrow{\text{two passages through AM}} r_3^2\, r_c\, a^4 \exp\left(i\varphi_c\right)\varepsilon_i^* \xrightarrow{CPM} r_3^2\, r_c^2\, a^4\, \varepsilon_i$$

The self-oscillating condition is expressed by $r_c^2\, r_3^2\, a^4 = 1$. There is no longer any conditions on the phases, so that no longitudinal mode can be defined [see Section IV.2.2.2(d)]. The frequency of the self-oscillating wave is determined by the phase conjugated mirror: It is that frequency for which r_c is greatest. The intensity of the wave is such that we have $a^2 = 1/r_c\, r_3$.

2 (a) We have (see Eq. IV.2.65)

$$\begin{aligned}\varepsilon_i &= E_0\left[\omega(0)/\omega(z)\right]\exp\left[-r^2/\omega(z)^2\right]\exp\left\{-i\left[kz+kr^2/2R(z)\right]-\phi(z)\right\}\\ &= E_0\left[\omega(0)/\omega(z)\right]\exp\left\{-i\left[kz+\phi(z)\right]\right\}\exp\left[-ikr^2/2q_i\right]\end{aligned}$$

with $q_i^{-1} = (1/R) - i\lambda/\pi\,\omega^2(z)$, and in the same way we obtain

$$\begin{aligned}\varepsilon'_r &= r_c\exp\left(i\varphi_c\right)E_0\left[\omega(0)/\omega(z)\right]\exp\left[-r^2/\omega^2(z)\right]\\ &\quad\times\exp\left\{+i\left[kz+kr^2/2R(z)\right]+\phi(z)\right\}\\ &= r_c\exp\left(i\varphi_c\right)E_0\left[\omega(0)/\omega(z)\right]\exp\left\{+i\left[kz+\phi(z)\right]\right\}\\ &\quad\times\exp\left[-ikr^2/2q_r\right]\end{aligned}$$

with $q_r^{-1} = -(1/R) - i\lambda/\pi\,\omega^2(z)$

Therefore we have $q_r^{-1} = -(q_i^{-1})^*$; i.e. $q_r = -q_i^*$, and the matrix representing the CPM is indeed $\begin{pmatrix}1 & 0\\ 0 & -1\end{pmatrix}$ (see Inset IV.2).

(b) Mirror M'_2 is located at a distance L from the CPM. It has a focal length of $f' = R'/2$. So for two back-and-forth trips:

$$q_0 \xrightarrow{\text{trip through air}} q_0 + \mathcal{L} \xrightarrow{\text{CPM}} -q_0^* - \mathcal{L} \xrightarrow{\text{trip through air}} -q_0^*$$

$$\xrightarrow{\text{mirror } M'_2} -q_0^* f'/(f' + q_0^*) \xrightarrow{\text{trip through air}} -q_0^* f'/(f' + q_0^*) + \mathcal{L}$$

$$\xrightarrow{\text{CPM}} q_0 f'/(f' + q_0) - \mathcal{L} \xrightarrow{\text{trip through air}} q_0 f'/(f' + q_0) \xrightarrow{\text{mirror } M'_2} q_0$$

(Use the fact that mirror M'_2 transforms the beam parameters as would a lens with focal length f, i.e. following expression V.2.3). As you can see, a self-consistent solution is obtained no matter what the initial shape of the beam. The cavity does not select any specific transverse structure. The shape of the beam is determined only by the amplifying medium and by the conjugated phase mirror.

VIII.3.6 Problem VIII.6: Ring cavity with phase conjugation elements

Let us take up again the diagram of Problem IV.6 and let us excite the amplifying medium by two pump waves, $E_p^{(1)}$ and $E_p^{(2)}$ propagating in opposite directions. This makes the amplifying medium behave like a conjugated phase mirror. It reflects waves $u_+(0)$ and $u_-(d)$ and thereby conjugates their spatial phases. The amplifying medium is no longer characterized by amplifications a_+ and a_-, but by the following relations:

$$v_+(d) = \mathcal{T}(c)\exp(i\phi_c)\,u_+(0) + r_c \exp(i\phi'_c)\,u_-^*(d)$$
$$v_-(0) = \mathcal{T}(c)\exp(i\phi_c)\,u_-(d) + r_c \exp(i\phi'_c)\,u_+^*(0)$$

Assume that r_c and $\mathcal{T}_c$ are constants characteristic of the amplifying medium and that the intensity of the pump waves are such that $\mathcal{T}_{(c)}^2 = 1 + r_c^2$.

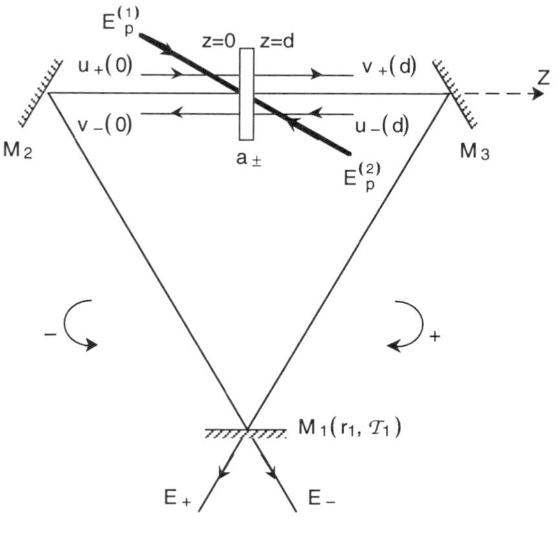

1. In this question, assume that M_1 is a plane mirror characterized by r_1 and $\mathcal{T}_1$, and use the plane wave approximation for the waves propagating inside the cavity.

(a) As in Problem IV.6, write the condition leading to a steady state of the two waves + and –, which are coupled to each other in this cavity, by writing that the two waves are self-consistent. Show that this condition can be expressed by a single real equation, and give this equation. Simplify this condition for the case where r_1 is very close to unity.

(b) Show that the output laser field E_+ can be expressed as a function of E_-^*. Compare the absolute values $|E_+|$ and $|E_-|$ of these fields. Was this predictable? What are the physical consequences of these results ? Compare them to the results obtained in Problem IV.6.

(c) Assume that the intensity transmission coefficient of the medium $\mathcal{T}_c^2$ saturates in intensity according to the law $\mathcal{T}_c^2 = \mathcal{T}_{c0}^2 / [1 + I / I_{sat}]$, where $\mathcal{T}_{c0}^2$ represents the nonsaturated transmission, I_{sat} the characteristic saturation intensity of the medium, and I the sum of the intensities of the two waves inside the medium expressed by $I = |u_+(0)|^2 + |u_-(d)|^2$. The phase ϕ_c is assumed to be constant and independent of I. Does the self-oscillation condition found in 1 (a) determine the frequencies of the eigenmodes of self-oscillation of the cavity, and does it give the intensities of these modes? Does the word "resonant modes" of the cavity keep its usual meaning?

Assume now that the self-oscillation frequency settles at ν_p, the imposed frequency of the pump lasers $E_p^{(1)}$ and $E_p^{(2)}$. Calculate the intensities of the two waves I_+ and I_- inside the medium as a function of I_{sat}, $\mathcal{T}_{c0}^2$, ϕ_c, L, and ν_p.

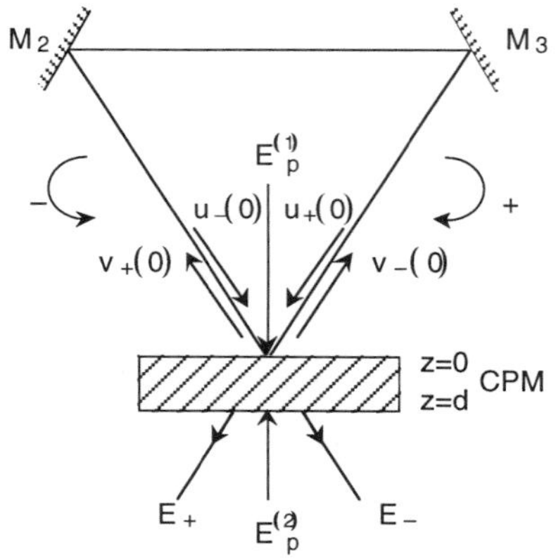

2. Let us take a new cavity, similar to the cavity of question 1 but with a conjugated phase mirror instead of mirror M_1. In this case, the relations between the complex conjugated amplitudes are

$$v_+(0) = r_c \exp(i\phi'_c)\, u_-^*(0)$$
$$v_-(0) = r_c \exp(i\phi'_c)\, u_+^*(0)$$

$$E_+ = \exp(i\phi_c)\, u_+(0)$$
$$E_- = \exp(i\phi_c)\, u_-(0)$$

(a) Use the plane-wave approximation and express the condition of self-oscillation of the two waves in this new cavity. Can we still speak of "longitudinal modes"? What determines the frequency of the self-oscillating wave? What is the relation between E_+ and E_-? What determines the intensity of the wave in the cavity? What is the advantage of this cavity as compared to the one of question 1?

(b) In order to shorten the length of the beams inside the cavity, a lens L with focal length f is placed between M_2 and M_3 in such a way that its distance from the conjugated mirror is the same, namely L/2, for both wave directions. Determine the characteristics of the fundamental Gaussian self-oscillating mode of the cavity.

Show that, by conjugating the complex amplitude of the wave, the conjugated mirror changes a Gaussian beam in the following way: If q_i represents the complex beam parameter of the incident beam, the beam parameter of the reflected beam q_r is given by

$$q_r = \frac{Aq_i^* + B}{Cq_i^* + D} \quad \text{where the matrix} \quad \begin{pmatrix} A & B \\ C & D \end{pmatrix} \quad \text{is equal to} \quad \begin{pmatrix} 1 & 0 \\ 0 & -1 \end{pmatrix}$$

Let q_0 be the complex beam parameter of the beam as it leaves lens L. Follow the evolution of this parameter as the beam travels through the cavity and express the self-consistency of q_0 in a steady-state regime. Can you determine the characteristics of the self-oscillating Gaussian laser-wave from the characteristics of the cavity? Give a qualitative explanation of the influence of the conjugation mirror on the transverse properties of the self-oscillating laser wave. What do you think is the purpose of the lens?

1.(a) Let L be the length of the cavity. We can therefore write

$$u_+(0) = r_1 \exp(ikL)\, v_+(d) = r_1 \exp(ikL)\left[\mathcal{T}_c \exp(i\phi_c)\, u_+(0) + r_c \exp(i\phi'_c)\, u_-^*(d)\right]$$
$$u_-(d) = r_1 \exp(ikL)\, v_-(0) = r_1 \exp(ikL)\left[\mathcal{T}_c \exp(i\phi_c)\, u_-(d) + r_c \exp(i\phi'_c)\, u_+^*(0)\right]$$

Taking the conjugate of the second equation yields

$$u_-^*(d)\left[1 - r_c\, \mathcal{T}_c \exp(-ikL)\exp(-i\phi_c)\right] = r_1 r_c \exp(-ikL)\exp(-i\phi'_c)\, U_+(0)$$

or again

$$u_-^*(d) = u_+(0)\; r_1 r_c \exp(-ikL)\exp(-i\phi_c) / \left[1 - r_c\, \mathcal{T}_c \exp(-ikL)\exp(-i\phi_c)\right]$$

Now the first equation yields

$$u_-^*(d) = u_+(0)\left[1 - r_1\, \mathcal{T}_c \exp(ikL)\exp(i\phi_c)\right] / \left[r_1 r_c \exp(ikL)\exp(i\phi'_c)\right]$$

Identification gives us

$$\left\{1 - r_1\, \mathcal{T}_c \exp\left[i(kL + \phi_c)\right]\right\}\left\{1 - r_1\, \mathcal{T}_c \exp\left[-i(kL + \phi_c)\right]\right\} = r_1^2\, r_c^2$$

In other words,

$$1 + r_1^2\, \mathcal{T}_c^2 - 2 r_1 \mathcal{T}_c \cos(kL + \phi_c) = r_1^2\, r_c^2$$

We obtain the same equation if we start by conjugating the first equation. When r_1 is close to unity, we can simplify it to

$$\cos(kL + \phi_c) = \left(1 + r_1^2\, \mathcal{T}_c^2 - r_1^2\, r_c^2\right) / 2 r_1 \mathcal{T}_c \approx 1/\mathcal{T}_c$$

b) $E_+ = \mathcal{T}_1\, v_+(d)\exp(ikL/2)$

$$= \mathcal{T}_1\left[\mathcal{T}_c \exp(i\phi_c)\, u_+(0) + r_c \exp(i\phi')\, u_-^*(d)\right]\exp(ikL/2)$$
$$= \mathcal{T}_1\left[\mathcal{T}_c \exp(i\phi_c)\, r_1 \exp(ikL)\, v_+(d) + r_c \exp(i\phi'_c)\, r_1 \right.$$
$$\left. \times \exp(-ikL)\, v_-^*(0)\right]\exp(ikL/2)$$

yielding $v_+(d) = v_-^*(0)\, r_1 r_c \exp\left[i(\phi'_c - kL)\right] / \left\{1 - r_1\, \mathcal{T}_c \exp\left[i(\phi_c + kL)\right]\right\}$

But $E_- = \mathcal{T}_1 \exp(ikL/2)\, v_-(0)$, therefore $v_-^*(0) = E_-^* \exp(-ikL/2) / \mathcal{T}_1$

and $E_+ / E_-^* = r_1 r_c \exp i(\phi'_c) / \left\{1 - r_1\, \mathcal{T}_c \exp\left[i(\phi_c + kL)\right]\right\}$

so $|E_+| / |E_-^*| = r_1 r_c / \left|1 - r_1\, \mathcal{T}_c \exp\left[i(\phi_c + kL)\right]\right|$

Since the second member of this equation is equal to one, we conclude that $|E_+| = |E_-^*|$. This equality means that the dextrogyre and the levogyre waves of the laser are symmetrical. The next-to-last equation shows that fields E_+ and E_- have coupled phases. The corresponding laser emissions are no longer independent, as they would be with an ordinary amplifying medium (without phase conjugation), like in Problem IV.6.

(c) We have $\cos(kL + \phi_c) = 1 / \mathcal{T}_c = \sqrt{1 + I/I_{sat}} / \mathcal{T}_{c0}$.
And $I = |u_+(0)|^2 + |u_-(d)|^2 = 2|u_+(0)|^2 = 2\,I_+ = 2|u_-(d)|^2 = 2\,I_-$. It is therefore *impossible to calculate independently* the frequency ν_q and the corresponding intensity I_q of a self-oscillating mode. There is only one equation relating these two quantities. Here, the frequency ν_q is determined by the frequency of the pump waves. The cavity no longer does the frequency selecting it usually does, and cavity "modes" cannot be defined in the traditional way. We have

$$I_+ = I_- = (I_{sat} / 2)\left\{\mathcal{T}_{c0}^2 \cos^2\left[(2\pi\nu_q L/c) + \phi_c\right] - 1\right\}$$

2.(a) Let us calculate $u_+(0)$ $u_+(0) = v_+(0)\exp(ikL) = r_c \exp(i\phi'_c)\, u_-^*(0)\exp(ikL)$
$= r_c \exp(i\phi'_c)\, v_-^*(0)\exp(-ikL)\exp(ikL) = r_c\exp(i\phi'_c)\, v_-^*(0)$
$= r_c\exp(i\phi'_c)\, r_c \exp(-i\phi'_c)\, u_+(0) = r_c^2\, u_+(0)$

So self-oscillation imposes that $r_c = 1$. Since the condition does not include any condition on the length of the cavity, we can no longer speak about longitudinal modes. The oscillation frequency is determined by the phase conjugation mirror. We can write

$$\begin{aligned} E_+ &= \mathcal{T}_c \exp(i\phi_c)\, u_+(0) = \mathcal{T}_c \exp(i\phi_c)\, v_+(0)\exp(ikL) \\ &= \mathcal{T}_c \exp(i\phi_c)\, r_c \exp(i\phi'_c)\exp(ikL)\, u_-^*(0) \\ &= r_c \exp(ikL)\exp(i2\phi_c)\exp(i\phi'_c)\, E_-^* \end{aligned}$$

And we find $|E_+| / |E_-| = r_c$. At steady state, $r_c = 1$; so in that case $|E_+| = |E_-|$ and $\mathcal{T}_c^2 = 1 + r_c^2 = 2$. Since $\mathcal{T}_c$ is a function of the ratio I/I_{sat}, $\mathcal{T}_c^2 = 2$ determines the value of the intensity. We conclude that the two output waves have the same intensity and their phases are coupled. The ratio $|E_+| / |E_-|$ no longer depends on L, as opposed to case 1 (b). The coupling between the output intensity and the cavity length therefore vanishes and the cavity is more stable.

b) The transfer matrix associated to the conjugated phase mirror is $\begin{pmatrix} 1 & 0 \\ 0 & -1 \end{pmatrix}$ (see Exercise VIII.5). The self-consistency of parameter q_0 can be determined after two complete transits through the ring and two reflections on the conjugated mirror, no matter what the values of L or of f. Indeed, we have

$$q_0 \xrightarrow{\text{travel of L/2}} q_0 + L/2 \xrightarrow{\text{CPM}} -(q_0^* + L/2) \xrightarrow{\text{travel of L/2}} -q_0^*$$

$$\xrightarrow{\text{lens}} -q_0^* f/(q_0^* + f) \xrightarrow{\text{travel of L/2}} -q_0^* f/(q_0^* + f) + L/2$$

$$\xrightarrow{\text{CPM}} q_0 f/(q_0 + f) - L/2 \xrightarrow{\text{travel of L/2}} q_0 f/(q_0 + f) \xrightarrow{\text{lens}} q_0$$

It is impossible to derive the characteristics of the oscillating mode from those of the cavity. In fact, it is the spatial distribution of the beam reflected off the conjugated phase mirror which determines the transverse structure of the laser beam. The purpose of the lens is to reduce the diameter of the laser beam so as to put to use most efficiently the active part of the conjugated mirror.

VIII.4 Bibliography

VIII.4.1 Nonlinear optics can be studied in:

BLOEMBERGEN, N. *Nonlinear Optics*, W. A. Benjamin, New York, 1965.

YARIV, A. *Introduction to Optical Electronics*, Holt, Rinehart and Winston, New York, 1976.

SHEN, Y. R. *The Principles of Nonlinear Optics*, Jonh Wiley & Sons, New York, 1984.

BUTCHER, P., AND COTTER, D. *The Elements of Nonlinear Optics*, Cambridge University Press, Cambridge MA, 1990.

BOYD, R. *Nonlinear Optics*, Academic Press, New York, 1992.

NEWELL, A. C, AND MOLONEY, J. V. *Nonlinear Optics*, Addison-Wesley, New York, 1992.

VIII.4.2 Optical Kerr effect and some of its applications to molecular physics (and more especially to the physics of liquid crystals) are found in:

LALANNE, J. R., BUCHERT, J., AND KIELICH, S. Fast Reorientations in Liquid Crystals probed by Nonlinear Optics, in EVANS, M., *Modern Nonlinear Optics*, Part 2, John Wiley & Sons, New York, 1993.

KHOO, I. C., AND WU, S. T. *Optics and Nonlinear Optics of Liquid Crystals*, World Scientific, Singapore, 1993.

VIII.4.3 Phase conjugation can be studied in:

DUCLOY, M. Nonlinear Optical Phase Conjugation in TREUSCH, J., *Advances in Solid State Physics*, Vol. XXII, VIEWEG, Braunschweig, 1982.

DUCLOY, M., AND BLOCH, D. Spectroscopy and Phase Conjugation by Resonant Four-Wave Mixing, in PROCH, D., AND GOWER, M .C. *Optical Phase Conjugation*, Springer Verlag, Berlin, 1993

VIII.4.4 A very simple presentation of the generation of harmonic waves is found in:

HIGGINS, T. V. Nonlinear Crystals: Where the Colors of the Rainbow Begin, *Laser Focus World*, 28 (1), 1992, p. 125.

And a very thorough textbook is:

YARIV, A. *Optical Waves in Crystals*, Jonh Wiley & Sons, New York, 1984

VIII.4.5 Photon echoes and, more generally, transient and coherent optical effects can be studied in the references given in the bibliography of Chapter IV, Section IV.5.3.

Index